AEROSPATIAL REMOTE SENSING IN GEOLOGY

Aerospatial Remote Sensing in Geology

JEAN-YVES SCANVIC

Section pertaining to radar was coauthored
by Jean-Paul Deroin

Translator
N. Venkat Rao
Centre for Exploration Geophysics
Osmania University
Hyderabad, India

Technical Editor
A. Bhattacharya
National Remote Sensing Agency
Hyderabad, India

A.A. BALKEMA/ROTTERDAM/BROOKFIELD/1997

Published by arrangement with Éditions BRGM, Orléans Cedex, France

Published with the assistance of the Office of the Counsellor for Cultural, Scientific and Technical Cooperation, Embassy of France, New Delhi.

Translation of: *TÉLÉDÉTECTION AÉROSPATIALE ET INFORMATIONS GÉOLOGIQUES*, BRGM. Updates provided by the author for the English edition in 1995.

Cover photograph: Spot image of the Damavand volcano, north-east of Teheran (Iran). © CNES, 1986. Distribution Spot Image: BRGM Processing

A.A. Balkema, P.O.Box 1675, 3000 BR Rotterdam, Netherlands
Fax: +31.10.4135947; E-mail: balkema@balkema.nl;
Internet site: http://www.balkema.nl

Distributed in USA and Canada by
A.A. Balkema Publishers, Old Post Road, Brookfield, VT 05036-9704, USA.
Fax: 802.276.3837; E-mail: info@ashgate.com

ISBN 90 5410 725 1

CONTENTS

The process of prospecting: to place oneself in front of nature and to interrogate it with the help of more and more powerful tools of investigation (C. Bernard).

Chapter 1

REMOTE SENSING IN GEOLOGY

Satellite-based remote sensing is considered in relation to developments in geological mapping over the last forty years to its present position within the panoply of exploration technics. This historical account, which clearly establishes the relationship between aerial photography and remote sensing and places the former within the modern classification of the latter, is followed by a simple but progressive methodology-the result of extensive team experience.

The acquisition characteristics of the major Earth Resource satellites, past, present and future, both experimental and operational, are reviewed and various criteria are proposed for selecting the most appropriate accessible data in view of a defined objective.

Finally, various methods of image analysis (e.g. image analysis with or without stereoscopy, classification, computer-aided photointerpretation) are discussed and various types of evaluation are proposed.

1.1 EVOLUTION OF GEOLOGICAL MAPPING

Geological mapping forms the basis for any surface or subsurface prospecting. It has evolved considerably over the last forty years through successive stages.

Prior to 1950, the initial stage, unguided mapping was the rule. For want of other preparatory tools, the field geologist planned his exploration programme according to a systematic network which he modified during the course of observations.

Since 1950 geological mapping has been guided by photointerpretation methods. Aerial photography, the first technic of remote sensing, was gradually systematised throughout the world. Later, or in parallel, new tools emerged which have enabled geologists to refine their knowledge of the surface and subsurface. Geophysics and geochemistry, originally conceived as exploration tools, have also become the tools of geological mapping.

Around 1972 multiband space-borne remote sensing found a place in this panoply of technics. Its integration with geological investigations has been facilitated since 1984 by the launching of second-generation satellites, such as Landsat TM, Spot HRV, into orbit.

Since 1984 geological mapping has undergone dramatic transformation, entering the era of digital technology. Computer-aided mapping, multicriteria analysis, three-dimensional representation and geographic information systems have displaced conventional technics. A new method of mapping has emerged in which the use of observations made by various tools, synergistic research and formulation according to the needs of management, and control of the environment have become indispensable.

1.2 INTRODUCTION TO PROSPECTING TECHNICS

Subsurface economic resources are concentrated due to certain specific palaeogeographic and tectonic conditions. Preparation of geological maps to facilitate their discovery thus forms the basis for prospecting for minerals, petroleum and groundwater. Such maps have enabled us to pass from chance discoveries, already exhausted, to intelligent exploration. However, as most deposits are concealed in the subsurface, ground observations alone do not suffice. To interpret conditions occurring at depth, geologists have introduced various probing technics or observations that precede exploration drilling. But since each deposit has its own specific characteristics the choice of technics to be employed becomes complicated. That is why the procedures and methods leading to the discovery of a deposit or water resource cannot be stereotyped.

To understand these technics it is necessary to keep in mind their evolution in mineral prospecting concomitant with the increase in demand for primary materials.

> Up to the industrial revolution of the 19th century and even until the 1940s, the approach remained essentially centrifugal, i.e., tracing the mineralised structure starting from an indicator or a deposit. Later, the procedures became more and more implicit and centripetal, starting from the known favourable regional-geological situation and advancing towards the final target whose areal extent often is not more than a few hectares (Bertraneu, 1973).

This statement summarises the historical evolution of mining geology; it is equally valid for petroleum geology and even for hydrogeology.

The centrifugal approach is based on the search for surface indicators and owes its success to metallogenic analysis or the synthesis of sedimentary basin studies and detection of indirect indicators. Geophysical and geochemical anomalies are known to have led to hidden subsurface resources. They are better understood because, after the era of chance discoveries, such as Texas oil, Shaba copper and Transvaal diamonds, geologists developed and introduced technics of subsurface exploration, having realised that surface data alone often does not suffice.

Some of these technics can be qualified as direct methods. They primarily include surface observations of rock outcrops of altered mineralised bodies and iron hats or gossans which help in mineral exploration. They also include precise measurement of core samples, which enables the petroleum geologist to reconstruct the geological history of a basin. Various petrographic and mineralogical indicators surround deposits. Finally, one may add to this panoply of direct technics the scintillometric method, which indicates the radioactivity of rocks and its variations which, if anomalous, lead to the discovery of radioactive deposits.

Geochemistry, or detection of detrital or chemical contamination produced by a rocky substratum or alteration of a suboutcropping deposit for identifying indicator elements in the soil or in stream sediments, is a hybrid technic and not readily qualified as a direct or an indirect method. It may be considered a direct method in that it measures the 'contamination' of soils precisely and reflects quite faithfully, in most cases, the geological substratum. However, in terms of prospecting its application is indirect. The metallic traces transported to the surface, which constitute evidence for deeper mineralisation, are rarely distinct and are difficult to distinguish from other geochemical components of soils and sediments. The associated geochemical signature need to be isolated and the anomalies they produce enhanced.

The other technics used in prospecting can be considered indirect because based on measurement of certain physical parameters. The major ones are magnetic susceptibility, electrical resistivity, velocity of elastic wave propagation (seismics) and density (gravity). Magnetic or aeromagnetic and gravity methods are used in mineral prospecting, electrical resistivity in hydrogeology, and gravity and seismics in petroleum exploration. Their utilisation has led to numerous discoveries. Sometimes they directly indicate the presence of a resource but more often enable the geologist to interpret the geological nature of the subsurface and thereby to locate structures that are likely to host a resource.

Remote sensing is an indirect technic for mapping and prospecting. It differs from the geophysical method in that it gives, *a priori*, a single image of the surface or the zone very close to the surface. In other words, except in the case of high-frequency detectors, the information gathered in remote sensing does not pertain to the deeper parts of the landscape since the waves received by the sensors originate only from the surface. However, experience has shown that in certain cases the response of the surface may be the result of integration of several geological variables of the landscape, which are directly induced by deeper phenomena. This is the effect of transparency or transport, which will be discussed later.

However, based on a good map, exploration drilling and geological and metallogenic ratiocination (economic geology) offer major advantages. That is why remote sensing, even though its place in prospecting is still quite modest, is a tool widely employed in the reconnaissance stage; its application offers 'the possibility of obtaining the necessary feedback and of interrogating nature in another manner' (Pélissonnier, 1974). This possibility emanates from three major factors, viz., the spectrum, texture and morphology of the responses, which can be interpreted in geological terms.

1.3 HISTORY OF APPLICATION OF REMOTE SENSING IN GEOLOGICAL EXPLORATION

> ARCHAEOLOGY
>
> Bonaparte's balloon found at Aboukir
>
> The French archaeologists who excavated the wreck of 'The Patriot', sunk near Aboukir in 1798, have found a fragment of the balloon which Bonaparte carried in the invasion of Egypt to obtain an 'aerial' view of the battlefields. The divers also brought out Provencal tiles, some bottles, pieces of plates and a brick.
>
> The last item may have been from the furnace used for producing the hydrogen required for inflating the balloon. (UPI)
>
> Extract from the journal *Le Monde*

Among the group of aerospace technics for acquiring information characteristic of surface objects, remote sensing is the recent outcome of developments in photography, in particular aerial photography, which constitute the basis of photogrammetry and a new concept of mapping. After the initial failures of Icare, military scientists and engineers explored the concept of aerial and spatial manoeuvres. On attaining this

Fig. 1. The Richard stereoscope employed in France during the war of 1914–1918. This 'taxiphoto' was one of the first to be used in the world (Library of A. and M.; Scanvic, 1991).

objective, they sought to benefit from it by reducing the difficulties man encounters on the Earth whose imaging had been achieved for the first time. Alas, one of the primary motives was exploitation for military purposes (Fig. 1).

Subsequently, aerial photography and space imagery (remote sensing) have always accompanied, if not preceded, thematic mapping. Geologists have always been one of the promoters of these technics since their profession prompts them more than others to exploration of the Earth, which is often virginal and occasionally hostile and needs to be better understood before penetration. For the last forty years geologists have had only their enthusiasm and hammer to confront a given region on foot or on horseback, guided by the blind itinerary of a systematic grid for which the topographic maps were very approximate and aerial photos often did not exist. When rock outcrops were rare in an exploration area, they plodded kilometres without encountering one. When aerial photographs became available, many geologists of this period discovered much later that they had passed very close to exposures whose existence they could not guess. The poor workmanship of the maps prepared then arose from the ineptness of the means employed.

Sometimes the exploration programme had a tragic outcome. Such occurred in Gabon, Africa during the course of a geological and mineral exploration mission undertaken in 1955 in the Abeilles region, in the still unknown interior of the Chaillu Massif. In spite of excellent preparation, given the absence of aerial photographs, the risks inherent in the difficulties of penetrating rock outcrops could not be fully appreciated. Unforeseen obstacles present in the topography slowed drilling progress. Provisions ran short and two men lost their lives. The lesson of this tragic incident was quickly learnt. During the next mission two novel technics were successfully employed: preparation of photogeological maps and helicopter reconnaissance. These methods were rapidly improved during the second half of the 1950s. The first systematic aerial photographs of a major part of the country were obtained in this period, which marked the advent of mapping based on the data of photogrammetry and photointerpretation. Thus the quality of maps significantly improved and geological operations could be accelerated, especially in regions showing hydrocarbon potential.

The history of large petroleum discoveries (Perrodon, 1985) indicates that exploration for black gold benefited considerably from the first airborne experiments. Aerial photos were used by geologists for the first time in California (1920); the chance and flair of prospectors proved inadequate in Texas and the search for surface anticlines became the rule; aerial photos have been used in Sumatra since 1936; and in Iran and Saudi Arabia analysis of landforms has acquired maximum importance. Geomorphological studies have resulted in the discovery of vast petroliferous anticlinal structures: Santa Fe (85 Mt) in 1902 and Ventura (170 Mt) in 1926 in the USA; Pars (1800 Mm^3), Kangan (600 Mm^3) and Nar el Har in Iran.

The Sahara region is another example. In 1950, when geologists were few in number, to ensure quick geological mapping of millions of square kilometres, the French Bureau of Petroleum Research, which was responsible for exploration policy, despite the scarcely known potential of aerial photography, gave top priority to aerial and geophysical coverage of the region. These two technics became the preferred tools of an exploration programme and yielded significant results. In Ahnet province, for example, after discovery of the Berga, a dozen surface anticlines were drilled on the basis of surface and photogeological data; most of the wells revealed the presence of gas. In 1955, photogeological and field studies aided location of the first well on the Edjeleh anticline in the Polignac basin (Algeria). It proved productive, as did many subsequent wells.

The French Institute of Petroleum played a key role in demonstrating the efficacy of aerial photography and the promotion of this new technic throughout the world. Since 1958 it has facilitated the application of this technic to all the problems encountered by geologists.

Such developments have helped to widen the potential scope of this technic. Nevertheless, in spite of certain improvements, such as the Kodak infra-red false colour images, this technic has barely been explored. The transition from aerial photography to multiband digital images has been rather rapid.

Commencing in 1965, the National Centre of Space Studies in France initiated experiments for potential users of the first available technics, such as multispectral remote sensing, thermography and radar,[1] as forerunners to the already planned programme of 1972. The first space colour photographs, collected in 1965 by the cosmonauts of the 'Gemini' and 'Apollo', very quickly whetted the interest of geologists. Later, satellites circling the Moon provided an observation of the Earth. When on 21 July 1969 Aldrin was photographed by Armstrong treading on lunar soil for the first time, the 'LM' mission was preceded by acquisition of photographs over 38 million km^2 by the Lunar Orbiter satellites. These photographs, obtained on high-resolution Kodak film with an ultraminute objective of variable focal lengths (the very first) from the French firm Angenieux, had a ground resolution of 1-2 m. Subsequently, a ground reconnaissance was done and geologists carried out analysis of some 30 kg of samples collected on the lunar surface. While returning from the Moon the cosmonauts discovered a new vision of the planet Earth. They expressed surprise that it differed markedly from their vision as obtained from conventional maps. Perhaps it was this that spurred the birth of various programmes of remote sensing of earth resources, in particular the Earth Resource Technological Satellite, the forerunner of the Landsat (in 1972) and Skylab series.

The images furnished by the first-generation satellites are characterised by a spatial-resolution to ground-coverage ratio. They have provided geologists with extensive structural information whose importance in the exploration of minerals, petroleum and groundwater resources as well as their management is increasingly being appreciated. The phenomenal growth of structural geology is largely due to the development of remote sensing and the introduction of an appropriate methodology. 'Lineament' maps, important inputs for structural analysis, are not only appreciated, but critically reviewed and progressively succeeded by regional composite maps in which the value of integrating ground and space approaches has been proven. Thus remote sensing has gradually become a logical component of the methods and procedures used in structural geology. It has been demonstrated that the megafracturing of rock masses, often indicated by remote sensing in covered terrains, is an indispensable complement of structural analysis at the scale of outcrops.

---

[1]Using the Daedalus, Cyclope and Supercyclope sensors of the SAT, Vigie de Thomson and Hasselblad

At the same time, the limits of utilisation of this data are being delineated since geologists in changing from one mode of observation to another, demand higher precision and hence better resolution while nevertheless preserving the synoptic effect whose potential has been realised. This new exigency in structural geology also applies to geological mapping. Between 1972 and 1984 remote sensing provided some of the information missing in the domain of synthesised (composite) geological maps, an essential tool for the development of a country. It soon became evident, however, several exceptions notwithstanding, that the remaining essential information could not be obtained without improvement of the systems. That improvement became possible with second-generation satellites, such as Landsat V (Thematic Mapper) and Spots 1 and 2 (HRV).

Efforts to verify the geological features observed in remote-sensing data have rapidly increased but their identification by conventional methods of ground study has sometimes proven difficult. The contributions of geophysics and geochemistry to geology and the degree of their correlation with remote-sensing data ascertained through simple analogue methods, led to research in digital processing technics to promote synergistic approaches. Methods of statistical processing of interpreted data of linear objects came first, followed by methods for assessing the products of digital extraction of information. Computer programs of this type, such as Linéame-Dense (Dutartre, 1984), were developed at the BRGM. Several other programs exist but all of them suffer the common limitation of not being able to distinguish geological linear objects from non-linear ones and hence human intervention is still necessary in this regard. The methods of identifying such forms based on digital processing technics ought to compensate this handicap.

Attempts have also been made to develop methods for classifying geological features. Initially it appeared that such classification could readily be done on the basis of spectral characteristics alone, as demonstrated in laboratory studies. The number of spectral bands available with the Thematic Mapper made the level of such classification acceptable under certain conditions of geological setting of terrains, but regions satisfying these norms do not constitute a majority. Vegetation, soils and surficial cover over outcrops are more common and the spectral characteristics recorded do not represent the rocks. Research is presently underway on other methods that take into consideration the overall perception of environment. This approach, logical in the mode of spatial observation, necessitates an understanding of the mechanisms that contribute to the final result and makes it obligatory to remove the natural barriers between various disciplines of Earth and environmental sciences. Remote sensing has now become the driving force for a multidisciplinary and interdisciplinary science whose objective is to prove the reality of this concept. Only this approach can provide a geological key to classification based on the methods of mathematical morphology and offer a solution to the problem of integration of remote sensing into geographic information systems.

1.4 PRESENT STATUS OF REMOTE SENSING IN GEOLOGICAL EXPLORATION

Geologists of all specialisations involved in mapping have progressively adopted remote-sensing technology as a complementary tool for ground investigations. As a paradox of common accessibility which took long to occur, reservations are greater relative to the usage of aerial photography.

All the major projects of mineral or oil exploration have generally been based on remote sensing including aerial photography, especially since 1983 when recommendations were issued in the name of UNESCO by the Service of the Geological Map of the World. This resulted in a significant increase in use of remote-sensing data in regions of Africa and Asia for which small and medium-scale maps meeting high-quality standards have been prepared.

In the case of large-scale maps field geologists are more reserved about the importance of remote sensing. The resolution to scale ratio of the first-generation satellite data was considered very good relative to observations made in the field in a detailed exploration. Even though improvements in radiometric performances of the Thematic Mapper of Landsat 5 and High Resolution Visible of Spot have gradually attracted them, aerial photography has not been totally abandoned because it is simpler to use for proven results and is particularly amenable given its stereoscopic vision versus that of the space mode which is still restricted and cumbersome to use.

Generally speaking, the graph of utilisation of remote sensing in geology follows a regular progression, which can be explained as due to certain characteristics as well as its known qualities.

First of all, geologists are familiar with using the properties of electromagnetic spectrum advantageously. The latter is a powerful exploration tool used in geophysics-the electromagnetic method, scintillometry etc. Secondly, the operational space programmes are based on several favourable factors common to the Landsat and Spot series, e.g., liberalisation of dissemination policy, continuity in time (1972-1990 for Landsat, 1986-1995 with possible further extension for Spot) and universality of acquisition except when encountering meteorological problems.

On the other hand, certain limiting factors are also strongly felt which hinder expansion of application of remote sensing in geology. The network of receiving stations, though constantly being enlarged, still has many shadow zones for real-time acquisition, particularly in Western Africa. The quality of the data, collected in a more random manner, is sometimes inadequate as meteorological conditions still remain a perennial obstacle in certain equatorial regions. Image-processing systems that permit optimum utilisation of remote-sensing data are still expensive and often difficult to maintain even though the generation of microsystems has introduced a significant improvement in this field.

Certain performance characteristics are considered inadequate-spatial or spectral resolution, stereoscopic effect etc.-but developments are taking place with reference to these aspects and choice is presently possible. However, certain specialised detectors, such as 'Radar' or 'Thermal' satellites, whose importance was appreciated during experimental missions, are yet to be fully exploited. The satellite data of Radar

ERS-1, launched in 1991, is presently under avid study by the international geological community.

1.5 METHODOLOGY DEFINED

The methodology we propose for optimum utilisation of remote sensing in geology has emerged from experience in photointerpretation and analysis of problems confronting the geologist. Geological exploration is a complex enterprise involving large outlays and inputs. In the preparatory stage of his mission, the geologist works out a prospecting strategy, exploring means for expediting future mapping, especially in difficult regions, such as deserts, equatorial forests, glacial regions, and envisages hypotheses that differ from the already established and often penalising ones: 'that a mineral discovery which reorients the prospecting has been made semi-fortuitously by a borehole badly located (relative to the initial hypothesis) or prolonged for reasons other than geological' (Pélissonnier, 1974).

Remote sensing has tremendous capability for aiding the geologist in this preliminary analysis, since normally one refers to the digital and multispectral character of the data and the extent of area observed. Though very valuable in remote-sensing analysis, these are not the only advantages. It is less often remembered that through remote sensing we obtain the state of the landscape, its evolution, its variations induced by geological processes and especially an irreplaceable synoptic view of the terrestrial ecosystem which associates, in a single response, the complex vegetation, soils and their water content, rocks and their altered products. These characteristics of remote sensing represent a major and rich potential yet barely explored, which space observational platforms enable us to exploit.

Based on the experience of the remote-sensing team of BRGM, we have defined a specific procedure for space geology. It demarcates the present boundaries between photointerpretation and digital classification. In usage, furthermore, it can be adopted to totally new situations.

1.5.1 Technological Aspects

Remote sensing is a group of technics differentiated from one another by the type of sensors, aircraft or spacecraft, mode of acquisition, such as analogue or digital, passive or active, spatial resolution, spectral range utilised and the area covered.

The technics most widely used today are aerial photography, multiband[2] space remote sensing with visible, near and mid infra-red and thermal bands, and airborne or space-borne radar. The numerous radiometers or radar systems presently available permit airborne acquisition on request.

The earth resource satellites generally in vogue, or those that could be used as such, are of two types: experimental satellites and operational satellites. Experimental satellites (or shuttles) have a limited time duration of acquisition. The shuttle may function totally independently or may be integrated into the orbital platform system which enables it to make regular visits. The Soviet space station Mir, regularly visited by the

[2] It is recommended that the term 'multispectral' be replaced by 'multiband'.

shuttle Soyuz, is a typical example.

Operational satellites obey three simple rules:

1. Ensure total coverage of the Earth.
2. Form part of a series providing continuous service for a certain duration (8 to 10 years, for example).
3. Provide for dissemination of data on a commercial and non-exclusive basis.

1.5.1.1 Examples of experimental satellites

SKYLAB

This consists of three manned missions launched in 1973. Data acquisition was principally effected by means of six multiband 70 mm cameras. The spectral bands were: 0.5-0.6 μm, 0.6-0.7 μm, 0.7-0.83 μm, 0.7-0.9 μm, 0.5-0.88 μm, 0.4-0.7 μm. The data acquired is available.

HCMM (Heat Capacity Mapping Mission)

Launched in April, 1978 in a polar orbit, this satellite recorded data up to September 1980. The onboard radiometer measured the reflected solar radiation and the thermal emissivity of the surface between wavelengths of 10.4 and 12.5 μm, with a resolution of 600 m. Data acquisition was carried out from a subpolar orbit at 620 km altitude twice a day, at 12 h and 24 h. The data facilitated calculation for selected regions of temperature variation and thermal inertia, which furnished interesting geological information. The data acquired is available.

SEASAT

Launched in 1978, this satellite functioned for 4 months in an orbit at 770 km altitude, with a repetition cycle of 152 days. It had several onboard systems of measurement and acquisition of data, of which one corresponded to synthetic aperture radar (SAR). The SAR operated in a 23 cm wavelength with a resolution of 25 m and a ground coverage of 100 km width, along the track of the orbit. The limited data, especially over land, is available in digital form.

SIR A

This is one of the experiments of the space shuttle conducted in November 1981. In just 4 days 10 million square kilometres were covered by means of SAR, 23.5 cm wavelength, with a ground resolution of 40 m, at an angle of incidence of 50°. The limited data collected, especially over land, is available in analogue form.

This experiment was repeated in a second mission known as SIR B.

SPACELAB - Metric Camera

This was also a space shuttle flight conducted in 1983. The foremost instrument on board was the 'Metric Camera' operating in panchromatic and false-colour infra-red bands, incorporating stereoscopy. Resolution was 20 to 30 m.

The data of these experimental satellites is distributed in France by the GDTA at Toulouse.

1.5.1.2 Operational satellites

NOAA

The NOAA has launched a number of satellites for meteorological determinations over the last several years. The data acquired is also used in Earth sciences, however. The latest satellite in this series was launched in September 1986 into an orbit situated at 820 km altitude.

The AVHRR radiometer aboard the satellite provides for acquisition of data with a resolution of 1100 m in 4 spectral bands, such as visible (0.58-0.68 µm), near infra-red (0.725-1.0 µm), thermal (2 bands between 10.5 and 12.5 µm).

This data is very rarely used in geology because of its coarser resolution and for want of receiving stations. It is distributed in France by the Lannion meteorological station and the GDTA (SATMOS).

Plate 1 shows some of the geological features of France mapped from the AVHRR data.

LANDSAT 1-2-3

This represents a series of earth resource satellites. The first was launched in July 1972 by the USA under the name ERTS-1 (Earth Resource Technology Satellite) and revealed to the international scientific community the potential of remote-sensing applications.

Data is acquired from a 915-km orbit using a multi-spectral scanner (MSS) radiometer in four spectral bands: 0.5-0.6 µm (green), 0.6-0.7 µm (orange), 0.7-0.8 µm (red) and 0.8 µm (near infra-red). The resolution is 79 m and each scene measures a square 185 km per side or about 34,000 km^2. The repetition cycle is 18 days.

LANDSAT 4-5

These satellites were included in the preceding programme in July 1982 and March 1984 with different acquisition modes. These new space vehicles mark the beginning of the second generation of remote-sensing satellites characterised by two innovations, one spectral and the other spatial. The first refers to the greater capability for data acquisition in the infra-red mode and the second to an improved resolution of 30 m.

Orbiting at an altitude of 905 km the radiometer, named Thematic Mapper, provides data characteristic of seven spectral bands: 0.45-0.52 µm (blue), 0.52-0.60 µm (green), 0.63-0.69 µm (red), 0.76-0.90 µm (near infra-red), 1.55-1.75 µm and 2.08-2.35 µm (middle infra-red) and 10.4-12.5 µm (thermal). The resolution for the thermal band is 120 m and the repetition cycle 16 days.

All the Landsat data is distributed in France by the GDTA, Toulouse.

SPOT 1

The first non-American earth resource satellite, Spot 1 was launched by the CNES (France) in February 1986 in an 836-km orbit. It provides data in the panchromatic band with 10 m resolution and multiband data with a resolution of 20 m. Multiband imagery is acquired by an HRV radiometer in three spectral bands: 0.50-0.59 µm, (green), 0.61-0.68 μm (orangish-red) and 0.79-0.89 μm (near infra-red). Each image measures a square of 60 km per side or 3600 km^2 area. Panchromatic and multiband images can be obtained in the stereoscopic mode since the system is flexible, permitting an inclined orientation, and offers a theoretical repetitive cycle from 26 to 4 days on average.

Spots 1 and 2, the latter launched in January 1990, represent a technological advancement. The data is distributed in France by Spot Image (Toulouse).

1.5.1.3 Other satellites

MOS 1

Launched by the Japanese in February 1987 in a 909-km orbit, this satellite is primarily specialised for marine observations. Its repetitive cycle is 17 days. The MESSR system, one of the onboard radiometers, acquires data between 0.51 and 0.59 μm (green), at 0.61 μm (orange), between 0.72 and 0.80 μm (near infra-red) and 0.80-1.10 μm (near infra-red) with a resolution of 50 m over a surface area of 3600 km^2.

IRS

This is an Indian remote-sensing satellite launched in 1988 in a 904-km orbit with a repetitive cycle of 22 days. Its acquisition system consists of three 'scanning cameras', one LISS I and two LISS II, operating in four spectral bands: 0.45-0.52 μm (blue), 0.52-0.59 μm (green), 0.62-0.68 μm (red) and 0.71-0.86 μm (near infra-red). Its resolution is 73 m for the LISS I camera and 36.50 m for the LISS II camera. The ground swath width is 148 km for LISS I and 74 km for LISS II.

SOVIET SPACE SYSTEMS

The Russian space organisation, inherited from the Soviet regime, is quite complex and highly diversified. Until recently, mainly the Salyut programme of earth observation, commenced in November 1975 for meteorological purposes, was known. The orbiting station Salyut is occasionally visited by the space shuttle Soyuz.

Three major series exist for earth resource studies: 'Resource F-1', 'Resource 01' and Almaz.

Series 'Resource F-1': The sensors of this series (for photography) have been placed aboard the Cosmos satellites.

The first Cosmos satellite was launched in March 1962 and the last in July 1987 in a polar orbit. Some of the Cosmos satellites have a specific mission of earth observation: Cosmos 149 is equipped with a scanning telephotometer which allows measurement with a resolution of 6 m of radiations from the Earth in the visible, near infra-red and thermal bands between 8-12 μm and 10.5-11.5 μm. Cosmos 384 is mounted with a high-frequency radiometer specialised for snow studies. The precise characteristics of the data presently available on sale are as follows:

1. Kate 200 is a multiband camera allowing data acquisition in three spectral bands of 500-600, 600-700 and 700-850 nm. The format of the film is 180 × 180 mm with a scale factor of 1:1,000,000 and a resolution of 15 to 30 m. It has a longitudinal coverage of 60%.

2. MK 4 is a multiband camera with six spectral bands between 400-700, 460-505, 515-565, 580-800, 810-890 and 635-690 nm. The film format is 56 × 81 mm with a scale factor of 1:2,500,000 and a resolution of 15 to 30 m. The longitudinal coverage varies according to the spectral bands (20 to 80%).
3. Kate 140 is a camera with a single spectral band between 500 and 700 nm, with a resolution of 60 m. The format of the film is 180 × 180 mm with a scale factor of 1:1,500,000. The longitudinal coverage varies from 20 to 80%.
4. KFA 1000 is a colour camera which allows data acquisition in two spectral bands, between 570 and 680 nm and 670 and 810 nm. The resolution is 5 to 10 m. The format of the film is 300 × 300 mm with a scale factor of 1:250,000. The longitudinal coverage is 60%.

- Series 'Resource 0' (0 for digital)

The information acquisition system of this series consists of two multiband radiometers (MCY-3) with a spatial resolution of 45 m and two others (MCY-CK) with resolutions of 170 m and 600 m according to the spectral bands. It is placed in an orbit with a mean altitude of 650 km. Radiometers MCY-CK acquire 600 × 600-km images in the bands 0.5-0.6 μm, 0.6-0.7 μm and 0.8-0.11 μm, with a resolution of 170 m, and 10.4-12.6 μm with a resolution of 600 m.

The radiometers MCY-3 provide images of 90 km (2 × 45) × 45 km in bands 0.5-0.6 m, 0.6-0.7 μm and 0.8-0.9 μm with a spatial resolution of 45 m for one of the two systems and 27 m for the other in future (Resource 02).

The importance of this series undoubtedly lies in its multiresolution approach, which allows or will allow (Resource 02) in the visible band 170, 45, 27 m, and its thermal component which will be accessible day and night. The distributor of data is Planeta, Moscow.

- Almaz series

This is an SAR system placed in a low orbit of 300 km, operating in the S-band of 9.6 cm wavelength. Ground resolution is 10-15 m and the swath width 45 km.

Almaz 2 was launched in 1991 and furnishes digital information acquired over the entire globe.

ERS 1

This is an European remote-sensing satellite launched 17 July 1991 in a sun-synchronous orbit, with a repetitive cycle of 3 or 35 days.

The principal payload consists of two instruments: (i) an active microwave system, the AMI, operating with an angle of incidence of 23° in the C band of 5.6 cm, restores digital images with an average spatial resolution of 30 m and (ii) a radar altimeter.

At present, four receiving stations, viz., Kiruna (Sweden), Gatineau (Canada), Maspalomas (Spain) and Fucino (Italy) collect data in a real-time mode. Others are mostly operational in America and Asia.

CHINESE EARTH OBSERVATION SATELLITE

The purpose of this satellite is to acquire remote-sensing data to support the general inventory of resources in the national territory of China. It is presently in orbit and data is collected by means of a pressurised capsule separated from the satellite similar to a parachute.

The characteristics of this Earth observation satellite are as follows:

— altitude of orbit: 200 km;

— scanning panoramic view, angle of incidence 50°;

— scale factor: 1:220,000;

— ground resolution: 10 m;

— spectral range: 0.5 to 8.5 μm.

Geometric corrections are applied to the image after acquisition.

Wind and Cloud no. 1 is another Chinese meteorological satellite. Launched on 7 September 1988 in a geosynchronous orbit of the sun and situated at 900 km altitude, this satellite is similar to AVHRR, with a resolution of 1.1 km, five channels between 0.48-0.53 μm, 0.53-0.60 μm, 0.58-0.68 μm, 0.725-1.1 μm and 10.5-12.5 μm.

JAPANESE SATELLITE JERS-1

JERS-1, the Japanese Earth observation satellite, was launched in February 1992. It has a payload of a radiometer in the visible, near and mid infra-red bands and a radar.

The spectral characteristics of the radiometer, which comprises seven bands, are as follows (in μm): 0.52-0.60, 0.63-0.69, 0.76-0.86 (this channel generates stereoscopic data due to the specific configuration of the radiometer), 1.60-1.71, 2.01-2.12, 2.13-2.25 and 2.27-2.40. Spatial resolution, 18 μm.

The radar is of the L-band type of 23 cm wavelength and resolution of 18 m. Thus JERS-1 presents three major advantages: spectral bands responding to the principal minerals of hydrothermal alteration, concomitant stereoscopic view and simultaneous acquisition of visible and radar bands.

1.5.1.4 Some near-future satellites

RADARSAT

This is a Canadian satellite specialised in the active microwave domain. Projected launching, 1992-1994. The instrument is expected to have a frequency of 5.3 GH (C-band) and a swath width of 130 km. Resolution will be 28 m.

NEW GENERATION SPOT

Configuration of the onboard system is still under development; however, it has been more or less decided that it will incorporate at least two important characteristics:

— forward-backward stereoscopic acquisition and

— ground resolution of 5 m.

The NG Spot is expected to be launched around 1998.

1.5.2 How to Choose Remote-sensing Data?

The information contained in an image is spectral, spatial and temporal in nature. These different criteria determine the selection of data when choice is a possibility.

1.5.2.1 Spectral and spatial criteria

Considering the number of satellites in orbit, geologists have a wide scope for choosing the data best suited to resolve their various problems.

Several types of criteria will guide the selection. These criteria may be general-geographic, climatic, pertaining to the area of study and scale of operations. Other criteria concern the theme, type of application and exploration indicators. Finally, there are technological criteria related to adequacy of spectral and spatial characteristics of the available radiometers for the specific objectives of a given study.

Some simple, but not exclusive guidelines for the selection of data are given below.

- When the area to be studied is large and the scales of operation are of the order of 1:500,000 or 1:250,000, the Landsat MSSS data together with photointerpretation is the most suitable.
- When the area to be covered is small or medium and the scales of work of the order of 1:100,000, 1:50,000 and 1:25,000, it is reasonable to use the Spot XS data having a resolution of 20 m or TM data of 30 m resolution.
- If the type of investigation can take advantage of a large spectral range, extending from blue to thermal infra-red, the Landsat TM data is suitable. Depending on the nature of the objects under investigation, the specific characteristics of the spectral bands are as follows:
  - Band 1: 0.45-0.52 µm: penetration in water, material in suspension, differentiation between soil and vegetation.
  - Band 2: 0.52-0.60 µm: vigour of vegetation which is a secondary indicator of metal content in soils.
  - Band 3: 0.63-0.69 µm: structural analysis.
  - Band 4: 0.76-0.90 µm: stress in vegetation induced by geological phenomena. Detection of iron oxides.
  - Band 5: 1.55-1.75 µm: indication of water content in soils and in vegetation; mapping of clays, secondary indicators of hydrothermal alteration.
  - Band 6: 10.40-12.50 µm: discrimination between silicate and non-silicate rocks, determination of water content of soils, analysis of stress of vegetation.
  - Band 7: 2.08-2.35 µm: discrimination of rocks such as clays and micas in silicate formations, kaolinite and montmorillonite in soils.
- When the investigation requires a three-dimensional observation, the only present-day source of spatial information is the Spot satellite, which allows acquisition in a stereoscopic mode; however, mixed configurations such as Spot-Landsat can also be considered.

- When the study is localised in an equatorial or intensely clouded region, radar is the preferred instrument because of its ability to acquire data in any atmospheric condition. It allows the mapping of drainage networks, structure and micromorphological differences even under the difficult condition of dense forests. The radar wave is even more sensitive to micromorphological variations in arid regions, revealing differences in surficial formations on the basis of grain size and indicating geological phenomena even below a certain thickness of soil.
- When the phenomena to be studied are very localised and demand high spatial resolution with multiband data, three solutions can be recommended.
  - — Special airborne survey with one of the several radiometers available on a commercial basis.
  - — Special acquisition with combination of XS + P Spot.
  - — Combination of TM + Spot panchromatic or TM + Spot XS, choosing appropriate spectral bands.
- Finally, whenever necessary, aerial photography should always be included.

In all cases a precise costing must be done since ultimately the choice of data devolves upon this.

The approximate costs of acquisition of spatial data (1990) are given below.

Landsat MSS	34,000 km^2	6,000 F	(European zone)
Landsat TM	$\frac{1}{4}$ scene: 8500 km^2	13,000 F	(European zone)
entire scene:			
	34,000 km^2	35,000 F	(European zone)
Spot XS	3600 km^2	11,700 F	
Spot panchromatic	3600 km^2	14,000 F	
Spot panchromatic mode with programming	35,800 F		
IRC stereoscopic aerial photos, 1:50,000		15 F/km^2	(1988 price)
Cosmos: Price variable according to camera			
Example: 1 colour negative (KFA)		7,000 F	

Numerous calculations of costs based on area to be surveyed, number of spectral bands, spectral and spatial resolution (Cervelle, 1989) and digital parameters are available for comparison. However, quantification of certain important but undecipherable parameters is difficult, such as programming of data acquired and the likelihood that cost of data might be reduced per useful area for the user. This individual analysis, possible with Spot, is particularly important for those studies that require acquisition of data on specific dates under difficult meteorological conditions.

1.5.2.2 Temporal criteria

Given the feasibility of programming and accumulation of remote-sensing data representative of various climatic conditions, it is presently possible to choose from the available images those most suitable for a well-defined thematic problem.

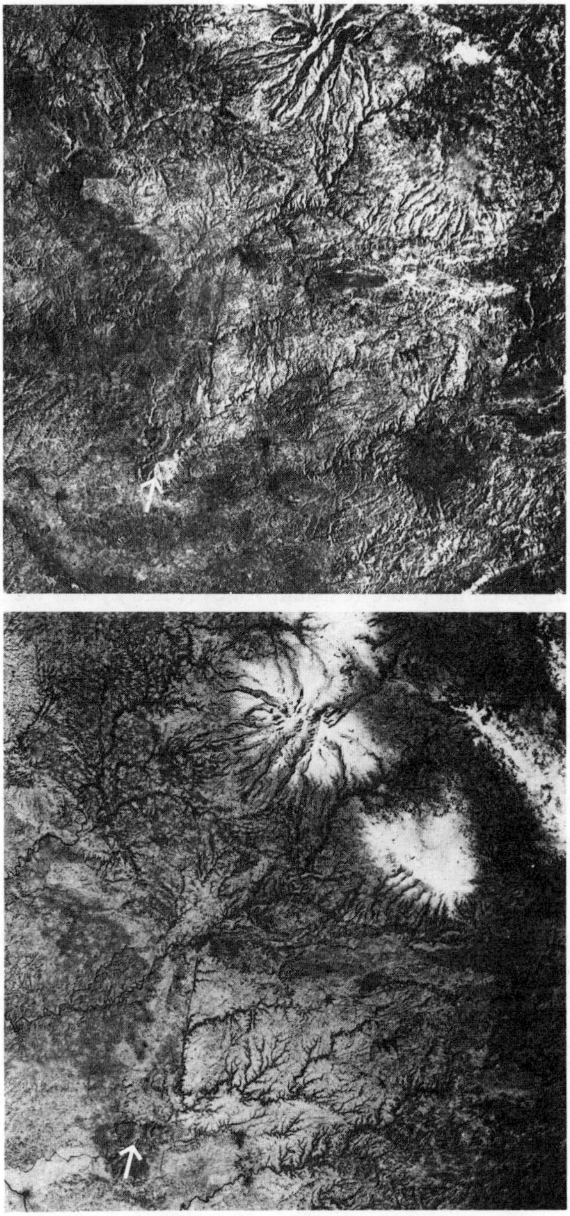

Fig. 2. Permian formation of Lodeve in northern Gaillac, confused with its Triassic core on Landsat MSS image of March 1973, distinctly revealed due to the effect of drought in September 1976.

Temporal information has a dual importance: it enables understanding the optimum radiometric response for a given region and provides scope for dynamic studies of the objects.

The radiometric qualities of the acquired data are of importance to all users. In Sahel (Africa) for example, even in the absence of cloud cover, several factors such as intensity of atmospheric haze, sand or laterite blows that obscure the terrain and dense forest fires which complicate responses, considerably affect restoration of the information desired.

Climatic conditions, which vary with seasons, also play a significant role in influencing the quality of data, particularly affecting the first and second-order features whose signatures are restored in the images. This effect, well known and very important in the field of agricultural studies, although not clear in the case of geological applications, is nevertheless indicated from experience and exerts a strong influence on methodology.

The effect of climatic conditions is notable in the case of geological mapping of regions of temperate climate wherein extreme conditions, such as prolonged drought or excessive rainfall exist. Some chosen examples in France during the drought period (1976) illustrate the importance of the images acquired under this condition over the Massif Central for mapping purposes (Fig. 2) or for hydrogeological applications in fissured terrain, in which moisture-resistant zones (Plate 2) and water-bearing fractures are discernible.

1.5.3 Restoration of Data

1.5.3.1 Selection of scale

Depending on the characteristics of acquisition systems and the themes of application, the user has a certain latitude in choosing the scale of data desired.

Using the Landsat MSS data, the structural geologist was initially contented with analysis of lineaments of structure of a region based on images in the scales 1:1,000,000 and 1:500,000, which facilitated a synoptic perception, indispensable in this type of investigation. The progress achieved from results of digital processing of multiband data enabled selection of scales 1:250,000 or 1:200,000 or even 1:100,000 in ideal cases. The features in these scales are perfectly compatible with reconnaissance maps and were used to accelerate the mapping of a number of countries during the 1980s. This effort was based on the application of rigorous technics of data restoration for interpretation and map preparation. These operations include high-quality geometric corrections on the basis of resampling and calculation of deformation models of the image data.

Recent improvements in the second-generation satellite systems, such as the Thematic Mapper of Landsat 5 (30 m resolution) and High Resolution Visible of Spot (10 and 20 m resolution) allow better performances. Scales to 1:100,000 and 1:50,000 meet the needs of detailed mapping and management well. Supported by appropriate digital processing, it is also possible to obtain the scale of 1:25,000 and even to 1:10,000 using the Spot panchromatic data which is of higher resolution (Fig. 3).

Conversion from one scale to another using different sensors subsequently formed the basis for a better understanding of remote sensing data, one closer to the realities on the ground. It has been observed in structural geological studies, for example, that a

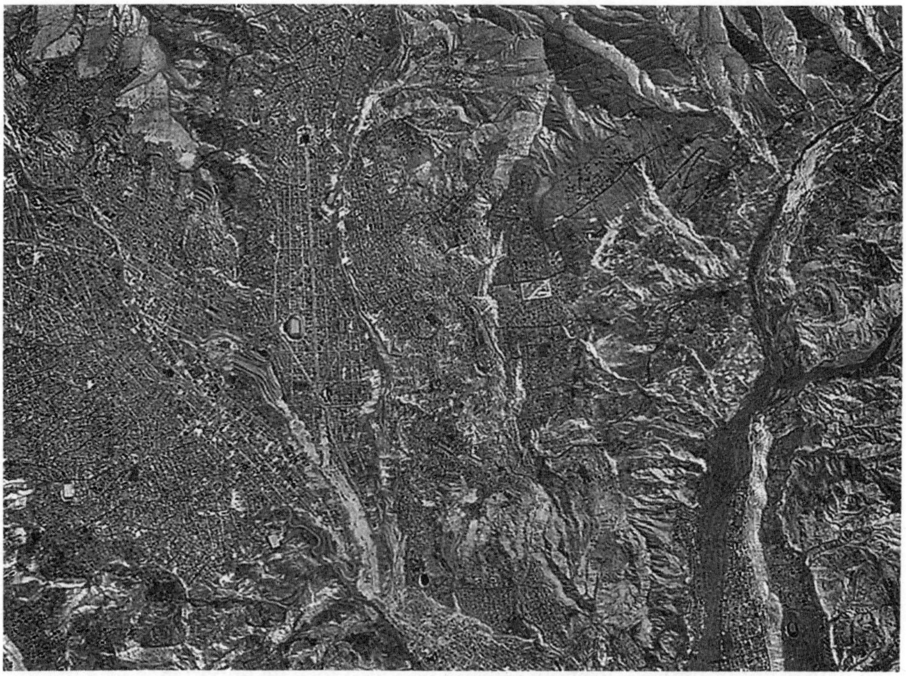

Fig. 3. City of La Paz, Bolivia; Spot view.
Restoration of a panchromatic image to 1:25,000 scale
(CNES, 1989. Distribution Spot Image. BRGM processing)

number of man-made features such as roads, high-tension lines etc., which are clearly visible and interpretable on Spot images, are not discernible in Landsat MSS images, thus eliminating the random nature earlier attributed to structural interpretation of linear discontinuities. Finally, this multiscale view of the ground allows application of fractal analysis, another mode of enhancing the information acquired.

1.5.3.2 Mode of data restoration

Restoration of remote-sensing data is presently carried out in two modes, i.e., a photographic mode which faithfully records the landscape and allows visual interpretation in optimum conditions, and a classification type of mode, resulting from digital processes, most often based on spectral criteria.

Due to the tradition of photointerpretation, 'photogeology' has strongly influenced geologists, especially since the results of digital methods are still inadequate or, leastwise, heterogeneous. Hence remote-sensing analysts still prefer the photographic restoration methods even in space geology.

However, the new subjects encompassed by remote sensing, such as geobotany and study of alterations, appearance of new sensors such as spectral imagers and thermal probes, which call for the use of digital technics, the newly emerging geographic information systems and especially multicriteria analysis-all these constitute new developments in this field. These developments are quite sensitive to the improvements in spectral classification and textural characterisation. Further, taking into account the progress yet to be made to perfect the transmission of information, it would be premature to offer an evaluation of the improvements that will flow from the employment of expert systems and digital methods for analysing image morphology.

It may nevertheless be said that these developments will play an important role in the restoration of data.

1.5.4 Methods of Interpretation

The methods of information extraction and investigation from remote-sensing data are derived from integrated photointerpretation technics in the case of visual approaches and from analytical interpretation in the case of digital approaches. They are supported by stereoscopic analysis technics wherever possible.

The visual methods are presently recognised as superior in geology since lithology and tectonics cannot usually be observed directly on the images because of soil and vegetation cover, and are interpreted through the help of second-order indicators by the combined action of eye and brain which simultaneously process information of diverse nature in the images. The result of interpretation is presented in graphic form, which may be digitised if necessary. However, while the digital format of remote-sensing data is one of the major characteristics of this source of information, it may be stated that beyond processing for improving visualisation, the original structure of the data is not optimised. Remote sensing will also lose its importance as a mapping tool in time if appropriate methodology is not developed. A decisive approach to this problem is necessary for several reasons, such as increase in number of spectral channels, for example 228 in the Earth Observation Satellite (EOS) programme to be launched in 1995, and improvement in spatial resolution to 5 m under development for Spot, which might lead to 'blurred images' (Durand, 1986) that would be more difficult to interpret visually. One might add development of combination systems and geographic information systems and their orientation towards the preparation of decision-making maps, for which remote sensing has a proper place.

Certain aspects of the foregoing can be summarised from recent publications (Flouzat, 1983, 1988). As a general rule, the stratiform configuration of the Earth's surface and traditional partitioning into different disciplines have restricted mapping to a single type of information without attempting to play with the effects mutually induced by one on the other, which represent a definite value added. This shortcoming is partially rectified in synergistic photointerpretation, wherein the image is fully exploited by invoking knowledge external to the theme studied, on the basis of combinations that may not always be explicit or formalised in logical terms. But the latter are rarely adaptable to digital classification methods, which thus remain imperfect. It is better to await an analytical interpretation technic based on the concept of image considered as a space view described by its specific characteristics, without appealing to external information. The combination and distribution of these descriptive parameters would then form the entities and image structures to be interpreted through comparison with external knowledge.

The methodological basis for analytical interpretation can be described by three components.

— Analysis of the problem: Identification of the relations that exist between external information concerning the phenomenon to be investigated and its representation, direct or induced, on the image.

— Description of characteristics or attributes of the object studied, that are apparent in the image, and their spectral response. These consist of spatial attributes, such as textural or geometric characteristics, and spectral attributes. This description allows deciphering of textural elements in the image which enables generation of a continuous scene, simplifying the image into homogeneous entities. Two major groups of elements generally appear, viz., random elements and linear elements.

The above two stages facilitate characterisation of image structures on the basis of external knowledge, which emerges from the know-how and experience of the interpreter.

— Modelling: This allows determination of the nature and organisation of digital processing necessary for quantitative extraction of the information indicated by the visual analysis.

This formulation places the experience gained, mainly from synergistic photointerpretation, and recent research in their proper perspective. Though obtained by different methods, experience, if considered from the new logic, represents an appreciable contribution to description of the nature of the objects in the image, both from a spatial as well as spectral point of view.

Proceeding along this constructive path, the thematic plan adopted in this book describes the specific image structures as defined by other methods, but the principal characteristics can be identified as follows:

— Spectral response, which determines the shade or colour of the objects seen, directly or artificially.

— Morphology of image visualised in two or three dimensions, which refers to the form of objects and their dominant mode, geometric for example, their type, plain or cellular, their size, their orientation and their distribution.

— Texture, defined by the mode of arrangement of the objects and their local variability, such as homogeneity, granularity, mixing, local contrast, amplitude and local density.

The role of the above in each response will be discussed in so far as possible in the chapters that follow.

1.5.5 Methods of Evaluating Interpretation

These are the methods that enable us to determine the geological significance of the information extracted from interpretation of remote-sensing data.

These methods pertain to concepts regarding topographic forms, utilisation of existing data, strategy for ground follow-up, i.e., terrain-image-terrain correspondence, and finally synergistic studies.

1.5.5.1 Photogrammetric aspects

Experience in recent years has shown that the scale of presentation, quality of supporting data and accuracy of restoration influence the validity of interpretation of remote-sensing data. Attempts are made to improve these aspects by generalising the processes of geometric corrections of digital data on the one hand, while systematising the utilisation of computer-aided mapping on the other. This second approach facilitates:

- Execution of interpretation.
- Correction of certain geometric deformations, thereby enabling interpretaion that accords with a given topographic base, using an appropriate programme.
- Outputs of maps of any scale on a given topographic base, for an accurate area of the ground, in black and white or in colour, with or without plotting, on transparencies or on paper. As the maps obtained take into account (through judicious graphic choice) the variations associated with deviations between the resolution of the sensor and the scale of restoration, geologists can precisely locate observations from the image, investigate their ground truth and assess the geological significance.

This approach is further improved if the study is conducted in a stereoscopic mode, calculated from a digital model of the terrain. Combined with the image this enables preparation of a new dataset, i.e., an orthophoto and a virtual complementary image calculated according to the optimum angle of view. Thus the interpretation originates from a geographic reference map compatible with the topographic maps.

1.5.5.2 Evaluation through available database

To optimise preparation of a map product from remote-sensing data, it is necessary to base interpretation on earlier (archival) data. This requirement, among others, justifies the creation of geographic information systems. Incorporating earlier data in the interpretation of data from remote sensing, either visual or digital, serves important functions:

- Optimising the preparatory map product by transforming it from a sketchy state to a semi-finished product, incorporating the information already authenticated on the ground.
- Verifying certain specific points.

Evaluation involves superimposing interactive modes on the archival map and the corresponding digital image, in order to authenticate a boundary in real time, correcting it through extrapolation or interpolation as required. Computer assistance is therefore essential for a true (validated) photointerpretation.

Evaluation also involves employment of combined systems that facilitate obtaining a more thematic value, defining and verifying the anomalies that have a potential importance for exploration of subsurface resources. In fact, the task of the remote-sensing analyst, in accordance with his mapping procedures, is to detect anomalies in the landscape he is analysing. Very often, relying on the principle of analogy, he assumes that the anomaly represents an indicator of resources. In reality, analogy alone does not suffice to establish an anomaly as an indicator; it might well be a red herring. Thus verification of anomalies is as necessary and as urgent as their detection; often their identification is less important. In this context programs of integration of multisource data, which characterise combination systems, constitute an instrument for qualifying or disqualifying remote-sensing anomalies.

1.5.5.3 Ground surveys—remote sensing and metrology

For some years the procedures for evaluation of remote-sensing data consisted of searching during conventional mapping operations in the field for significant elements that could explain the image features observed. This approach often led to an impasse.

Of late, the approach has reversed and special surveys are carried out to investigate only the observations drawn from remote-sensing data whose geological significance is presumed. This kind of field work utilises image criteria for conducting verification. Besides conventional geological observations, it may also rely on the measurement of certain physical properties or the study of significant relations between different components of the landscape. This approach proceeds in the direction of modern development in metrology, which is gradually being affirmed as a mapping technic.

Those physical properties which form the basis of remote-sensing methods are normally studied. Some are presently measurable, such as reflectance, temperature, roughness, emissivity and dielectric constant. These parameters were measured in the past mostly to guide the choice of spectral bands of onboard radiometers, to orient the digital processing of data or to verify the validity of certain hypotheses. For example, it was from laboratory measurements of reflectance variations in plants located on soils somewhat enriched in various minerals, such as lead, copper, arsenic etc., that the existence of metal stress was demonstrated (see Sec. 4.6).

But the uniqueness of the proposed method lies in attempting to understand and explain by *in situ* measurements the anomalous changes observed in remote-sensing data. For example, when a temperature anomaly is observed on an image, thermal measurements are taken in the soils and boreholes drilled to estimate the water content of surficial formations or the rate and nature of gaseous emanations, to establish the phenomena which induce them and to propose a geological explanation. Over time and under certain conditions, this approach could develop into a true prospecting tool on the ground, as is already the case with portable ground radars (Peragallo, 1978).

Investigation of the significant relations between various components of the landscape proceeds from a more naturalistic approach. It consists of explaining the variations in an image by detecting characteristic associations that largely imply geological causes, for example the associations of lithology-fracturing-vegetation, lithology-surficial cover-vegetation, altered lithology-water content, soil type-lithology etc. Each type of association defines a typical indicator (Dutartre and King, 1986).

1.5.5.4 Correspondence between terrain and imagery

The efficacy of remote sensing depends on the degree of its inclusion in the field operations of geological investigations and adherence of the geologist to it, while preserving the objectivity of exploration in a virgin area.

It is thus a point of methodology since apparently confusion exists between spontaneous analysis and supervised analysis. The former consists in identifying without embellishment the morphological characteristics of a region and depends on the experience of the remote-sensing analyst. This may form the basis for a new working hypothesis. The latter is based on prior knowledge of the terrain targeted for exploration and whose mineral potential is also known.

To preserve the originality of a discovery in a region for which exploration is launched by optimising all the available sources of information, analysis of remote-sensing data is best adopted, using the correspondence between ground truth and image, facilitated by interactive systems. This approach should be iterative and interactive and leads to inclusion of remote sensing in the logical sequence of exploration methods. The procedure for this approach can be outlined in six steps:

- Preparation of a basic image in the form of a colour composite suitable for lithological mapping or in a derived form which aids in the interpretation of lineaments.
- Spontaneous interpretation based on a first-order analysis. This is the first objective vision of a region before its exploration. It should be supported by existing data, without losing its objectivity, for improving the geological perception of the map.
- First phase of exploration, involving analysis of remote-sensing data with proper optimisation of the technic. This facilitates inclusion of remote sensing in the logical sequence of exploration methods.
- Supervised interactive analysis. This means investigating the specific responses of the components observed on the ground, improving their perception by digital processing technics, and extending the result to the entire zone of exploration. It involves all the concerned specialists-structural geologists, field geologists, hydrogeologists, economic geologists etc. It may lead to 'geological' classifications that are entirely digital or to an oriented interpretation based on a second-order, more detailed analysis, centred on definite targets and directed towards determining the geological importance of the image features, without being able to identify them with certitude. This interpretation, which may be supported by modelling, is subsequently digitised.
- Multicriteria analysis, a logical follow-up of the preceding phase. It relies on the study of correlations, complementary nature and synergistics between remote-sensing data and the results obtained from other technics, such as geological, structural, geochemical, geophysical, metallogenic, geomorphological, soil mapping etc. Based on digital processing of different variables, it allows, for example, selection of priority zones for planned prospecting.
- Synthesis of maps based on synergistic studies. This stage-wise sequence makes it possible to obtain a map product from remote-sensing data, whose perfection depends on the occurrence of rock outcrops and quality of existing data.

However, other types of programmes have been employed by the geologists of BRGM, utilising different conditions of study, such as technical, economic and geographic.

Thus, in Oman, the most recent example (see Chapter 2), there was no systematic interpretation of Spot images obtained for conducting the project prior to field surveys.

1.5.6 Use of Remote Sensing in Geological Exploration

The principle of utilisation of remote sensing for geological exploration is essentially based on analysis of the landscape using methods inherited from photogeology. It enables us to distinguish different geological objects and to detect certain peculiarities.

Depending on the condition of outcropping of the resources explored, the responses identified by various means of analysis may be of the following types:

- lithological,
- structural,
- morphological,
- mineralogical,
- geobotanical.

These are analysed in the following chapters, starting with studies that enable preparation of preliminary reconnaissance maps, improvement of knowledge about mineral potential, demarcating the areas for exploration or refining the targets, or discovering new indicators or new deposits.

Chapter 2

GEOLOGICAL MAPPING AND REMOTE SENSING

Examples of geological mapping, based to various degrees on digital processing and interpretation of remote-sensing data from satellites, are presented and analysed. The examples were selected from all latitudes and thus represent different conditions of outcrop observation. The relevant studies are based on a choice of data acquired by different satellites, such as Landsat MSS, Landsat TM and Spot, to enhance their specific characteristics: area covered, spectral bands, resolution and in particular stereoscopy, which is rediscovered since it appears to be essential for the geologist.

The advantages and limits of remote sensing are revealed by this presentation of different-scale mapping surveys: composite maps, exploration maps and large-scale (1:50,000, 1:100,000) survey maps. The problems involved in integrating remote sensing into the logical sequence of mapping methods are defined in relation to the type of map, the mapping concept, the type of rocks and environmental conditions.

2.1 INTRODUCTION

Geological mapping constitutes the basis for any surface or subsurface prospecting. It is also required for studies oriented towards civil engineering and natural hazards. Identification of rock types, understanding the mechanisms of emplacement and the phenomena that have affected them, dating of rocks and establishing the chronology of geological events are the principal tasks of mapping. Geological mapping is accomplished using various technics and different scales, viz., microscopic, ground, space or subsurface exploration, drilling and geophysical methods.

Remote sensing a priori offers various possibilities, such as discrimination, demarcation, mapping of landforms, their identification and quantification etc., but not all of them are presently operational with the same degree of efficiency, due in particular to the occurrence of rock outcrops.

The methods of geological mapping by remote sensing can be classified into three groups: photointerpretation of images, digital/spectral analysis and computer-aided 3-D photointerpretation. They are not presently equal in level of efficacy and have not attained the same degree of operation.

Photointerpretation of images

Depending on the occurrence of rock outcrops and the extent of soil cover, vegetation and surficial deposits, two methods can be defined for extraction of geological information from visual interpretation of images.

— Direct reading method based on the thematic comprehension of the interpreter who compares the objects observed on the image with the geological models familiar to him from experience.

— General interpretation method, which consists of an integral analysis of the basic attributes—reflectance, landform and texture. Based on the experience of the interpreter, this method enables differentiation of morpho-landscapes whose geological nature is presumed.

Image interpretation is optimised in both methods by interactive application of digital processing technics based on specific algorithms. The use of stereoscopy, introduced in the Spot generation, improves the perception of forms and enhances the quality of maps.

Digital image processing

In spite of improvements in the digital image processing algorithms, the mapping technic gives results that are inferior to those obtained from photointerpretation. This is particularly so in the case of absence of rock outcrops since classification programmes, which are sensitive only to spectral variations, remain ineffective for textural analysis. The results are better in exposed regions but nevertheless hindered by the oxidation cover which frequently homogenises the responses. Finally, the methods of extraction or identification of forms developed from digital processing technics, require further visual interpretation to become operational.

Computer-aided photointerpretation

This method, presently under development, combines the knowledge and experience of the photointerpreter with the expertise of information technology, facilitating the preparation of essentially digital maps. It is based on the acquisition of stereoscopic data, application of digital processing technics, geomorphological interpretation on an interactive screen and direct evaluation by existing databases. Handicapped earlier by its analogue form, which required cumbersome and constrained transformation for quality, photointerpretation has regained its importance in problems that can be processed by simpler digital formulations.

Whatever be the method employed, remote sensing makes possible the preparation of maps of different scales for different purposes. Remote sensing orients fieldwork preparation, aids homogenisation of maps drawn at different times by different persons, varying in perception of geological history, extrapolation and interpolation of boundaries of formations, verification and marking of observations.

Examples of map preparations by computer-aided interpretation cover large areas. They are representative of varied climatic and outcrop conditions, which, depending on the nature of the predominant rocks and the tectonics affecting them, influence the quality and utility of the results.

The cases presented in this chapter focus on the role of remote sensing in various stages of geological mapping, as applied to prospecting for subsurface resources. The results obtained in preparing maps of different scales, i.e., composite maps, reconnaissance maps and 1:50,000 maps prepared by means of various remote-sensing data for regions of highly diverse outcrop conditions are evaluated.

2.2 COMPOSITE MAPS

Composite maps are small-scale regional geological maps compiled from various sources. Such geological maps are the documents used for orienting exploration of subsurface resources. They aid in preparation of the mineral plan of a country. Several countries have been covered by such maps, which vary in usefulness, especially those that are very old. Prior to space-borne remote sensing, these maps were prepared from the data available and subjective interpolations. Since 1972 Landsat data, whose scale is perfectly adaptable to this type of mapping, has enabled preparation of maps in which the shortcomings of field observations are eliminated by the extrapolation and interpolation method based on morphological analysis.

One can cite as an example the map prepared by the Geological Survey of Sudan in collaboration with the BRGM (1981). This was the first time that a composite map of such a vast country (extent approximately 3,000,000 km^2) was prepared by means of space images (Landsat). The remote-sensing technic, supported by existing geological data of different scales, contributed greatly to the production of a high-quality map. Without this technic, progress beyond the level of the old composite map would not have been possible.

Maps of this type have also been prepared for Mozambique, Zaire, Togo, Madagascar etc., largely supported by remote-sensing data. In the case of equatorial regions, e.g., Gabon, Brazil, composite maps have been produced using radar data (Radam project). Small-scale geological mapping in Africa has witnessed significant improvements over the last 10 years under the impetus of remote sensing.

2.3 RECONNAISSANCE MAPS AND REGULAR GEOLOGICAL MAPS

Production of reconnaissance maps and regular geological maps, of scales varying between 1:200,000 and 1:250,000, is the objective of the second phase of mineral or petroleum exploration. Aerial photography has long been the remote-sensing technic most widely used in the preparation of these maps, according to an itinerary commencing from sedimentary terrains of arid regions to basement formations in all climes and latitudes. Now the possibilities have been enlarged and a number of regular mapping programmes are supported by space remote-sensing data, used alone or together with aerial photographs.

Some examples, selected because unique, are described below.

2.3.1 Geological Mapping and Mineral Exploration of Gabon Craton and Its Environs—Exploration across Borders

These projects of the Directorate of Mines and Geology of Cameroon, Equatorial Guinea and Gabon respectively were co-ordinated by the BRGM (France).

The projects envisaged reconnaissance mineral mapping around the Gabon Craton, making use of the geological and geochemical prospecting data available in the three African countries, namely Cameroon, Equatorial Guinea and Gabon. The programmes were carried out autonomously in view of the sovereignty of each country but the target of exploration was the same, namely the craton and its environs, which extend over the three countries. Remote sensing has since become a major tool in such programmes since crossing borders poses no obstacle.

In southern Cameroon, despite a vast equatorial forest, photogeology constituted the base for preparing 1:200,000 reconnaissance maps (Scanvic, 1980). These maps were then reinterpreted on the basis of ground observations and used to update the old 1:500,000 geological maps. The importance of circular structures superposed on magnetic anomalies and associated with abnormal concentrations of Ni, Cr and Co in terms of economic mineral deposits could then be reassessed. Analysis of microwave data provided by the space shuttle of the SIR A mission enabled re-finement of certain aspects of these maps (Champetier de Ribes and Scanvic, 1983).

In Equatorial Guinea (Brohan, 1981), in the absence of aerial photographs, ground surveys were preceded by an analysis of Landsat images of average quality and by interpretation of radar data (Motorola), ensuring total coverage of the country along two flights of complementary orientation. Supported by ground observations, interpretation of the remote-sensing data facilitated preparation of a geological map that is entirely different from the older maps. With regard to mineral exploration, the radar images played a significant role in covering this region for the first time in spite of a thick cloud cover, and in demonstrating the continuation of certain formations known in Cameroon into Equatorial Guinea. This study refers in particular to the iron-bearing formations whose regional structure was revealed from remote-sensing data. This structure, the result of tectonic activity, the effects of which can be observed from the north-west at the border with Cameroon up to the south-east at the Gabon border and even beyond, confirms the earlier suspected continuity across the three countries. The study of radar images also led to the discovery of diapirs in the adjacent sedimentary terrain, which until then were not known to petroleum geologists.

Ground surveys were preceded by geological mapping in Gabon using the existing aerial photos and radar data ('Goodyear') in the X band of 3 cm wavelength, recorded throughout the territory under the aegis of the Sodeteg society (Baelz-Manière and Gilli, 1988). The BRGM study revealed the importance of these maps in mineral investigations at different levels, which included orientation surveys in regions of difficult access in which altered formations predominate, mapping iron-bearing series and detection of NNW faults which control the emplacement of kimberlites. Finally, radar observations aided the choice of geochemical sampling methods locally as this data distinctly revealed zones of intensive peneplanation in which appropriate methods of sampling and analysis had to be adopted.

2.3.2 Mineral Exploration in Northern Mozambique—Importance of Mapping Support and Multicriteria Mapping

This project was carried out by the BRGM for the National Directorate of Geology of the People's Republic of Mozambique. The province studied extends over 140,000 km^2 and was covered by 7000 aerial photos but only 7 complete or partial Landsat MSS images. The special feature of this project was that, in the absence of topographic maps, an efficient alternative method had to be proposed for a rapid and economic survey, which would concomitantly ensure precision compatible with geophysical data (radiometrics and magnetics) and with aerial photos. As remote sensing satisfies these demands, the Compagnie Générale de Géophysique (France) prepared a controlled mosaic of Landsat images supported on a UTM projection and presented in the scale of 1:100,000 or a regional basis.

The quality and number of planimetric details visible on the Landsat MSS images and the precision of the method employed, the lack of regular topographic maps notwithstanding, ensured good integration of geophysical and photogeological data. This expedited ground verification of anomalies, control of geological surveys and preparation of geological maps which were subsequently preserved in the databank.

A second project was undertaken in 1983 in the northernmost part of the country, a virgin area for systematic survey. Its significance lay in demonstrating how the use of remote sensing could facilitate optimisation of geological mapping, in other words how a photogeological map prepared from aerial photographs and Landsat MSS images enables planning of further programmes, reduces duration of ground operations and expedites correlation of geophysical anomalies with morphological features. Thus geological surveys actually programmed on the basis of remote-sensing data enabled verification of the actuality of 'morphological' groups defined by remote sensing, assessment of their geological nature and determination of the ultimate significance of geophysical anomalies on the ground.

During the operation of this project the interpreted results of aerial data were compared with those of space remote sensing. These interpretations, each with its own specific characteristics, underscored the importance of certain geological and structural details, i.e., better perception of large units and their spatial evolution, categorisation of major faults etc. Their integration with radiometric and aeromagnetic maps, prepared on a common reference of Landsat MSS mosaics, revealed interesting correlations with precise locations on the ground. The accuracy of maps, in conformity with the estimates, was of the order of 200 to 300 m and all the anomalies were rapidly located on the ground.

Photological interpretation and analysis of Landsat MSS images have been used for planning geophysical surveys and verification of anomalies and, accordingly, for preliminary demarcation of the zones for surface prospecting based on certain criteria (Pinna, 1983, Pinna *et al.*, 1988), namely:

- favourable geological contacts,
- extended structures of other anomalies,
- circular structures,
- dykes or veins.

They could also be used during subsequent preparation and definition of programmes for detailed prospecting, focussed on multiple geophysical anomalies grouped in a narrow zone, which present significant photogeological characteristics. Interpretation of remote-sensing data has already constituted the basis for presentation of maps in different scales, 1:250,000 and 1:500,000.

In conclusion, completion of the Mozambique project, as detailed above, demonstrated the role of remote sensing in the planimetric study, for which it represented a unique and original solution in geological investigations, in which its real efficacy was established by the reduction, as envisaged, in groundwork without jeopardising the quality of mapping; and in mineral investigations in which it proved valuable along with geophysical data in the selection of areas for detailed prospecting. So it is now possible to integrate various databases, an operation which in the case of prospecting alkaline rocks for example, enables us to focus on small areas.

These objectives were achieved in Mozambique because the results obtained from remote sensing were compatible with its potential due to the fact that use of the technic was conscientiously planned during the project and the end-goals clearly stated.

2.3.3 Mineral Prospecting in the Red Sea Hills Province, Republic of Sudan—Role of Remote Sensing in Exploration of Arid Regions

Between 1978 and 1982 the BRGM carried out geological and mineral exploration in the Red Sea Hills province in the north-eastern part of Sudan for the Geological Service of Sudan. This programme was preceded by the preparation of a compiled geological map in the scale of 1:2,000,000 for the entire country of 3 million km^2, based on earlier data and interpretation of nearly 100 Landsat MSS images, which led to the selection of this province for launching prospecting work.

Geological mapping in the 1:250,000 scale of the Red Sea Hills was based on the preparation of photogeological maps for those zones for which aerial photographs were available. Ground surveys supported by these maps indicated the presence of a volcanosedimentary sequence, the Ariab sequence, which looked promising as a metallogenic province, especially for gold and copper mineralisations. Space remote sensing was then employed to demarcate the extension of the Ariab sequence throughout the Red Sea Hills. A mosaic of black and white Landsat images, Channel 7, was prepared in the 1:500,000 scale. Interpretation of this mosaic, supported by the mapping observations of the first phase of fieldwork, facilitated demarcation of the extension of this sequence up to the Red Sea, a zone of major faults which affect the sequence and complicate its mapping on the ground. This interpretation confirmed the continuation of the sequence on either side of the Red Sea, with identical formations known in Saudi Arabia.

Application of this group of methods had several advantageous economic consequences. It quickly reduced the area to be mapped by accurately tracing the extension of the metallotect and enabled us to modify the prospecting strategy by focussing closer to the sea coasts and slopes. The overall time duration between the beginning of reconnaissance mapping and the discovery of the deposit was reduced by half.

Subsequently, the methodology was improved and the surveys in this region were planned based on the digital analysis of Landsat MSS data with the interactive participation of field geologists. The maps resulting from this work presented in the 1:250,000 scale were used for the ground surveys. They facilitated extrapolations between the points of observation and location of controlling zones or facies, which saved more time for surveys and prospecting operations. These maps place the various sequences and facies in their regional context, refine observations made on the ground at isolated sites, and indicate the extension of geological groups, units and complexes outside the area of investigation or beyond the zone of access. The gossans were mostly identified by their shape rather than spectral response, since iron is present everywhere as an oxidation layer.

The application of remote sensing in this previously unexplored region resulted in the discovery a few years later of a copper-bearing belt containing a number of economic sulphide accumulations. Reserves of gold have been estimated and a deposit opened for exploitation at Hassai. The geological and structural environment of the prospected area is now well understood but still continues to be the target of study since management of the exploitation process necessitates additional investigation, particularly to ensure water supply to the processing plant. In view of the specific geological and climatic conditions, viz., acidic and basic igneous rocks, gossans, hydrothermal alterations, desert covers and temporary water flows in wadi, the Thematic Mapper data of Landsat V, which has a resolution of 30 m and provides access to mid and thermal infra-red spectral bands, was preferred for this unique study. Oriented towards the development of methodology, this study led to the following precise conclusions in a regional context:

- Utility of mid-infra-red spectral band: In arid regions, where the degree of outcropping is high, this band enhances the possibility of lithological discrimination based on the sole criterion of reflectance, and the water content strongly influences certain responses in the 1.52 μm band.
- Thermal response. Free of the effect of oxidation cover characteristic of rocks in arid regions, its discriminating capacity is greatly improved.
- Various photographic restoration technics, such as false colour (4, 3, 2) mid-infra-red (7, 4, 2), thermal infra-red (6, 7, 4) and principal component analysis, indicate that the discrimination possibility can be increased if the spectral criteria and textural criteria are combined, as is conventionally done in photointerpretation. This is the case, for example, of acidic and basic varieties of volcanosedimentary formations which have a similar spectral response but differ in texture. The acidic horizons are elongated because their schistosity is well developed, whereas the basic horizons are thick.
- Application in Hydrogeology: The importance of remote sensing (also see Sec. 3.4.3) lies in lithological and structural studies and in detection of groundwater movement in wadis. The latter is clearly visible in infra-red, especially thermal infra-red images. Investigated in detail on the ground by structural analysis, these observations facilitated site locations and the boreholes drilled yielded highly positive results (Berard *et al.*, 1990).

2.3.4 Preparation for Geochemical Prospecting in Miskah Province, Saudi Arabia—Problem of Surficial Deposits

This project was carried out by the BRGM for the Ministry of Petroleum and Mineral Resources (Fourniguet and Motti, 1983).

The project was unique in terms of objective, namely to define the characteristics of the surficial deposits for undertaking geochemical prospecting, and in methods of investigation. It involved development of a new approach based on the study of quaternary geology supported by remote-sensing data.

One of the problems faced in geochemical prospecting was that the area of dispersion of elements or detectable traces is strongly influenced by morphoclimatic conditions. The methodology employed had essentially to negate the effects of this dilution in order to enlarge the sampling grid without obliterating the results. In desert terrains the study of surficial deposits is a means of solving these problems and this project demonstrated how remote sensing optimises the achievement of this objective.

The study proceeded in three phases:

— Processing of Landsat images optimised for the interpretation of surficial deposits (1:250,000 scale). These images were obtained from a selection of processing technics on the interactive console. The interpretation resulted in a map consisting of a drainage network represented up to the 5th order and photofeatures representing 16 units or complexes.

— Ground verification carried out by a helicopter mission.

— Collation of ground data and preliminary interpretation, supported by new simulations on the interactive system, and preparation of a final map consisting of 29 units or complexes.

This type of map made possible the selection of deposits to be explored depending on the nature of their occurrence: allochthon, autochthon, area of influence etc. Its reliability proved good in arid regions where the method becomes operational when the integrity of correspondence between groundwork and image is maintained.

2.3.5 Regular Geological Maps

These maps are termed 'regular' because prepared from a strategy of mapping an entire country according to more rigid norms than those adopted for reconnaissance mineral maps. The two examples presented below are representative of the two quite different methodologies employed in a semi-arid zone.

2.3.5.1 Mapping in Saudi Arabia

This involved regular geological mapping of the eastern sedimentary margin of the Arabian Shield. Map scale: 1:250,000.

Preparation of the preliminary maps and their editing was based on Landsat MSS data which, under the situation of occurrence of rock outcrops and climatic conditions of Saudi Arabia, are considered quite high quality, enabling preparation of exceptionally large-scale maps (1:100,000).

Since 1979, 20 maps of this type have been prepared by the geologists of BRGM for the Saudi Ministry of Petroleum.

Methodology was based on the combined interpretation of aerial photographs and Landsat images in a preliminary phase, preparatory to ground surveys.

This interpretation guided the ground surveys conducted by a remote sensing geologist, a stratigrapher and a sedimentologist, who together prepared the lithostratigraphic sections suitable for use in a regular map.

Finally, the sections and point observations were compiled after analysis of remote-sensing data and the map prepared from Landsat images.

We shall consider three important aspects of this example for the quality of results: utilisation of stereoscopic data, systematic and preliminary photointerpretation and integration of remote-sensing analysis with ground mapping.

2.3.5.2 Mapping of the Sultanate of Oman

Systematic geological mapping of the entire country was done in the scale of 1:250,000. One hundred space maps prepared from Spot XS data acquired under different meteorological conditions constituted the base for the preparation of maps and their finalisation.

Another 28 maps have been prepared for the Ministry of Petroleum and Minerals by the geologists of BRGM since 1990.

Methodology was based on a group of optimised, integrated approaches, which included aerial photographs, Spot images of 1:100,000 scale, heli-borne surveys and location by space navigation using the GPS system.

Initially, as in Saudi Arabia, a team of geological specialists (stratigraphers, sedimentologists and remote-sensing analysts) mapped the reference profiles without preliminary photointerpretation during helicopter missions, for which navigation was ensured through Spot images and ground control by means of space geodetic observations. These reference sections later served as the basis for extrapolation of lithostratigraphic boundaries that were interpreted from aerial photos and stereoscopic Spot images.

This methodology was observed to be especially suitable for mapping vast inaccessible territories with a simple structural set-up. In certain complex zones the efficacy of remote sensing was found to be low and the number of ground observation points had to be increased in order to ensure precise mapping of lithostratigraphic boundaries.

The obvious operational advantages of this methodology can be identified as optimisation of helicopter missions by the double navigational approach—Spot images and GPS; usage of stereoscopic data; integration of remote-sensing analysis with ground observations; and selective and *a posteriori* interpretation of images for the purpose of contouring.

2.4 LARGE-SCALE MAPS

In view of the poor resolution of multispectral radiometers in Landsat 1 to 3, this type of space data was generally not used for the preparation of detailed maps, except for a few examples of 1:100,000 scale.

Use of aerial surveys or the space data of Landsat TM and Spot HRV has now become possible for large-scale mapping in 1:50,000 or even a 1:25,000 scale. However, the scope of interpretation remains limited due to lack of stereoscopic vision.

2.4.1 Copper Queen Region, Zimbabwe-Utility of Ground-image Correspondence

A remote-sensing study was carried out on the Lomagoundi basin in Zimbabwe (Scanvic *et al.*, 1986) in which copper and gold mineralisation are known to occur, to ascertain whether exploration should be renewed in this region. The processing and interpretation of Landsat MSS data augmented the earlier regional metallogenic information and guided the new prospecting activities. Interpretation of Spot data revealed a complex tectonics and led to the discovery of new lead-copper occurrences, thereby enabling refinement of the 1:100,000 scale map of this metallotect, a classic volcanosedimentary formation.

New observations in the environs of the Emerald gold mine, which had been abandoned, also revealed that the jasper horizons represent the metallotect of this metal at their intersection with the dolerite veins; geochemical prospecting could thus be renewed. Application of the ratio type of digital processing on Spot data facilitated mapping of iron hats (see Sec. 5.2.2.1).

2.4.2 Example of Hydrogeological Survey-Kanye Region

Exploration of aquifers necessitates mapping in particular the structural elements that determine their shape and in some cases even their recharging. With this information in hand, those areas requiring more detailed geophysical surveys can be demarcated and drilling sites determined. Mapping is more and more often being carried out using Spot data and structural analysis in the field. The former provides information about the regional tectonic frame and the latter offers a dynamic picture of the faults. Before the launching of Spot only a few studies (Bérard, 1982) had been conducted using Landsat MSS data because its spatial resolution did not enable sufficiently accurate ground location of the phenomena observed in the images.

In 1986-1987 the BRGM carried out a study with Spot imagery for the purpose of improving the water supply to the capital (Gaborone) of Botswana. It is presented here with the permission of the Ministry of Mineral and Water Resources of the Republic of Botswana.

The study region is situated between Kanye and Mmathete where the bedrock consists of fissured dolomite formations of the 'Transvaal group', intersected by syenite intrusions and dolerite dykes (Brosse *et al.*, 1987).

The major objective of the study was to define the geometry of the network of fractures that might constitute recharge conduits favourable to the existence of a groundwater aquifer, and to search in particular for those fractures not affected by hydrothermal activity as this network would be more recent than the intrusive events in the region. The structural study was based on interpretation of Spot imagery and aerial photographs, ground reconnaissance of the area and quantitative analysis of fractures. The Spot imagery was processed and restored in standard false-colour composites in the scale of 1:70,000. Interpretation facilitated the mapping of principal lithological units and major tectonic structures.

The major lithological units distinguished by the specific and contrasting reflectances of the surficial deposits accompanying them (Plate 3) are listed below:

— Vendersdorp group and Kanye felsite in the east;

— 'Transvaal group' dolomites and Masoke-Segwagwa syenite intrusions in the centre;

— Synclinal-folded Pretoria formation in the west;

— Mmathete granite in the south.

Spot imagery aided in distinguishing different facies within the 'Transvaal group' which were subsequently identified on the ground as quartzites, chert bands, ferruginous formations and dolerite dykes.

The principal tectonic structures are essentially the fractures, the only folded structure is the syncline of the Pretoria formation.

These are the major groups of fractures:

— NE trending faults: This is the most important group observed throughout the Spot imagery. It consists of faults with distinct horizontal extension of large amplitude (5 to 10 km) in the dolomite formations, which cut across the Mmathete granite and the syenite without significant displacements. Thus the dextral movement of these faults occurred earlier than the intrusions and these old structures were rejuvenated subsequent to these intrusions.

— Faults of E-W orientation are less numerous and control the dolerite sills.

— Faults of N-S and NW orientation are few and disappear at the contact with the dolerites.

Analysis of the Spot imagery thus aided in the preparation of a base geological map that is more exhaustive than the earlier maps, and simple but reliable. It also helped in mapping and identifying the various groups of fractures, some of which cut across the marker geological formations. The chronological sequence of tectonic events affecting the region was derived from these observations, facilitating selection of zones in which fracturing is more recent, at least later than hydrothermal sealing. Based on this map a structural study was conducted in the field for selecting zones for geophysical surveys, in which ultimately boreholes were located for exploration of the aquifer. These investigations demonstrated the richness of the geological and structural information derived from the Spot data and the need for thorough knowledge of the lithology, alterations, chronostratigraphy and tectonics of a region when selecting zones favourable for groundwater. It may be noted that this type of study based on processing and interpretation of Spot digital data, experimentally successful in intertropical zones, could be extended to other similar regions in which the vegetation cover and surficial deposits do not alter the perception of geology from space.

2.4.3 Mineral Prospecting in Yemen—Updating Geological Maps Using Remote-sensing Data

This study was carried out to evaluate the mapping potential of Spot imagery in an arid and well-exposed region, Nihm, in the Arab Republic of Yemen (Christmann and Scanvic, 1987).

The Spot imagery, restored in the form of an optimised colour composite after digital processing, facilitated excellent mapping on a 1:100,000 scale and its lithological confirmation of earlier ground surveys in the same scale. The new map is a considerable improvement in several respects-shape of contours, new lithological identifications and tectonics. It resulted from improvements in spatial resolution and its correspondence with ground observations, which underscores the importance of a synoptic view. This is particularly so with regard to the following (Plates 4A, B and C):

— Recent volcanic domes in which the diameter varies from 30 to 100 m. Those already known were clearly visible on the image and new ones discernible.

— Differentiations within a complex formation consisting of amphibolite gneisses. Lower horizons up to 40 m thick are distinctly identifiable and indicate the disjointed nature of the faults.

— Multiphase character of the Haris Nihm granodiorite intrusion, whose successive contacts could not be seen on the ground due to inaccessibility, were distinctly visible in the Spot imagery.

— Regional faults and in particular a N-S trending strike-slip fault whose effect on the structural disposition of some formations is perfectly identifiable on the image.

A detailed interpretation was carried out in a limited area using an optimised image restored to 1:50,000 scale, which proved comparable to the surveys conducted on the ground on a 1:20,000 scale. The proposed mapping technic refined or confirmed the extension of two metallotects of Pb, Cu and Zn-known in the sedimentary formations.

— Dolomites associated with the surface of unconformity below which mineralisation occurs. These dolomites are readily differentiated from the stratified dolomites by their spectral response (Fig. 4).

— Faults striking 130° N and 45° N and especially their intersections which constitute favourable controls for the development of high-grade mineralisation.

It appeared in the course of this study that monoscopic interpretation in this type of stratified sedimentary formation is disadvantageous since it does not allow perfect restoration of chronostratigraphic information.

2.4.4 Petroleum Exploration—Study of a Sedimentary Basin in Madagascar

The 1:100,000 geological map of the Karroo and post-Karroo basin of Morondava on the western side of Madagascar, was prepared over a long period of time and accordingly the surveys conducted by different geologists who varied in procedures and ideas. Consequently, the combined map lacked coherence and the geologists of the basin desired its reinterpretation on the basis of a uniform mapping procedure for the purpose of petroleum exploration.

For example, with the picture emerging from the feedback and from integrated perception, it appeared in particular that sea transgression started very early in the north, viz., in the Jurassic, and much later in the south, in the Upper Jurassic.

Before launching directional radiometers of high spatial resolution on the satellites Landsat TM and Spot HRV, a study was carried out in 1981 by the Hunting Society using MSS data. These images, whose spatial resolution was fair though not optimal, aided in the preparation of a composite map of the basin. This study, the only means available at that time, was inadequate for detailed investigation, which is reflected in the composite map.

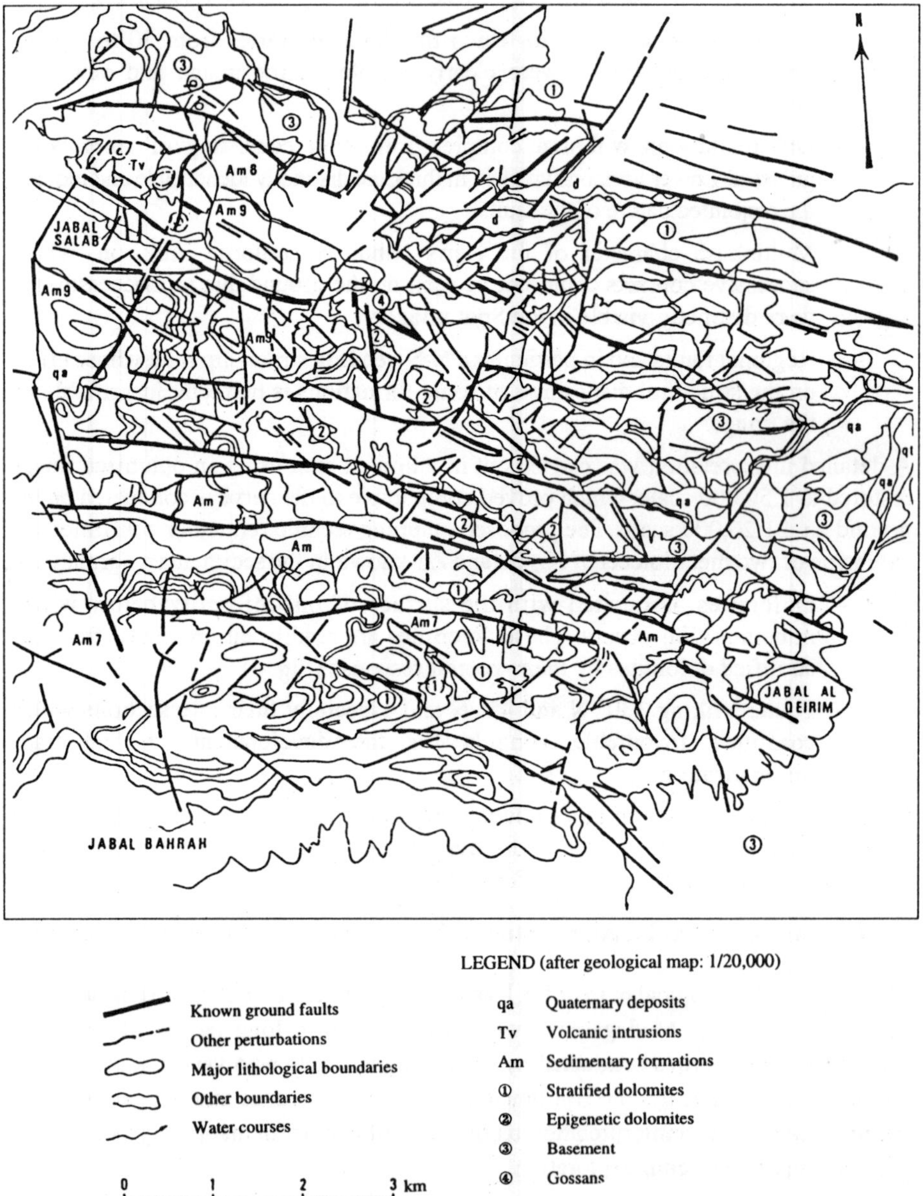

Fig. 4. Spot image of Yemen (Plate 4 , point 1): differentiation of (1) primary and (2) secondary dolomites.

That is why OMNIS, the Madagascan organisation responsible for prospecting, initiated a new study with Spot imagery and asked the BRGM to be the operator for interpretation and geological synthesis (Berthiaux *et al.*, 1988).

The study involved interpretation of 26 multispectral Spot images of 20-m resolution, ensuring complete coverage of all the spectral bands for the Bemahara, the promising part of the basin, and of the panchromatic stereoscopic mode of 10-m resolution, covering two specific zones. The study was expected to achieve, within the technological limitations of remote sensing, the accuracy desired in mapping the marine, continental and intermediate facies so as to reveal the evolution of littoral zones, ascertain the lithostratigraphic profile, emphasise the existence of eventual shoals, indicate facies variations and, lastly, facilitate geological and structural interpretation of certain gravity anomalies.

Based on earlier maps of different scales, interpretation of Spot images and the new data acquired on the ground by OMNIS geologists, 11 geological maps on a 1:100,000 scale were prepared in four months. They represent a unique mapping approach, supported by numerous new observations (Fig. 5).

These new maps proffer different levels of interest:

- They provide an updated compilation of old maps that lack uniformity. Updating pertains to accuracy of contours, characterisation of new lithological units and maintenance of uniformity in dissimilar data, but obviously does not deal with chronostratigraphic information as its parameters cannot be extracted from Spot data.
- Due to the spectral and spatial characteristics of Spot imagery, identification of new units within a formation and detection of large structures are possible. These images further enable integration of interpreted elements into a general context, as revealed by ground coverage of each image.
- Elements of palaeographic interest are also incorporated.

For example, the differentiations observed in the various horizons of Isalo constitute indications of the distribution of continental deposits along the eastern boundary of the basin. During Isalo I sedimentation may have been relatively more homogeneous in southern Miandrivazo than in the northern part where the formation shows considerable diversity of appearance. Just the opposite situation occurred in Isalo II; sedimentation appears to have been fairly homogeneous in northern Miandrivazo but diversified in the southern part, as evidenced in particular by the development of the Makay facies.

During the Jurassic the conditions that prevailed in Isalo II appear to have been maintained with a highly diversified sedimentation in the southern sector, consisting of continental facies, mixed facies and then marine facies in the Upper Jurassic. In the same epoch, in the central regions and in northern Miandrivazo marine sedimentation was initiated from the Bajocian and stabilised throughout the Jurassic.

Development and movement with respect to time and space in progradation zones in both parts of the Miandrivazo sector suggest that this zone played a significant role in the distribution of sediments in this part of the Morondava basin. It has been established, for example, that the Sakamena formation became thinner in the

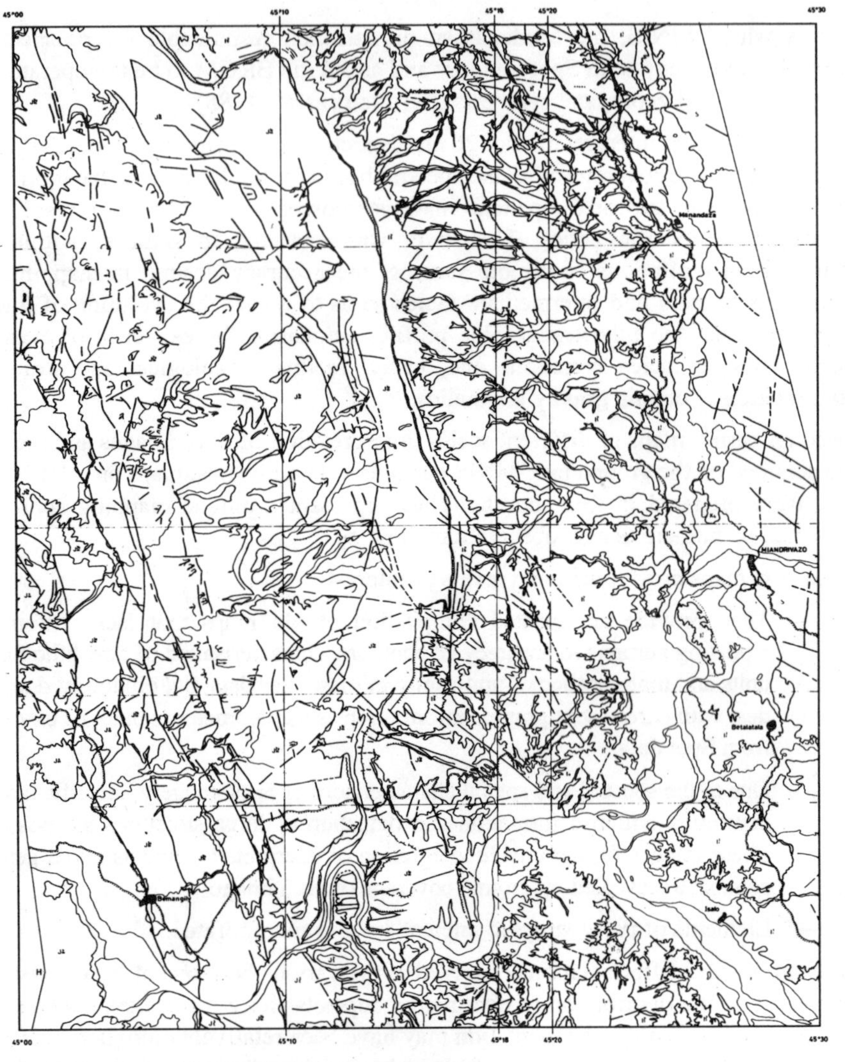

Fig. 5. Madagascar. Spot image of the Morondava basin and its geological interpretation (CNES, 1987. Spot Image Distribution and Processing).

southern part of the Morondava by losing its lower units and then disappeared, enabling another phase of sedimentation in the Isalo to settle directly and discordantly over the basement.

In the northern part of the basin the facies distribution confirms the existence of the Maneriverina and Bemolanga shoals where the Isalo facies rests directly over the basement. Contrarily, the Sakamena facies was brought to the surface by an uplift that affected the enclosing Fonjay and Amberemy structures:

— They (the new maps) reveal the parameters which confirm, complete and characterise the structural set-up of the Morondava. Examples: a series of large fractures parallel to the major axis of the basin, which were responsible for segments of subsidence; discontinuities in vinculas striking 120 to 130° N, some of which indicate shear displacement; basic intrusions which emphasise the 120 to 130° N and 0 to 30° N lineaments; slight deformations of low amplitude, which appear to be contemporary with segments that subsided.

— They furnish clues that will orient and facilitate new ground reconnaissance and prospecting programmes, which constitute the basis of objective mapping.

— Lastly, they give the basic parameters for an integrated approach which enables evaluation of all the data acquired by this mapping for an in-depth understanding of the history and structural geometry of the Morondava basin.

Thus the interpretation of 1:50,000 images, given their scope for stereoscopic vision, enabled accurate determination of the existence of basement structures trending 130 to 150° N. It also enabled a synergistic analysis of gravity and remote-sensing data wherein the hypothesis concerning the existence of shoals as potential structural traps along the direction of lineaments in the same trends could be affirmed.

Based on these maps and the integrated analysis of the basin, exploration permits have been sought by international petroleum companies.

2.4.5 Mapping in Velay

This mapping programme sought to ensure within the frame of systematic surveys total geological coverage of France in the scale of 1:50,000 and to execute a regional synthesis in the scale of 1:250,000.

With this in mind, a methodological study based on the scientific research programmes of BRGM was conducted at two levels: 1) regional, using a Landsat TM mosaic and 2) local, using the stereoscopic mode of Spot-Landsat TM images. From 1990 to 1992 four temporal images on a 1:50,000 scale were interpreted.

The methodology adopted for remote-sensing data was based on the application of digital processing technics and optimised by using the reference characteristics of certain formations investigated on the ground by structural geologists and petrographists.

The enhanced digital data was then restored in the form of photographic images in which the oblique Spot XS and the vertical Landsat TM data provide stereoscopy and facilitate geological interpretation of the landscape. This methodology demonstrated both its potential and shortcomings whereby the component of the photointerpreter is limited to the analysis of the image and identification of the equivalent between the demarcated features and the lithostructural units as marked on earlier maps (1:80,000 scale).

For example, one can confirm the significance of stereoscopic vision for details of lithostratigraphic reconstructions, especially in the case of sedimentary formations-structural information as contained in the image and the possible utilisation of certain spectral responses when induced by strongly mineralised media, such as phonolitic occurrences, periglacial formations of the Aubrac Mountains and Mont Lozere, and granites.

The apparent limitations of the method pertain to the mapping of basement rocks, which are structurally very complex and often obscured by a cover of alterite and vegetation, and whose petrographic differences may generate only feeble morphological or spectral variations.

Technical and financial limitations are also apparent, such as inadequacy of geometric corrections, calculated without the support of digital processing methods, which are introduced only when the altitude variations are large, and very high cost of the space remote-sensing technic compared to that of aerial photographs.

The important characteristics of the methodology of this example are: potential use of stereoscopic data of Spot-Landsat; performance of digital processing in an interactive mode and supervision with reference to the spectral response of certain geological formations.

2.4.6 Radar Geology

It was during the 1970s that the first airborne radar surveys were carried out (Westinghouse, Goodyear, Motorola). This imagery formed the basis for the 1:250,000 mosaics used in particular for geological purposes. The choice of this technic of remote sensing was frequently necessitated by the climatic conditions of the study areas (equatorial regions), which made the acquisition of aerial photographs difficult.

At the end of the 70s and during the first half of the 80s, the radars installed on satellites (SEASAT) or on space shuttles (SIR-A and SIR-B) have provided the first evaluations of the use of high-frequency region in spatial context. Applications in the field of earth sciences have been developed for desertic (Blom *et al.*, 1984; Elachi *et al.*, 1984), volcanic (Gaddis *et al.*, 1989) and equatorial (Sabins, 1983) terrains. The major results of such investigations pertained to the penetrating power of the radar signal (perception of structures by transparency principle), textural analysis (specific surface conditions that reflect the substratum) and structural analysis (major tectonic features).

More recently, radar studies were reactivated through the launchings of SAR (Synthetic Aperture Radar) of ERS-1 (European remote sensing satellite) in 1991, JERS (Japanese remote sensing satellite) in 1992, ALMAZ-1 (Russian satellite) in 1991, the flight of SIR-C on the space shuttle in April 1994 and the preparations of other missions (Radarsat, ERS-2). Numerous airborne experiments have come into vogue as support to spatial surveys: SAREX over South America (Deroin *et al.*, 1994), Canadian airborne SAR (Lowman, 1991) and others.

The BRGM (France), for its geological and mineral exploration needs, has conducted operational studies over different equatorial regions, which have enabled it to acquire a competence in the utilisation of aerospatial radar remote sensing: in 1980, the frontier zone between Guyana and Brazil (Goodyear image, 3 cm, Radam project, Brazil); in 1981, the diamondiferous zone of Roraima in Venezuela (Goodyear images, 3 cm); in 1983, over entire Equatorial Guinea (Motorola image, 3 cm, two look directions, north and south); between 1984 and 1990, in Gabon, covering the 1:200,000 map sheets of Mitzic Boué, Mouila, Koulamotou, Lastourville and Franceville (Goodyear images, 3 cm); in 1986, over the entire Togo territory (Motorola image, 3 cm).

In addition, the BRGM has participated in the analysis of experimental data acquired over different sites in France and abroad, in the course of its diverse programmes of applied research: in 1977, 1979 and 1984, over the Vans zone, Ardèche region (France), with the airborne images of Goodyear (3 cm), JPL (25 cm) and SAR 580 (X, C, L) (Cluzeau *et al.*, 1983; Scanvic *et al.*, 1985), with the SIR-A data in Mali, Saudi Arabia and Cameroon, with Seasat data in Wight Island (Deroin et al., 1991) and in the Alps (Rebillard, 1980) and with the data of satellites ERS-1, JERS-1, ALMAZ-1 in France, Guyana and Colombia.

Finally, as a member of the All Time Action Group (GATT) formed by the CNES for defining the thematic requirements in radar remote sensing and the associated experimental projects, and as a part of the programme Prosper designed to study the integration of radar and optical data, the BRGM has actively participated, co-ordinating with the University of Paris VI, in different investigations on the methodological exploitation of results.

Based on this experience, as also on the results published elsewhere, we present here some of the acquisition technics and develop ideas on the present application potential of radar data in the field of geological mapping.

These results were presented at the meeting 'Cartographie Radar' of the 'Sociéte Française de Photogrammetrie et Télédétection' (Nice, France, 24-25 November 1994) and published in its bulletin (No. 138), from which they are taken by the authors, J.P. Deroin and J.Y. Scanvic (1995).

2.4.6.1 SAR of ERS-1, an aid to mapping and mineral exploration in an equatorial region (coastal forest and savanna of Guyana)

The radar remote sensing investigations conducted for geological applications over Guyana form a part of the mission Pilot Project PPF-12, 'Littoral and Fluvial Environment of Guyana', undertaken by the European Space Agency (ESA) during its first call for offers (1991) aimed at encouraging proposals on experiments using radar data of ERS-1 launched in July 1991. Guyana, a French overseas territory (South America), is covered by a dense equatorial forest over almost its entire territory.

Since 1986 several attempts to acquire optical data have met with failure, but the SAR of ERS-1, since its launching, has facilitated acquisition of digital images even during rainy seasons. Their study (Rudant *et al.*, 1993 and 1994; Lointier *et al.*, 1993; Deroin *et al.*, 1993) has led to precise conclusions regarding the geological potential of SAR band C data for mapping in these latitudes.

The SAR imagery has been used in these studies. It consists of an active signal in the high-frequency domain, here band C of 6 cm wavelength. The incidence angle is 23° and the spatial resolution is 30 m. The principle of this sensor is different from the passive mode of acquisition of Spot or Landsat. This type of imagery offers a special feature of representing the time taken by the signal to be backscattered to the sensor after it encounters a ground object. Thus, in montane regions, a peak is imaged before the valley in its proximity. The interpretation of radar requires experimentation before optimising the instrumental parameters (wavelength, incidence angle, polarisation, pixel size, degradation of raw image, etc.) and the terrain parameters (morphological conditions, soil types, surface conditions etc.).

The ERS-1 radar data acquired during the Pilot Project has been used for methodological investigations aimed at improving the signal: elimination or attenuation of speckle, testing different filters, geometric corrections, superposition of radar image on a digital terrain model, analysis of multi-temporal images etc. The best result was obtained for a mosaic of radar images that furnished the first complete mapping reference over a tropical zone. This mosaic, however, is not georeferenced.

The analysis of ERS-1 image shows a high capability of recognising the lithologies when a sufficient geomorphological contrast exists. The lithologies are most often represented by a characteristic image-facies. Some identifications can be made on the basis of structural criteria (foliation in migmatites, basaltic dykes etc.). Differences within the same lithological nature are difficult to find. Also, differentiations in a granitic or gneissic basement are difficult to make. The lithostructural map shows about a dozen image-facies that can be correlated with known geological units in Guyana. It should be remembered, however, that the active nature of the radar signal introduces an anisotropy in the representation of structural orientations detected, as the trends situated in the axis of the incident signal, i.e., perpendicular to the direction of flight, are not observable. From the structural point of view, the ERS-1 image provides significant information about dyke systems, major regional fractures and about the criteria of folding, foliation, stratification and schistosity. However, detailed mapping of tectonic contacts between different units appears to be difficult because of the angular distortion and the lay-over observed in the high-relief zones. Lineament analysis enhances specific trends in areas of gold prospects (the Kaw Mountain, Tortue Mountain). Fractures that host basalt veins are observed throughout the scene.

The ancient coastal bar is clearly seen on the ERS-1 image. On the ground, it corresponds to a continuous sand body between Cayenne and Sinnamary. The disposition of Quaternary low plains can be seen on either side of the Kourou River, where the outlines of sandy bars stand out prominently over a black background of hydromorphic soil. It may be noted that the bars occur as fans, indicating variation in space and time. The distribution of Quaternary high plains is quite distinct on the ERS-1 image. The downstream boundary of a Quaternary high plain is marked on the image by a reduction in the density of black hydromorphic areas and by the continuity of the first sand bar. The upstream boundary is less distinct except when the hills of ancient basement form a topographic contrast of several tens of metres. The altered basement can be clearly identified on the image when circular hills are numerous and contrasting in relief. On the other hand, isolated mounds that are poorly delevelled relative to Quaternary alluvial plain are not indicated distinctly. A hardpan can be identified if the morphology is favourable. In other cases, lacking morphological contrast, the hardpan is not discernible.

Some radar images, acquired over French Guyana during the airborne survey SAREX-92 (TREES programme), have been studied to evaluate the importance in earth science applications (geomorphology, lithology, structure) of the airborne SAR sensor, as a complement to ERS-1 satellite data. The results confirm the significance of airborne SAR, whatever be the band (X or C) or mode of polarisation, for tracing the time and space evolution of coastal geomorphology, such as, for example, mapping the littoral bars that indicate the palaeocoasts. On the contrary, the lithologic and structural differentiations, illustrated by analysis of ERS-1 satellite data, are rarely revealed in the radar images because of the small size of the images available (a few km^2) and the littoral sites of acquisition (Quaternary formations), and consequently are not favourable for tectonic, metallogenic and geodynamic studies. The sites of geological or mineral importance have been more difficult to analyse due to geometric problems associated with the topographic variations, even of small magnitude.

Another example of using radar data, in Gabon in equatorial forest zone, can be found in section 7.6.2.2 (Plate 34).

2.4.6.2 SAR ERS-1: Results of mapping in a weathered terrain in temperate climate

A methodological study was carried out over the Mauges Plateau (Loire Atlantique) (Scanvic *et al.*, 1993), a region of metallogenic importance (Braux *et al.*, 1991). The objective of the study was to interpret the multidate image lineaments for the purpose of mapping the basement below a cover of weathered layer (average thickness 10 m) and to evaluate the effect of different acquisition parameters as indicators of geological phenomena: look angle, humidity of soils, vegetation, etc. The preliminary images acquired had the following parameters (see Fig. 11):

Image 1

30/09/1991, ascending mode, look direction eastward, acquisition after 10 days of rain (54 mm);

Image 2

21/05/1992, descending mode, look direction westward, acquisition after 10 days of dryness (0 mm).

The geological interpretation of these images resulted in maps of very dense image lineaments on image 1, and very scarce image lineaments on image 2. Two other images were hence chosen from among the data available in the catalogue:

Image 3

22/09/1992, descending mode, westward look, acquisition after a dry period.

Image 4

27/10/1992, descending mode, westward look, acquisition after a spell of rain (35.4 mm).

Interpretation of images 3 and 4 resulted in two maps of image lineaments which, without attaining the density of image 1, were useful in terms of structural geology and provided a preliminary identification of principal indicators. In fact, as the slopes exposed to the west on Mauges plateau are affected by erosion (Wyns, personal communication) and the vegetation (pasturage and cereals) is in full growth in May, it can be inferred that in this region:

— the choice of look direction is of primary importance;

— the humidity of soils and the vegetation cover play a significant role; their individual contributions are difficult to estimate without additional data.

However, a comparison of interpretations of images 2 and 3 showed that the first factor has a lesser effect than the second.

From the preliminary map of image lineaments (image 1), the following documents were prepared:

— A map of major structural units, obtained from a reinterpretation based on the latest databank on geological mapping (Wyns, 1984; Wyns and Bechennec, 1992). It aided in finding all the structural elements in the Mauges sequence, in precisely demarcating or modifying the boundaries, thus suggesting new approaches of investigation on the ground.

— A series of data based on the gravity picture of the concealed Chemille granite, which confirmed and refined its structural setting determined from the radar images: the general form, since the correlation between the boundaries at depth determined from gravity (Corpel, 1990) and the image lineaments (from radar) is excellent; gradient is reproduced by fractures; the N 10-20° lineament, a transverse fault, confirms the gravity interpretation of a massif with two negative structures; geometric forms (hexagonal, subcircular) indicate exposed or concealed granite intrusives; the westward extension of the massif suggests a possible continuation of Pommeraie granite. It may be noted that this granite, as determined from gravity data, is not distinguishable from metamorphic country rock, either by density of Landsat lineaments or by fractures visible on the aerial photographs (Delpont and Dutartre, 1983; Berthiaux *et al.*, 1987; Berthiaux and Marot, 1989). The same is true for ground data in which points of observation or measurement are scarce.

In conclusion, this study showed that use of radar images has great potential for structural interpretation of bedrocks below a weathered cover or other surface formations, because of the 'skin effect', which is very sensitive to the conditions of acquisition. The observations made on radar images:

— Justify recent mappings made in difficult conditions of outcropping;

— provide more detailed structural information than that derived from ground surveys or from the interpretation of Landsat or Spot images, thus bringing in refinements, modifications and new orientations in ground investigations;

— demonstrate the integration of different methods, viz., ground surveys, optical imagery, radar imagery and gravity data, which enhances certain signatures.

The foregoing confirm the importance of repetitive function promoted by the all-time nature of radars and the need for multi-temporal analysis.

2.4.6.3 Comparison of ALMAZ-1 and ERS-1

A comparative study of ALMAZ-1 and ERS-1 data was undertaken for the northern Cévennes region (Langogne-Villefort axis) in collaboration with VNIIKAM (St. Petersburg, Russia) (Deroin and Rogov, 1994) (Tables 2.1 and 2.2).

SAR images were analysed by a method based on marking a certain number of geological and environmental objects on both images. Later, we evaluated the significance of three parameters (angle of incidence, resolution and frequency) on the recognition of these objects. Polarisation, totally different for the two satellites (HH and VV), was judged a difficult parameter to evaluate. This parameter, like direction of flight, was considered of minor importance by Marek and Schmidt (1994).

Table 2.1. Principal parameters of SAR of ALMAZ-1, ERS-1 and other conventional radars

	ALMAZ-1	ERS-1	JERS-1	SIR-B	SEASAT
Agency	Russia	Europe	Japan	USA	USA
Year of Launching	1991	1991	1992	1984	1978
Orbit (km)	300	785	568	225	800
Swath width (km)	40	100	75	30	100
Resolution (m)	15-30	30	18	25	25
Angle of incidence	30-60°	23°	35°	15-60°	22°
Frequency (GHz)	3 (S)	5.3 (C)	1.3 (L)	1.28 (L)	1.28 (L)
Wavelength (cm)	10	5.7	23	23.5	23.5
Polarisation	HH	VV	HH	HH	HH
Antenna (m)	15	10	12	10.7	10.74

In the case of mountainous regions, the ALMAZ data appears to be more significant than that of ERS. This is related to the angle of incidence, as the lay-over effect is very high in ERS. It has been noted, for example, that Lake Villefort situated in a structural setting controlled by the north-south trend is not seen on the ERS-1 image, as the signal is incident from the east (descending orbit), nor does it furnish any information on the lake situated in a high relief zone (the Borne massif).

Spatial resolution is another parameter of importance to be investigated, because it aids in detecting small size objects (rocks of Mount Lozere, for example, of the order of 1000 m^2). The morphology, some mineralogical features and drainage pattern are found to be more effective in ALMAZ.

Some of the parameters of ALMAZ also show a significant improvement compared to the visual interpretation of landscapes on ERS radar images. However, the satellite parameters of the path directions of ALMAZ, launched in a low orbit, still need to be perfected.

Table 2.2. Evaluation of three parameters of SAR of ALMAZ-1 and ERS-1

	Geological						Soil type			
	min.	mjtf.	mntf.	morph.	drn.	detl.	veg.	vill.	rbet.	
ALMAZ										
Angle of incidence	2	3	3	4	4	2	3	?	?	21
Resolution	3	4	2	2	3	2	3	3	4	26
Frequency	2	?	?	?	?	?	2	?	?	4
Total	7	7	5	6	7	4	8	3	4	51
ERS-1										
Angle of incidence	1	3	2	2	2	1	1	?	?	12
Resolution	2	4	2	1	2	1	2	2	2	18
Frequency	1	?	?	?	?	?	1	?	?	2
Total	4	7	4	3	4	2	4	2	2	32

Abbreviations: min.: mineralised area, mjtf.: major tectonic features, mntf: minor tectonic features, morph.: morphology, drn.: drainage pattern, detl.: details, veg.: natural vegetation, vill.: villages, rbet.: roads, bridges etc.

'Visual' quantification of perception: (4) Excellent, (3) Acceptable, (2) Sometimes significant, (1) Not significant, (?) No information.

2.4.6.4 Evaluation of two hyperfrequency bands (X and C) for lithological mapping (example of Larzac)

As a preparation for its future programmes, the CNES (Centre National d'Etudes Spatiales) investigated the space systems of earth observation that might be used operationally at the beginning of the twenty-first century. Among such systems, a spatial synthetic aperture radar, forming part of the payload of the optical system of Spot satellite, appears to be an efficient tool for exploration of natural resources and management of the territory. During the initial projects the CNES conducted E-SAR radar flights in France. Of the various important sites, the Larzac region was selected for evaluating SAR for geological mapping applications (Deroin and Rudant, 1991; Nezry *et al.*, 1993). The terrain is *a priori* favourable for differentiating the lithological variations. Using the E-SAR data, we attempted to improve the lithological identifications in carbonate formations (carbonates, dolomites) and their alteration products (decalcification clays). We tried to evaluate the utility of two bands.

The test-site of Larzac is located in the north of Herault region and extends east and south-east of Caylar town. The choice of an area is based on the possibility of observing surface variations associated with the conditions of the geological substratum. It should also have a plane topography and limited vegetation cover. The Larzac plateau offers these characteristics, since, in addition to pronounced aridity and plane topography, the sparse human occupation makes minimal contribution to the details of landscape. Panchromatic Spot images and infrared aerial photographs were used for these test studies.

The Institute of Hyperfrequency Technology of DLR (Deutsche Luft and Raumfahrt) has developed airborne radar sensors for study of the earth's surface, with different frequencies and polarisations. The main characteristics of E-SAR are: radiometric resolution (4 views) 2 db; geometric resolution 4 m; maximum flight elevation 3500 m; nominal swath width 4 km. The present study was conducted on the analysis of a couple of radar scenes, consisting of a band X (3 cm) image (VV polarisation) acquired on 25 April 1990, and a band C image (6 cm) (HH polarisation) acquired on 16 July 1990. The two images have the same angle of incidence (about 45°). Except for a few cultivated segments, the soil types were not changed between these two dates.

Data acquisition covered an area of about 12 km^2 centred around the hamlet of Vernède. The lithological types are: alluvial cones (C), karst plateau (R), residual siliceous deposits (A), gravelly and bioclastic limestones of backreefs (j^9AR), massive dolomites (j^8D), bioclastic limestones with reefal constructions (j^8R), light-beige sublithographic and bioclastic limestones (j^8b), beige limestones, sublithographic and sometimes bioclastic (j^8 a), and maroon, lithographic to sublithographic limestones, at times bioclastic (j^7).

In general, there is no correlation between backscatter signal and lithology, i.e., radiometric classification is not possible. Some structural features are observed forming the boundaries of belts that are homogeneous from the radiometric point of view. Topographic variations, even moderate, are responsible for the image characteristics of backscattering. *A priori*, it is needed in geological studies to mark objects of tens of metres or more in size; metre-sized objects (marker-beds) may also be required to be identified. Perception is excellent in band C for residual formations. Lithological variations, fractures, the effects of slope and morphology can be clearly observed. Perception is poor in band X, especially for morphology and residual formations. This band, less useful for geological studies, is better suited for soil type investigations. For example, soils with thickets and small plants are clearly identifiable. B and C reveals more significant information than band X because of its greater penetration (higher wavelength). It thus appears preferable for mapping surface conditions related to the nature of the bedrock.

A map was prepared using the data of band C, whose preliminary analysis showed that it forms the best data for geological applications (Fig. 6). The extension of dolomitic facies can be recognized from their relief (tower-like) and their associated vegetation (*garrigue*). The thick forest zones can also be very clearly identified as they correspond to the formations j^8D and j^9AR, with a strong backscatter signal. In the area of formation j^9AR, infra-red photographs do not reveal the bedrock because of the vegetation cover.

The objective of this study was to evaluate the utility of the two frequency bands (X and C) of E-SAR radar in geological investigations in a plane test site with varied lithologies (different facies and fault zones). The study of images obtained in bands X-VV and C-HH has shown that the latter is more effective for distinguishing geological or geomorphological objects; it is developed by others for aspects of image degradation (Rudant *et al.*, in press) and will be completed by evaluation of band L (about 25 cm), which might turn out to be more significant than band C (NASA,

1989). Visual interpretation provides an accurate estimation of the areal extent of the residual clay facies (controlled by fractures), distinctly identified within the dolomitised zones, yet indicating the elements of regional fractures. The presence of noise (Lopes et al., 1990) inherent in radar images makes it impossible to carry out digital classification of formations using statistical criteria of the first order. Variations in the backscatter signal are attributable to the effects of soil moisture, variations in angle of incidence, surface unevenness, vegetation cover etc.; it is not possible to identify and separate these causes. However, general tendencies may be recognised concerning, for example, the effects of slope observed in band C data. Degradations of source images (complex data) have facilitated identification of configurations that are best suited for geological studies. The configurations of 2 m/4 views and 4m/16 views have been found to be significant. The satellite configurations of ERS-1 (30 m/4 views) with a pixel size of about one metre do not carry the information detected in this case. Radar, which furnishes more specific information compared to the methods of ground investigations, as also relative to other image data, can thus be considered a necessary tool for regional geological studies.

Fig. 6. Band-C SAR Image (A) of the Larzac region and its geological interpretation (B).

1—Small, enclosed zones of very low and smooth backscatter signal, often along or aligned parallel to a fault. They represent areas of decalcification clays, 2—Single dark belt covered by zones of facies '1', with finer texture. Corresponds to no known geological boundary and appears to be a subgroup of '7', 3—Zone extending over 2 km, of dark shade, probably exaggerated by proximity of the edge of the scene, with a very fine texture, exactly corresponding to the outcropping zone of bioclastic limestones, with reefal constructions, j^8R; 4—Dark belts in the north of the scene, bounded on the south side by an E-W fault, of very fine texture, corresponding mainly to limestones of type j^7; 5—Coarse-textured belt in the SW boundary of the scene, darker than '11', of which it is the extension. Has no separate geological significance; 6—Elongated belt of very smooth texture, bounded by a fault in its western margin, corresponding to the area of outcropping limestones j^8; 7—Numerous zones in the southern half of the scene, mainly characterised by their coarse image-texture. They represent areas of outcrops of type j^7 limestones, much coarser in texture than that of '4' in the north of the scene; 8—Large number of zones, distributed throughout the scene. The dominant tone is clear, the image texture is coarse and often indicates the presence of bushes. They represent areas of extension of dolomitisation, a phenomenon that affects the limestone horizons; 9—Belt extending over 2.5 km, bounded by a fault in the east, very coarse texture (dense bushes), corresponding mainly to the outcropping area of gravelly limestones of type j^9AR; 10—Belt bounded by faults, only in the north-western corner of the scene, very coarse texture, with no specific geological importance. It represents an atypical zone of facies j^8; 11—Very distinct and very coarse-textured belt within zone '8' (dolomitic), corresponding to type j^8 a limestones; 12—Belts in contact with zones '3', '7' and '8', probably corresponding to an image differentiation of dolomitised areas. More distinct and less coarse than '8', they indicate thinner vegetation cover and higher mineral component.

(a)

(b)

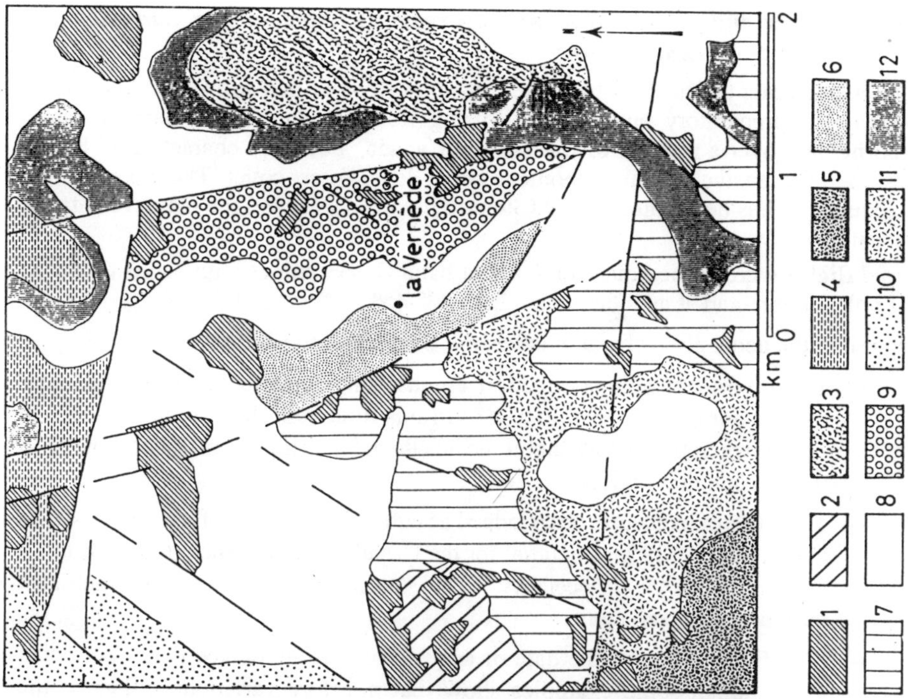

2.4.6.5 Radar and geology of permafrost regions: Experience of Canadian Remote Sensing Centre

For several years the geologists of the Canadian Remote Sensing Centre have been involved in preparatory studies for utilising Radarsat data. They have acquired airborne data over a number of test sites in Canada, a country characterised by the presence of intensive glacial phenomena and a dense forest cover. The methodology of acquisition has been summarised (Singhroy *et al.*, 1993) and the major results are presented below.

Aerial flights were carried out with SAR of band C, sometimes with bands C and X, HH polarisation, and at an incident angle of 23°-45°.

Regarding the acquisition characteristics, the Canadians concluded:

- The importance of look direction, emphasising in particular that, if the linear objects are parallel (within ± 20°) to this direction, they are not detectable except when the texture that indicates them has a high contrast compared to the surrounding region.
- The angle of incidence, often judged as unfavourable if it is of the same order as that of ERS-1 (23°), as optimal for the Canadian Shield when it is close to, or greater than 45°, especially if the geology introduces no significant morphological variations.
- Significant effect of the date of acquisition but the choice is not perfected even though some examples show that moisture variations influence the textural responses related to permeability of soils and glacial formations.

Concerning the geological information that can be gathered from the interpretation of radar images in the Canadian conditions of rock exposure, the following inferences are made:

- The scale of survey is regional, viz., 1:250,000, though in some cases 1:50,000 investigations have been carried out.
- The importance for structural geological studies is high, provided, however, the gaps in observation are compensated by additional radar scannings.
- The potential of radar for lithological mapping is judged as uneven, complementary to optical imagery, and sometimes specific, since its utility is dependent on the relation between geology and morphology below the forest cover, on the technical capability to extract structural information from the image lineaments (Roy and Boily, 1993), and on the clarity of textures, controlled by soil moisture, which facilitates a clear differentiation of Quaternary and glacial formations.

With regard to the methods, the Canadian geologists stress:

- The importance of visual interpretation of images and, hence, their digital optimisation (filtering).
- The combination of optical and radar remote sensing, though it may not be a final solution.
- The need for integrating radar data with geophysical and geological maps, particularly by means of IHS type of processing to balance the observations and to improve the existing maps.

— The integration of data into geographic information systems, to enhance their utility by statistical analysis, for example, in a mineral exploration perspective.

2.4.6.6 Conclusions

Based on the review of case studies representing diversified morphoclimatic environments, it is possible to comment on the quality of geological information that can be extracted from modern radar data and to propose some methodological developments which need to be investigated.

Compared to optical remote sensing-space imagery or aerial photography, most geologists agree on three aspects, viz.:

— Radar imagery cannot replace optical data and can only be complementary. The information it provides is often incomplete, but it may be very significant, for example, on a base of multidate acquisition data. Its area of excellence is that of structural geology and its optimal scale of survey is regional, such as 1:100,000 to 1:250,000.

— The best results are obtained in conditions of difficult climate and limited or obscure outcrops, in regions where sufficient knowledge about the terrain is available. One can expect useful data from the effect of higher penetration associated with the lower frequency radar bands (S, L or P) and from the 'skin effect' of bands X and C. Such is the case, for example, in equatorial dense forest zones where lithology is well indicated by topography, zones of alterites in temperate climates, semi-arid areas where lithological contrasts are distinct, permafrost regions, arid zones where images of bedrocks under a thin soil cover can be obtained, in volcanic areas, and so forth.

— The radar image is disturbed in its geometry (lay-over), in its radiometric characteristics (speckle) and by certain parameters of the system such as angle of incidence, look direction and perhaps polarisation.

Geologists who experience the need for diversifying their tools of mapping and the special significance of radar are looking for developments concerning in particular:

— Control over the geometry of radar images, i.e., they have to be incorporated into georeference databases;

— Optimisation of stereoscopic acquisition, which is *a priori* difficult due to the 'shadow' produced by the look direction and by the ratio between the angle of incidence and the relief (Deroin and Rouzeau, 1994);

— Integration of data of varied sources, such as ground observations, geophysical and geochemical surveys etc.;

— Evaluating the effects of season, moisture, texture and vegetation, which may be masking or enhancing geological information;

— Statistical processing of structural information and its optimisation: multidate analysis, digital interpolation and filtering, structural domain analysis, relation with topography, exposure of slopes etc.

When the above studies are undertaken, the geological utility of radar imagery will be considerably enhanced.

However, the radar wave is complex, with its parameters of amplitude, polarisation and phase. The last is not fully exploited in geological remote sensing and is highly promising. Studies are presently being carried out on radar interferometry (Massonet, 1992) which are based on the principle of measuring constructive or destructive interference of coherent waves recorded by satellite SAR sensors that facilitate observation of the same part of the ground surface under various geometric conditions of viewing using multiple orbital passes. Differential interferometry consists of detecting phase changes caused by variational effects and relating them to altimetric data. Studies are in progress (Carnec, 1993) for evaluating the possibilities of this technic to measure deformations associated with landslides and to ensure the monitoring of risk zones.

2.4.7 Mapping of Granite Massifs in Temperate Zones

This is a specific aspect of mapping that poses special problems duly considered here.

In a region of basement rocks and temperate climate, where outcrops are rare, often altered and covered by surficial or alterite products, medium and large-scale mapping from space by direct observation of outcrops, as in arid regions, appears *a priori* difficult and can be attempted only by analysis of surface characteristics.

Numerous studies have been carried out on ancient massifs using Landsat data, whereby the validity of the methods preceding or accompanying mapping could be established. In particular, for the Armorican Massif and the Massif Central (France), such studies facilitated evaluation of the criteria to be taken into account for this type of mapping and explanation of the phenomena generated.

Let us consider two situations: one, granite buried under surficial deposits of alterites and the second, granite occurring below other rock formations. Their mapping was envisaged based on the methodology of two concepts: the notion of orthotype and the phenomenon of transparency. The orthotype is a group of variables governed by definite functional relations between geology and vegetation, i.e., the state of the surface as induced by the geological substratum and its altered products. This deterministic principle leads to the search for significant surficial effects and explanation of the causes which govern them. These effects are spectral, structural and textural in nature. They determine the variations of colour, circular shapes, density and orientation of specific fractures, texture of drainage network and the characteristics of soil (Dutartre and King, 1986). The transparency phenomenon emerges from processes that are much more complex, since it aids (Marconnet, 1982; Marconnet *et al.*, 1985) in detecting granite bodies occurring under the cover of metamorphic rocks. It is thus based on the existence of contact metamorphism and its detection using spectral and textural criteria of the images.

2.4.7.1 Mapping of granites under surficial deposits

This is a common case in France. The criteria used in the interpretation vary according to the granite bodies and several are generally used to characterise and demarcate the facies on Landsat or Spot imagery.

- *Colour and textural* details of the image often constitute the basis for discriminating the surficial conditions induced by an underlying granite massif. The dominant colour represents the spectral response of the surficial deposits or altered products and is strongly influenced by their water content. The texture is induced in most cases by the soil characteristics or drainage network. Examples par excellent are the Pontivy and Villader granites in the Armorican Massif and Ussel-Meymac and Margeride granites in the Massif Central. In the case of granites of the Massif Central, it may be emphasised that the morphological differentiations appear at the contact between the Millevaches granite and the Velay granite whose convergence of various petrographic features is not *a priori* a favourable discriminating factor (Plate 5).
- *Colour and morphology* can be combined to facilitate excellent discrimination. In this case colour is the more common of the two criteria since it characterises a vegetation that often differs slightly from that in the actual context. Its response may be accentuated by variations in chlorophyll activity. Shape, whether circular or oval, is a better discriminating criterion. The significance of these structures is found primarily in the morphology, the underlying blocks being domal or giving rise to distinct relief. Such is the case of Quentin granite in the Armorican Massif, whose western part is perfectly identified in relief in the south-western part of the North Armorican trough, whereas the eastern part is not so clearly distinguished as its elevation is almost the same height as the surroundings (Plate 6).
- *Colour and geological structure* resulting from tectonic and petrographic effects often play a discriminating role. The responses then emanate from the specific type of fracture systems and are smoothened by the scale and resolution parameters, as in the case of Villefranche-de-Rouergue (Massif Central) granite (Dutartre, 1983) or by more complex phenomena, such as the direction of magmatic flows, as is the case of Margeride granite (Massif Central) in Thoras. Other phenomena may also be associated for inducing surficial characteristics. For example, in the Ambazac hills (Massif Central) the petrostructural distinctions of the Brame, Saint Sylvestre and Saint Goussaud granites are characterised by spectral variations induced by the orientation of fractures and magmatic fluids, e.g., N-S fractures in the Brame Massif (at least up to the Gartempe), E-W fluid flow in the Saint Sylvestre block and any fractures in the Saint-Goussaud Massif. Lastly, north of the Gartempe the Brame facies is considerably attenuated on Landsat images, tectonics and petrography once again playing a preponderant role in this variation of response (Plate 7).

In the preceding examples the responses are clear and simple photointerpretation suffices. Such is not always the case; it sometimes is necessary to employ more complex technics based on the digital processing of images and/or multivariate analysis. This was so for the eastern Marche granite in the Massif Central (France) (Royer and Laurent, 1982) where the multispectral approach aided differentiation between leucogranite and biotite-granite and identifying variations within the leucogranite, conforming to ground observation data. However, these differences in responses are small and their classification only approximate since it is very much dependent on vegetation, which in this case was an indicator of bedrock. That is why an attempt was made to enhance these differences by employing regrouping methods based on factor analysis of variables of remote sensing, major element geochemistry and trace element geochemistry. This type of approach aided in refining certain correlations indicated between geochemistry and remote-sensing data and in explaining them by a number of simple processes, such as:

— Variations in chemical composition observed within the granite blocks which are often controlled by the original structure (dykes, batholiths) and the various tectonic events to which they were subjected.

— Alteration phenomena largely dependent on the permeability of the rocks and the state of their fracturing.

— Nature of the vegetation cover, closely related to the state of alteration of the substrata and which may be considered an indicator of the geochemistry and mineralogy of the underlying rocks.

Given this perspective, it is possible to correlate the surface characteristics observed from remote sensing with the geology of the covered bedrock (see Sec. 6.4.2).

2.4.7.2 Mapping granites below other rock formations

This project was more ambitious since the signal received by the remote-sensing detectors emanates from the surface or very close to the surface. Experience nevertheless shows that it is possible, due to the transparency effect (the term 'transfer' would perhaps be preferable), to interpret certain surface signatures, for example the effect of buried domal structures. However, contrary to the case of granites below surficial deposits whose existence can be deduced in a simple and economic manner, identification of granites occurring below other rock formations requires the application of more exhaustive means of analysis. This type of information has been obtained in France and Spain using several surface criteria.

Such is the case of Saumur granite (France) (Scanvic, 1977) in the Cretaceous and Eocene horizons that cover it, indicated by a group of faults and spectral responses independent of the boundaries of these sedimentary formations. These faults and boundaries reproduce rather exactly the form of a significant gravity anomaly which correlates with a granite known from drilling. The real effect of transparency is observed here, which can be explained by the reactivation of deep faults and/or the differential subsidence phenomena that induce specific surficial manifestations.

At Saint-Pardoux (Massif Central, France) a circular structure visible on Landsat images has been correlated (Maroncle, 1985; Maroncle *et al.*, 1989) with an aeromagnetic anomaly (see Sec. 4.4.1.2) produced by a magmatic intrusion known to outcrop farther north and emplaced below a cover of metamorphic rocks. The stress effects resulting from this emplacement are probably the cause of the surficial manifestations observed on the Landsat images.

In the region of Redon (Armorican Massif, France) an interconnected chain of granites in the crystalline terrain, outcropping at places, has been found to occur below the crystalline schists. This is also indicated by the presence of a gravity anomaly as well as by a typical crystal structure of mica on the surface (Guigues, 1984). This granitic axis was identified in remote-sensing images because of the structural manifestations induced by it on the surface and indicated by a narrow lineament associated with the anomaly in the crystal structure of mica.

In Spain (Marconnet, 1985) remote-sensing data facilitated detection of a leucogranite body, often concealed, which is responsible for the mineralisation of tin and tungsten. These intrusions have developed a number of transformations in the enclosing metamorphic rocks (contact metamorphism, hydrothermal alteration, fracturing), which have been identified in remote-sensing data through the application of certain digital processing technics. Two exemplary cases of 'transparency' of a host rock vis-à-vis an underlying magmatic body have been observed in the Zamore province (Spain) and in Rehamna (Morocco).

2.4.7.3 Additional comments

Other methods of remote-sensing study have also been successfully employed for mapping concealed granite bodies in regions of temperate climate. One may mention, without going into detail, the multiscale approach employed by Brosse (1975) for the Villefranche-de-Rouergue granite (France) and by De Fraipont (1982) for Saint Sylvestre granite (Massif Central, France). Additional information on the application of thermal remote sensing can be found in the reports of Rousselin (1985) and in Section 7.2.2.1 of this volume (Baud granites). Most of these results were obtained from Landsat MSS data only, with a resolution of 80 m, on localised granites in a contrasting petrographic environment. These granites were mapped with notable precision in a medium-scale survey using photointerpretation of satellite images and specific application of certain digital processing technics.

These performance capabilities have now been improved with usage of the new spectral bands of Landsat TM, in particular the 1.6 µm band, and Spot data with a resolution of 20 m, enabling significant enhancement in quality and opening scope for geological mapping in the 1:50,000 scale.

2.5 CONCLUSIONS REGARDING GEOLOGICAL MAPPING

2.5.1 Geological Map and Remote Sensing

The examples described in this chapter illustrate the capability and efficacy of remote-sensing analysis in different stages of geological mapping, especially when based on the use of stereoscopy:

- To facilitate updating and improvement of existing data as well as its coherence;
- To aid in orientation before and during investigations;
- To secondarily fill the lacunae in observations;
- To detect specific features in each case in zones known to be homogeneous; and
- To accurately map specific geological features, metallotects and aquifers.

However, assessing the degree of efficacy of stereoscopic aerial photography led to certain concepts concerning the ratios between scale of the image and ground features. These ratios are governed by a number of factors whose effect on the results can be evaluated as follows.

- *Type of geological maps*

It is often necessary to know whether the given maps pertain to regular systematic surveys attempting total geological coverage of a country, the scales of these surveys and their ultimate application, or whether they refer to mineral exploration, hydrogeology or management (materials, natural hazards etc.). When the requisite scales are large and the acquired information is vast, use of stereoscopic aerial photography is still attractive to the geologist, particularly in view of the new generation of sensors with high geometric resolutions expected to be available by about 1998 (Spot NG, with a resolution of 5 m).

In general, the efficacy of remote sensing is considered to be good for small and medium-scale surveys, especially when these are carried out for a well-defined purpose.

- *Concept of mapping*

A geological map is the result of integration of ground observations and measurements made on outcrops that are often discontinuous. It is a complex geological document which combines the chronology and geometry of geological units on a topographic and drainage base. It is the final document resulting from interpretation of observed facts and hence is greatly influenced by the concept of mapping and the structural hypotheses adopted. Some of the factors that strongly affect the results of mapping are:

— Stratigraphy, which predominantly influenced, for example, the geological map of France, i.e., the effect of major stratigraphic boundaries on the chemical nature of rocks.

— Lithology: certain maps highlight differences in the nature of rocks while preserving the stratigraphic connotation.

— Petrography and structural hypotheses, which constitute the basis for the reconstruction of geological history but are very sensitive to the weight of various theories.

— Surficial deposits represented simply in some maps and in a complex manner in others.

The efficacy of remote-sensing interpretation, which is very sensitive to the relief, is enhanced when geomorphology is fully taken into account.

- *Nature of rocks*

As the morphological response exerts a dramatic effect on photointerpretation, the results of mapping vary with the nature of the rock. Photogeology has responded primarily to the demand of petroleum explorers since sedimentary rocks are clearly distinguishable in the landscape, often better than crystalline rocks.

- *Mode of emplacement and effects of tectonics*

The mode of emplacement of certain rocks, such as intrusion or extrusion, manifests the morphological expression.

Tectonic effects enhance the potential of interpretation when simple. When complex, interpretation becomes less reliable, as in the case of overthrusting.

- *Environmental conditions*

The effect of environmental conditions on the results can be significant and may also play a decisive role in the choice of methods.

Other factors may also be cited, such as occurrence of outcrops which facilitate or hamper observations; accessibility of study areas, which determines duration of fieldwork and choice of methods; climatic conditions, which constrain the periods of fieldwork and control the choice of remote-sensing technics; lastly, the restriction policy and costs involved. All these factors can lead to maps which are not state of the art.

It can be seen from this brief analysis that the relative weight of various factors on the results of mapping necessitate definition of several types of maps, depending on the mode of acquisition and validation of space information. It is also possible to propose an adaptive methodology for each type of map.

A recently published book (Prost, 1994) presents a comprehensive detailed current guide to small-scale photogeology of remote-sensing images, applicable primarily to petroleum exploration.

2.5.2 Role of Remote Sensing in Future Geological Mapping

Remote sensing has elicited a significant change in the concept of geological mapping today. Several types of investigation can be identified which would resolve such primary concerns as:

— Geological optimisation of the digital structure of remote-sensing data;

— Realisation of a detailed digital exploitation of such data by modelling the extraction of geological information derived from analytical interpretation based on a multi- and especially interdisciplinary integrated approach. While a number of natural processes are known, relatively speaking in detail, their mutual relations and the effects they produce are less understood.

— Analysis of multicriteria, i.e., choice of variables and expected function of mutual relations.

The answers to these questions would perforce determine the role and significance of remote sensing in each phase of mapping, since the latter undergoes development at different levels.

At the level of computer-aided mapping, studies are presently aimed at enlarging its usage primarily towards the tasks of conceptualisation and preparation. This approach must allow in future for conceptualisation of thematic maps wherein data collected at different stages of interpretation, of interpolation and locally during ground surveys is utilised concomitantly. This would necessarily involve some change in graphic indexes presently conceived for printing a single map, for regrouping information according to themes. The structure of geographic information systems is amenable to those changes and remote sensing must find a place amongst them. At present, most maps preliminarily prepared from field surveys are graphic in nature because obtained from visual interpretation. Rarely is this information digitised from the beginning for comparison with earlier data and preparation of a common map. To rectify this situation, at least in part, efforts must be made to develop a geological information system provided by remote sensing which can be digitised, and to critically evaluate the potential of different methodologies based on several approaches:

— Direct digital photointerpretation in 3 or 2 dimensions.

— Investigation of shapes and linear or circular lineaments.

— Spectral classification of geological objects.

— Spectral and textural classification based on the concept of landscape unit.

Direct digital synthetic interpretation is based on the principles derived from photogeology. It has sufficient scope for development in future since certain programmes capable of incorporating digital processing methods and generating images already exist. Modification of an existing database created on the basis of direct two-dimensional observations, such as an image or map, into a three-dimensional screen and extraction of certain information that allows precise determination of the thickness and dip of a sedimentary sequence or extrapolation of boundaries below surficial deposits are already standard procedures. Such programmes have been developed by the ISTAR Society, France. However, the limiting factors that hinder routine application of the method are the cost of operations and paucity of digital processing methods. The Spot system could ameliorate this situation through the acquisition of data in the stereoscopic mode.

Digital extraction of linear objects and their statistical analysis have been in vogue for several years, subsequent to the development of such programmes by Linéame (BRGM). They suffer certain common limitations, however, namely not being able to *a priori* distinguish linear lineaments of geological or structural origin from those of other causes, and their final classification requires the intervention of human experience. But were this information to be transformed into density maps, programs (such as Dense) are available whose statistical results are not distorted by the medley of automation, except when applied to certain types of land-use.

Important research on the identification of geological forms is presently in progress at the Pierre and Marie Curie University (Chorowicz *et al.*, 1989).

In general, geological interpretation of the features on satellite images is carried out by geologists who have specialised in remote sensing and are able to analyse their particular pattern. They employ diverse parameters not visually apparent and combinations thereof that are often not explicit, to arrive at an integrated interpretation. Such an interpretation for digitally classifying an object differs from the concept of analytical interpretation which considers the image as a space to be described by the characteristics constituting its entities and structures (Flouzat, 1988). This method, though simple, is not yet very reliable in geology albeit physicists have demonstrated in the laboratory that geological objects can be classified according to their spectral attributes alone. Under pressure of demand such digital procedures were mainly based on spectral classification initially to optimise the differences and later to tentatively identify the minerals and rocks. According to the ground conditions, whether arid, temperate or equatorial zones, the quality of results varied markedly, depending largely on the spectral range utilised. Even if geological, metallogenic or geobotanical responses were obtained, they were severely altered by those of the surficial deposits.

Thus the results obtained to date show that the spectral response of geological bodies is a necessary but insufficient condition of classification. Hence one should also use other compatible attributes along with analytical interpretation. Studies have been done (Flouzat and Merchoub, 1983) and are in progress using the methods of mathematical morphology applied to the definition of geological textures. In our opinion it is important to know whether the attributes utilised characterise geological phenomena and why. If so demonstrated, it will be possible to improve classification in regions where surficial deposits and vegetation cover predominate. Preliminary results have shown that one can distinguish 'landscape units' on satellite images. These are some extended areas that present constant relations between some of their components. The different components exhibit strong correlation between vegetation, morphology, fracturing and bedrock. The method thus consists of integrating these datasets, of choosing homogeneous and representative segments and then classifying them. An experimental study of the types described in Section 6.4 is under-way at the Cevennes (Deroin *et al.*, 1989).

The problem of multicriteria analysis in geology constitutes an important area of investigation in different fields, such as geological mapping, mining geology, hydrogeology, natural hazards and geotectonics. These studies are facilitated by the existence of modules of multicriteria analysis and geographic information systems capable of analysing, managing, processing and restoring the data localised in space.

In this perspective the remote-sensing technic presents a great potential for providing initial basic information. Multicriteria maps utilising remote sensing as a component have been and are being developed on this basis. One can cite several examples executed at the BRGM:

— Geological and mineral map of Mouila-Gabon;

— Hydrogeological map of southern Morocco;

— Groundwater map in the Nord-Pas-de-Calais;

— Map showing zones of sensitivity to landslides in Bolivia, Colombia and Taiwan;

— Map of possible erosion in Guadeloupe and China;

— Map of mineral occurrence in the Armorican Massif and hydrogeological prospects in the Sudan.

The specific type of information required from remote sensing for each example varies according to the target:

— Integrated visual and digital interpretation for Gabon.

— Lithological and structural interpretation, digitised and integrated with hydrological information on a space-image base in Morocco (Plate 8).

— Digitised structural interpretation, classification of soil types and surficial deposits, combined with hydrological information in a balanced approach in the case of Nord-Pas-de-Calais.

— Combinations of different classified or digital maps and the products derived from numerical modelling of the terrain based on Spot data or topographic maps for Bolivia, Taiwan (Plate 9), Colombia (landslide hazards), Loess plateau in China and Guadeloupe for erosion.

— Combination of different modes of observation and measurement.

In conclusion, several studies for improving the performance of digital and analytical interpretations can be foreseen. Such improvement will depend on the degree of integration of remote sensing into geological information models. Though visual interpretation presently gives good results, taken alone it would underutilise development of higher performance scanners, as it restores only part of the filtered information contained in the imagery.

Chapter 3

IMAGE LINEAMENTS

In order to remove ambiguities attached to the concept of lineaments, which was first used in geological remote sensing, the concept of image lineaments is proposed through analogy with geophysics. An image lineament, which is a linear object of a priori geological origin, is a structural expression detected by remote sensing. The examples presented show the importance of multidataset analysis in this field, which enables satellite-derived information to be validated in its use for applied geology: exploration for aquifers in areas of fractured rock, mineral exploration associated with shear zones and fracturing in general.

Original features that can be obtained through observations made on active microwave and thermal data are then examined in relation to structural geology. This is followed by a discussion of the principles of radar and the prospects that it offers since specialised satellites (ERS 1, JERS 1) have been put into orbit.

The conclusion stresses the need now felt for interpreting remotely sensed data before undertaking field studies so that spot information collected in situ can be located. It hints at the advantages that should eventually be obtained through modelling lineament networks, in particular when this is based on the fractal theory.

3.1 GENERAL

Images collected by radiometers of the first-generation satellites, especially the Landsat 'multispectral scanners', have revealed through the synoptic vision they provide, the existence and geological significance of linear discontinuities in the landscape. These discontinuities are termed 'linears' in English literature and 'alignments' and 'lineaments' in French.

Images obtained by second-generation satellites are characterised by improved spatial and spectral resolution. In some regions, where interpretation is based on second-order indicators (Fig. 7), these high-resolution images have facilitated differentiating image discontinuities, i.e., image lineaments, into two categories according to their morphological or radiometric expressions.

— A morphological type (Plate 10) characterised by lineaments that are emphasised by the shadow effect over valleys, slopes or cliffs. It should be noted that even major faults may not be indicated on the images by this phenomenon if they are parallel to the direction of illumination and, on the other hand, minor faults may have their effects enhanced or exaggerated if they are oblique or perpendicular relative to the same illumination.

— A radiometric type characterised by an abrupt change of colour. These changes represent a variation of vegetation population or the alignment of vegetation and have specific textural forms that reflect complex relations between geology and soil parameters. For example, in the Armorican Massif these radiometric lineaments, when observed on bare soils, were associated with distinct veins or large blocks of quartz (Braux *et al.,* 1990). In the Massif Central they represent a fault with a vertical throw of 3000 m, revealed from geophysical surveys (see Plate 2).

These composite lineaments, which may not always have an evident structural implication, have generally been identified with 'metallogenic lineaments'. This type of assumption has been made in the past, which may not have been correct since all the linear features did not necessarily have a proven structural origin, as validated by remote sensing. Nevertheless, the concept that lineament is associated with a metallogeny is largely held by those specifically concerned with studies of lineaments. If the overrating is removed, this notion is validated by remote sensing and its application to mineral exploration, even though it is now known from accumulated experience that lineaments are generally reflected only on the data of first-generation satellites due to their medium spatial resolution and wide synoptic view, which reduces the structural detail in favour of identification of composite lineaments.

In certain respects the high-resolution images of second-generation satellites are important because the lineaments are often more accurately observed, which expedites identification of faults, fractures and diverse structural phenomena that can be readily located on the ground. In fact, it is clear that the term lineament is perfectly adaptable to the structural study of regions with thick pedo-vegetation cover, in which ground verification by conventional means is more difficult and often deceptive if important methods of investigation such as geophysical surveys are not employed. In the case of arid or well-exposed regions in which lineaments in general

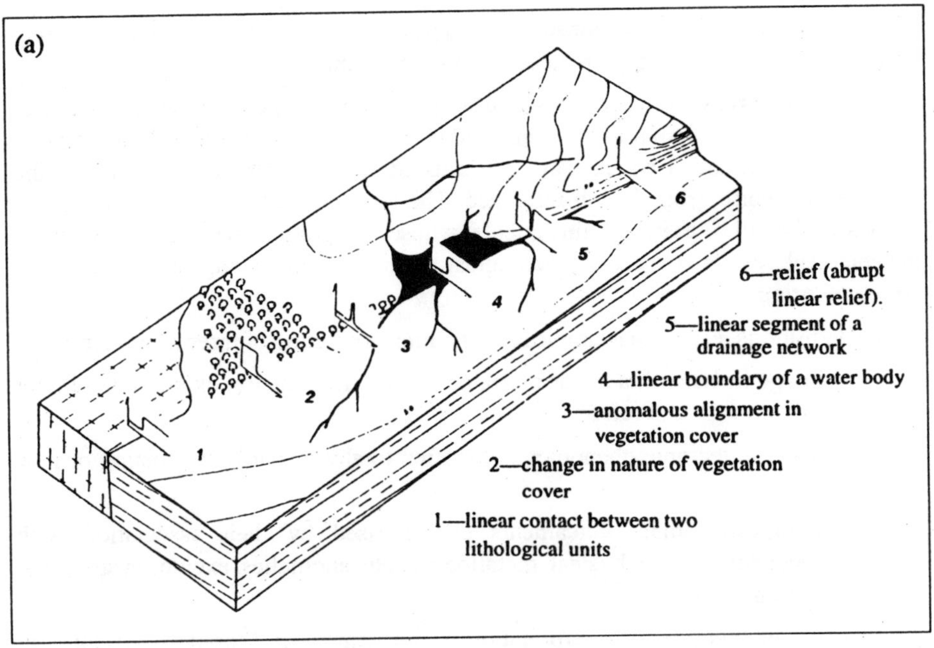

(a) Variation in signal of a radiometric profile across a lineament

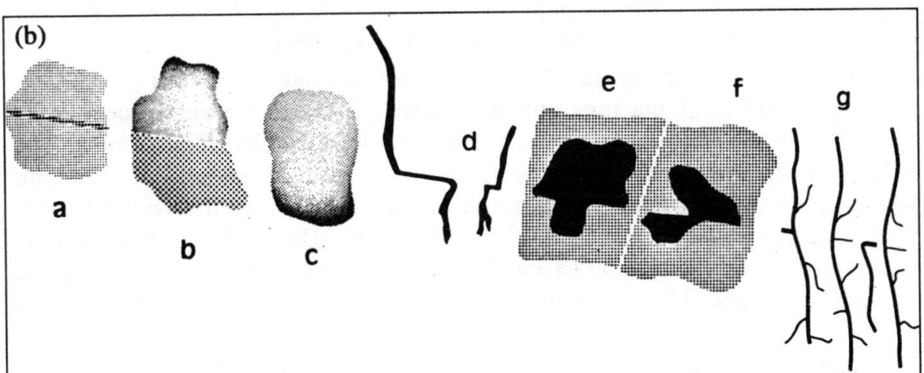

a—linear arrangement of pixels of a different shade in an otherwise homogeneous shade
b—drop in tonality
c—continuous variation of the surface
d—linear segment of a water course
e and f—lakes with linear banks
g—water courses having linear branches or tributaries.

Fig. 7. Some criteria for defining an image lineament (after Rouzeau, 1989).

have a clearly identifiable response on any type of image, the term lineament is rarely employed or used depending on the structural implications.

In view of these reasons, it is proposed to adopt the term image lineament (Castaing *et al.*, 1989), which is similar to the term used in geophysics, to qualify all the linear or curvilinear phenomena which can be presumed to be of structural origin, in the absence of precise ground truth reference. The term composite image lineament is proposed for all the image lineaments whose integration with geophysical or geochemical results reveals the geological nature without being able to confirm structural origin.

These definitions lead us to a simple classification consisting of three major terms:

— Image lineament, a linear morphological or radiometric expression without a precise ground reference;

— Specific tectonic phenomena identified totally or partly in existing or new maps;

— Composite image lineaments characterised by their association with geochemical, geophysical, metallogenic etc. anomalies, excluding structural geological features.

An example of this type of interpretation in the Armorican Massif is presented in Plate 11.

3.2 IMAGE LINEAMENTS OF STRUCTURAL ORIGIN

Mapping of structural phenomena often plays a major role in different phases of prospecting for subsurface resources. Structural geology has for several years been one of the preferred tools of exploration geology or applied geology.

It is in this domain that the contribution of remote sensing is the most perceptible. Structural analysis of outcrops and inventories of lineaments on space images perfectly complement each other, as the methodology developed has shown. The most evident approach for accelerating the pace of investigations is modelling because the features observed on images at a different scale from that on the ground, represent certain models defined by structural geology. The problem is thus to know what they represent on the ground and whether the model described for that specific area can be transposed to the one appearing large scale on the landscape viewed from space.

Remote sensing constitutes a tool for increasing our knowledge of rupture tectonics of the entire planet, which we shall attempt to demonstrate with examples.

3.3 COMPOSITE IMAGE LINEAMENTS: THEIR RELATION WITH METALLOGENIC AND GEOLOGICAL LINEAMENTS

We shall now discuss the evolution of the term lineament proposed by Hopkins (1841), well established in usage and its limitations in remote sensing, and show the sense in which we attribute importance to it in the strategic phase of prospecting for subsurface resources.

The relation between a tectonic structure and an alignment of geomorphological features affirmed by Hobbs (1902) is a significant element in the definition of lineaments. According to Hobbs, lineaments reveal the hidden structure of the geologic substratum without indicating the precise mechanism. That is why Sonder (1938) subsequently used the terms rhegmagenesis and rhegmatic faults characterised by structural phenomena of geometrical order, implying force mechanisms. These faults have a spatial continuity and are confirmed on the ground (Caire, 1975). Their trend is generally linear, slightly broken and barely curved. They may present a tensile nature in the whole or part of their trace, which facilitates intrusion of magma or localisation of mineralisation.

The term lineament evolved further in 1963 when Bardet associated the idea, which he took from Sonder, with the preferential localisation of diamondiferous kimberlites. According to Bardet (Fig. 8), control for emplacement of kimberlites is provided by faults or structures that are at times surficial but more often deeper. The alignment directions of the points of kimberlite intrusions emphasise these faults. In Africa they are associated with major basement faults, reflected in the pattern of coastlines and

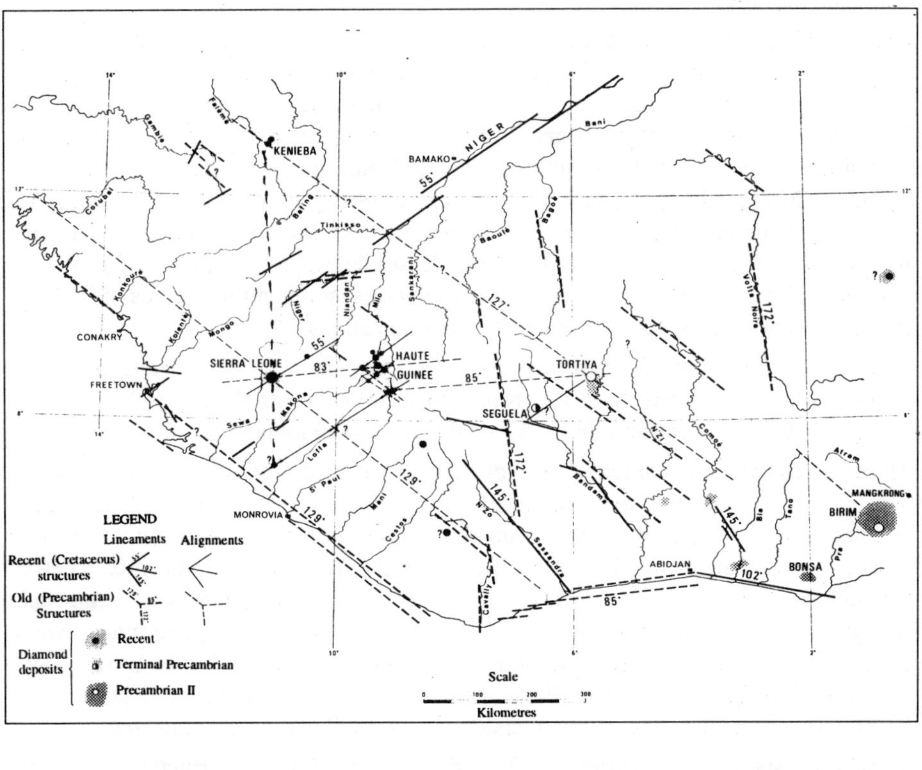

Fig. 8. Lineaments and diamond deposits of west Africa (after Bardet, 1963).

river courses. These structures are generally not marked by faults visible on the ground but some are distinctly indicated by geophysical discontinuities; that is why Bardet preferred the term lineaments for such structures.

Other geologists extended this concept to the process of mineralisation of a particular group of deposits and created a lineament and rhegmatic school of thought. Based on diverse investigations carried out in North America in particular, they tried to demonstrate that high mineral concentrations tend to be emplaced in the intersections of lineaments or mineralised nodes (Mayo, 1958; Kutina *et al.*, 1967; Wertz, 1963, 1971).

This concept was also extended to petroleum resources (Moore, 1971) of a region in the USA in which the major producing fields are aligned along the lineaments or along the flanks of curvilinear structures.

Elaborating this phenomenon, Pélissonnier (1974) defined a lineament as 'the boundary between two units of totally different geological histories'. More precisely, he affirms that 'copper deposits, without exception, appear to be associated with major lineaments on a regional scale and that in the case of large porphyry deposits, representing more than half the tonnage of metal reserves, these lineaments correspond to the boundaries of certain orthogeosynclines of the coast where the deep shear plane that characterises it (Benioff plane) is subducted'. This represents a very restricted definition.

Nicolini (1980) supplements it by separating the 'geological lineaments' from the 'metalliferous lineaments'. The geological lineaments are sometimes masked by more recent events and appear in the form of alignments emphasised by erosional boundaries, deviating river courses, continental slope margins, alignments and intrusions. They are linear, sometimes curved. Using them, one can investigate the origin of different features, such as ancient fractures, plate margins and boundaries between mobile and stable zones. He also remarks that the metalliferous lineaments, well known for tin, tungsten, gold and to a lesser degree for copper, on the African continent in particular, do not have evident relations amongst themselves or with the geological lineaments, at least for certain mineralisations.

These discordances amongst lineaments can be explained by the palaeogeographic, sedimentological and biological environments associated with the deposits, which are more recent than the causes that produced the geological lineaments.

Thus the concept of lineaments has evolved over a period of time and remote sensing aids in interrogating the significance of linear structures revealed by space images and the similarity they have with lineaments. To answer such questions, it is desirable to consider the data available to those who were the 'inventors' of lineaments before the development of space images. The data primarily comprised small-scale topographic maps (Daubree, 1879) and composite regional geological maps, which facilitated the first inventory of lineaments, particularly in the USA. In 1958, Huppé thus prepared a world map of lineaments based on the rhegmatic hypothesis of Sonder. But evidently these maps revealed only a small part of the lineaments.

Metallogenists such as Bardet (1963) and Nicolini (1980) constructed lineaments from maps of known occurrences of mineral deposits. Robert (1968) included in this construction the preferred directions developed by orogenesis, seismicity, volcanism, hydrothermal activity-all the elements which contribute to the surface manifestation of lines of weakness due to which the primary metallogenic emplacements would take place.

Finally, certain geologists (Duplan, 1973) attempted to extract lineaments from large-scale topographic maps. This approach is effective only when the topographic maps are uniform in representation of features, which is not always the case; often they reflect only a part of the lineaments, those that have a morphological expression marked in the elevation contours.

In every case the underlying principle for preparation of rhegmatic network maps, which turned out to be lineament pattern maps, was to detect by superposition of various maps the longest geofractures, probably also the deepest, corresponding to shear zones along which crustal blocks are displaced.

Since 1972 remote sensing has become an essential tool for mapping lineaments because it is the only economic and universal means of discovering, without preliminary constructions, all the characteristics of the landscape that are responsible for the architecture of lineaments. It facilitates the preparation of a relatively detailed inventory of homogeneous though imperfect conditions (problem of scanning, for example).

3.4 IMAGE LINEAMENTS AND PROSPECTING

The keen interest shown by certain schools of metallogeny for the concept of lineaments and fractures makes remote sensing of image lineaments a supporting technic in the prospecting of subsurface resources. However, while remote sensing allows quite an exhaustive inventory of linear structures, their metallogenic significance is not always evident and varies with the scale of presentation. The problem has been approached in different ways and we shall present and discuss it on the basis of specific examples.

3.4.1 Evaluation of Image Lineaments and Multicriteria Analysis

In order to utilise remote sensing data for prospecting, it is necessary to evaluate it in terms of structural and lineament characteristics by all complementary tools of analysis on the ground. One effective method is to compare visual, statistical or digital maps of lineaments with existing geological, tectonic, geophysical or geochemical maps. This approach facilitates precise estimation of the nature and geological significance of these lineaments on the whole or part of their trace, to classify them and to characterise the uncorrelated elements by continuity or analogue. Several methods are in vogue.

3.4.1.1 Comparison of remote sensing and tectonic maps

This exercise has been done for France in different scales. Using a small scale, the whole of France was recently covered by a geological map (Dottin, 1980) at 1:1,500,000 and a tectonic map at 1:1,000,000 scale (Chiron, 1980). A comparison of these maps with the linear structures interpreted from Landsat MSS images revealed some common features and faults whose tectonic nature was ascertained by subsequent ground surveys.

- One of the common features in the two maps is that all the known major faults are indicated in them, as in the case of the Armorican trough, Sillon Houiller, the pre-Cevenole boundary and a number of others of lesser importance. Allow us a slight digression here to examine a statement made by Routhier (1980): 'If *a priori* too little is known about the geology and metallogeny of a region, they [the Cevenole faults] attract no more attention in remote sensing data than any others.' This statement is true when only the lineament map of France is considered (Scanvic *et al.*, 1980) but not so if we look at the image which forms the basis of interpretation. In Bretagne the Armorican trough, the faults of Montagne Noire and the Arres anticlinal ridges have a distinct expression. Similarly, Sillon Houiller, or the faults of the Cevenole margin are strongly indicated by a characteristic morphology which facilitates their identification. It is also remarkable that these major faults are conspicuous on the thermal HCMM and NOAA data whose resolutions are much poorer than that of Landsat images, thus offering an additional means of evaluating their significance through the analysis of data of different scales.

Accordingly, even though the existence of deposits near these faults cannot be foreseen, the stage-wise mapping of linear structures aids in effectively planning the prospecting strategy in the reconnaissance phase itself. The example of Cevenole boundary is of significance in this regard since the faults which brought the shallow Triassic formations to the surface and which are responsible for emplacement of important mineralisations, are strongly reflected in the morphology, making them distinguishable from other regional faults and fractures (point A, Fig. 9).

Some lineaments whose tectonic significance was not known at the time of publication of geological or tectonic maps but which were visible on Landsat images, have been subsequently identified on the ground.

An example of such identification is the south-westward extension of the group of Metz-Luxembourg faults (France) observed on Landsat images in 1977. Their tectonic origin has been established in the Champagne chalk formation by a detailed study of microfauna (Labourguigne *et al.*, 1978). A more recent publication (Coulon *et al.*, 1988) has revealed the significance of the response essentially recorded in band 7. The chalks associated with the lineament are 'split' under the effect of brecciation induced by hydraulic fracturing in a tensile medium.

The Sillon Houiller (Massif Central, France) is another interesting example. After 1975, following the preliminary band-by-band interpretations of the black-and-white images of 'ERTS-1', it appeared (Scanvic, 1975) that the Sillon Houiller is not a singular fault, as represented in various geological and tectonic maps, including recent (1981) ones. Space observations showed it to be a group of parallel faults defining a more or less continuous valley. Additional studies subsequently carried out on colour composite images enlarged to a 1:100,000 scale after resampling type of digital processing, revealed the complex structural setting of the Sillon Houiller (Fig. 10), broken or at least intersected by transverse faults of Armorican or Variscan orogeny (Scanvic, 1986). These original observations are confirmed by 1:50,000 scale geological surveys in the Pontgibaud region and in the south (Feybesse, 1981), where they have a remarkable expression in thermal images (see Fig. 14).

Fig. 9. The Cevenole boundary seen from Spot (after Deroin, 1990) (CNES, 1986. Spot Image Distribution. Processing by BRGM)

- Mapping of lineaments in the Arabian Shield and its sedimentary surroundings constitutes another significant example outside France. This project of the Ministry of Petroleum and Mineral Resources of Saudi Arabia was carried out under the aegis of the Minatome Society (1982) by a unit of BRGM. It covered an area of about 920,000 km^2 using 1:1,000,000 Landsat images, with the objective of tracing the structures favourable for the occurrence of uranium. The inventory was supported by lithostructural maps when such existed, which facilitated identification of the geological nature of certain image lineaments, demonstrating the reliability of the technic and validation of extrapolations or interpolations made. The outcome of this study has been the preparation, within a very short period, of a complete structural map of the Arabian Shield, until then covered only by segment-wise interpretations.

As far as mineral exploration is concerned, this map provided new ideas about the location of certain potential traps of mineralisation and in particular:

— The existence of Red Sea type tensile faults, some of which constitute meridional continuations of the phosphatic structures of Jordan and are potentially significant for uranium.

— A volcanosedimentary sequence whose metallogenic potential is established elsewhere.

— Structural control of recent volcanism by the N-S oriented fractures, which appear to correspond to old basement faults reactivated subsequently.

— Tectonic control of intrusive structures that are regularly associated with N-S and NW-SE faults. These intrusive structures are considered potential exploration targets.

This study could not have been possible without the support of remote sensing because the state of progress of geological mapping of the Arabian Shield would not have enabled its execution within a short period and with the required quality.

- In the case of petroleum exploration the project report by the Bureau of Industrial Studies and Co-operation of the French Institute of Petroleum (BEICIP) regarding the Arc of Gibraltar and the Alboran Sea illustrates the application potential of this method (Biju-Duval *et al.*, 1975).

This study was carried out using Landsat MSS images by adopting the following procedure for interpretation:

— Attributing a geological weightage to the linear structures according to the criteria of identification (definite, probable and possible geological origin).

— Classification of linear structures according to tectonic criteria, interpreted either directly from the image or after the preparation of geological maps: shear faults, tensile faults, strike-slip faults etc.

— Identification of tectonic provinces by demarcating zones of homogeneous fractures.

— Developing a regional tectonic model.

The results of this study in the offshore region were compared with bathymetric and geophysical data. The comparison helped in developing a uniform regional structural model, including onshore and offshore zones, and proposal of a model for the evolution of the region consisting of the Arc of Gibraltar and the Alboran Sea. This interpretation provided a reasonable answer to certain questions posed by the petroleum geologists engaged in prospecting in this region.

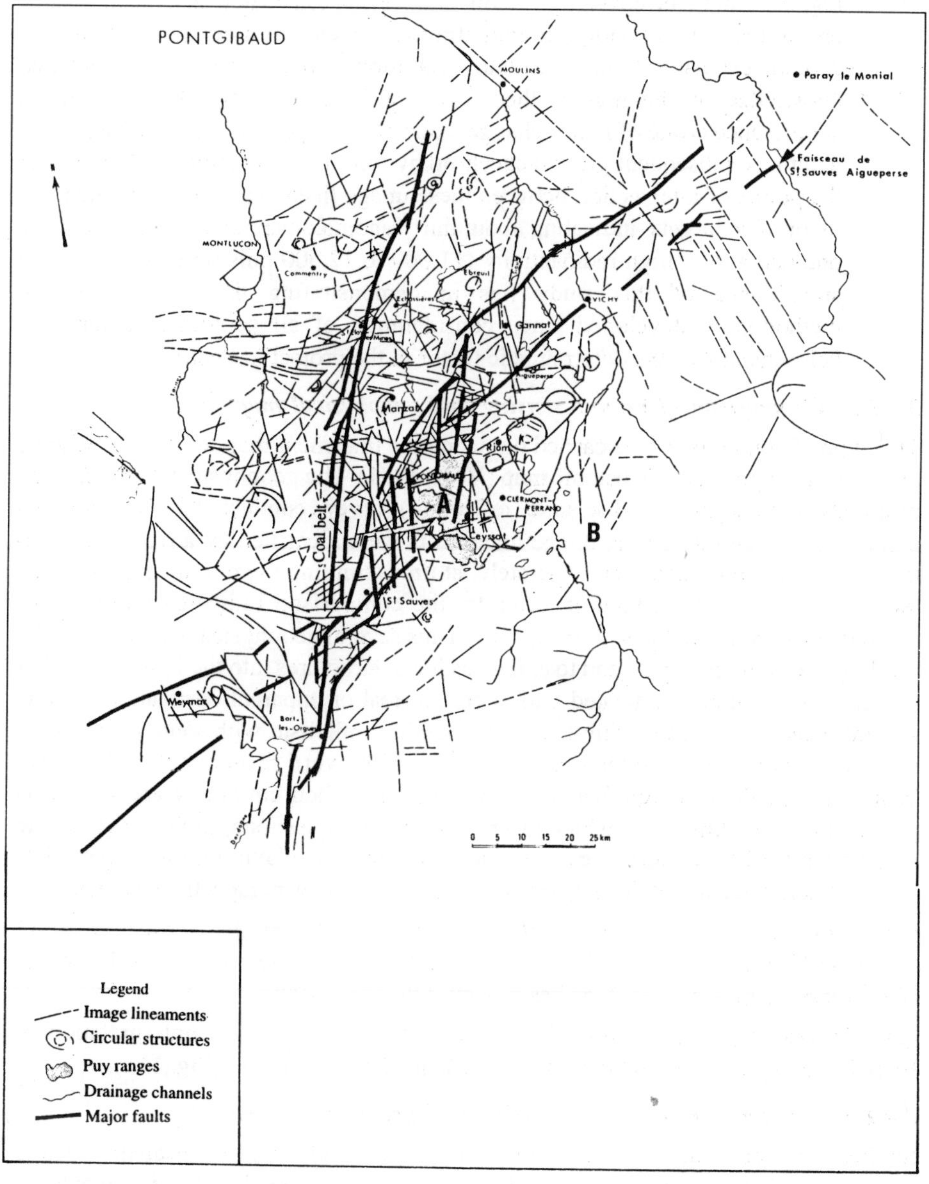

Fig. 10. Pontgibaud fault system (interpretation of Landsat MSS images).
A—Puy ranges; B—Limagne.

- Another example of application of a similar approach in mineral exploration pertains to uranium exploration in the Montmarault region (France). In this case (Sossa, 1980) the exploration was planned using an image lineament map based on Landsat images and a photogeological map of fractures. These together provided a satisfactory compilation of structural geology of this region, which is covered by soil and vegetation. They also facilitated a preliminary classification, highlighting alignments of a certain continuity, a certain extension and a certain repetition. Thus, based on a detailed knowledge of the massif, the author selected some for their geological significance (zone of facies change, zone of foliation, fractures etc.) and their length. A subsequent ground check-up by scintillometer surveys showed that the points of intersection between these image lineaments represent the most favourable targets, indicating about thirty anomalous points, of which at least one had a very high radioactivity (11,000 to 12,500 pulses per second). The significance of this study lies in demonstrating the efficacy of the methodology developed since consideration of all geological parameters beforehand was the deterministic factor in the project's success.

3.4.1.2 Comparison of remote sensing and geophysical maps

This type of comparison was carried out using the lineament map of France (Scanvic *et al.*, 1981) in preparing a tectonic map (Vogt, 1981). The geophysical data consisted of gravity, aeromagnetic and seismic (reflection) surveys conducted for hydrocarbon exploration. Even though the image lineaments were identified by a rigorous selection and were represented very discretely, at times limiting them to the most consistent portions so as to reduce uncertainty, the comparison indicated a number of coincidences with geophysical lineaments. Thus several geophysical structures of the pre-Pyrenees region were found to extend as linear structures into the Pyrenees. Other associations were also indicated with a certain real or apparent correlation. Lastly, several image lineaments exhibited a remarkable parallelism with some geophysical structures. Thus some major trends such as 70° N were subsequently identified. While their geological significance is known locally, their continuity and repetition give them a new dimension which prompts numerous questions. In this case remote sensing tends to pose more questions than it solves, thus playing a role expected of it-stimulating the curiosity of geologists by suggesting new perspectives of tectonics.

Other investigations on a regional scale have led to similar results in Berry (Debrand-Passard *et al.*, 1977), Limagne (Debeglia and Gerard, 1973), Limousin (Alsac *et al.*, 1980) and the Paris basin (Rouzeau, 1985, 1989).

Another example, chosen from the Cholet region, compares the interpretation of ERS-1 radar image with the gravity anomaly of Chemille granite (Fig. 11).

3.4.1.3 Comparison of remote sensing and geochemical maps

Studies have been done on the geochemical nature of certain lineaments. These studies are of interest as they indicate the possibility of evaluating lineaments by methods other than conventional structural analysis.

Limited to a few regions where geochemical surveys have been made and not generalised, these investigations have yielded significant results. Thus based on statistical analysis of mineral resource inventories of France by the autocorrelation method, Roquin (1984) detected the existence of linear geochemical anomalies that correlated perfectly with linear structures. This type of correlation reveals the possibility of reinforcing one piece of information with another and facilitates the selection even when the structures thus confirmed do not reflect the faults visible on the ground. Such a situation is especially observed in the Bodennec region (Roquin, 1979), Nort-Sur-Erdre (Scanvic, 1984), the Armorican Massif and in the Iberian peninsula (Cottard, 1982). In Pontgibaud district of the Massif Central (France) this type of comparison revealed a strong correlation between the geochemical anomalies (As, Ba) and the image lineaments visible on Landsat MSS images. Thus it indicated a new metallotect at the intersection of NNE and NNW faults suspected from geological and geophysical data (Maroncle, 1985, 1989).

Valencon and Berger (1980) reported the demarcation of zones of circulation of mineralised fluids from geochemical and remote-sensing studies. Using the 'swath width' map, a special mode of representing geochemical data, they characterised at each point the local variations in the concentrations of the most soluble elements. Thus they detected lineaments along which anomalous concentrations were observed for barium, strontium, uranium, sodium and silicon. These anomalies were interpreted as corresponding to the fracture zones that channelise mineralised solutions. They also investigated whether these fractures are indicated by lineaments on Landsat MSS images or on geological maps and found excellent correlation, especially in the case of remote sensing, even though the images chosen corresponded to the unfavourable season and the processings done were limited. In this case remote sensing aided in locating the leaching zones and in mapping the zones of fluid circulation.

Ahmadzadeh (1984) investigated the same phenomenon in a slightly different manner; his study was based on geochemical indicators and known deposits. At different scales, either regional (Massif Central, France) or local (gold-bearing district of Saint-Yrieix), he observed a specific distribution of gold-antimony mineralisation along the NE-SW lineaments and lead-barium mineralisation along NW-SE lineaments on Landsat images.

3.4.1.4 Multicriteria analysis

The study of gold mineralisation in the Saint-Yrieix (France) district constitutes an example of multicriteria analysis. These mineralisations are concentrated in quartz lenses along the shear zones (Bouchot *et al.*, 1988; Bouchot, 1989). Geological mapping helped to pinpoint the location of these quartz lenses on the surface. However, the auriferous shear zones could only be traced approximately and hence other means of locating them were investigated.

Landsat Thematic Mapper imagery has thus been used in association with various themes, such as economic geology, structural geology, geomorphology, soil mapping etc. This peneplain region is characterised by scarcity of exposures, dense vegetation cover and surficial alterations; so the direct approach to mineralisation is limited to the rare rock outcrop and ancient 'gold' workings. As these phenomena are aligned

(a)

Fig. 11. ERS-1 radar image of Mauges (France) (a) and its structural interpretation (b).

(b)

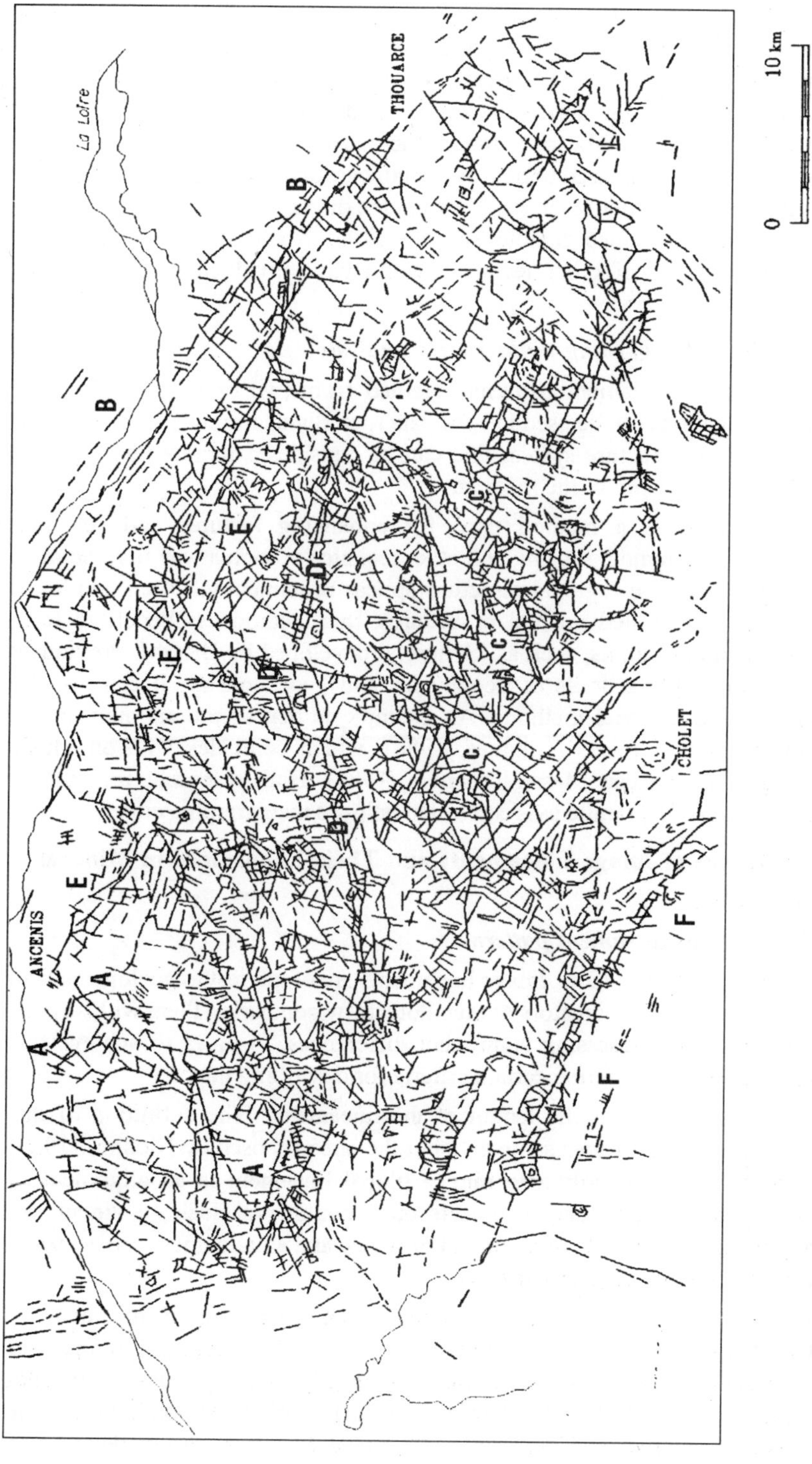

A—Champion series
B—Layon fault
C—Chemille granite
D—Mauges series
E—Plessé Ligné fault
F—Choletais

along structures that often show a random disposition, they would at most be useful for an approximate and even simplistic tracing of the shear zones. In such a context, quite often observed, multicriteria analysis aids continuous and reliable mapping of the shear zones, selecting the auriferous zones in different basement structures and characterising the later clay-bearing faults and barren quartz veins. The results obtained are presently being used to plan systematic ground investigations for tracing the underlying auriferous quartz lenses. They are the direct outcome of the integrated methodology employed since remote sensing alone would not have helped in obtaining them and use of other technics would have involved much time and uncertainty.

3.4.1.5 Conclusions regarding multicriteria analysis

The superposition of lineament patterns obtained from two or more methods of geological investigations can be used to eliminate spatially uncorrelated elements and to define a composite fracture system that can be considered representative of the most probable heterogeneity of the medium.

Multicriteria analysis is a method used for enhancing features and selecting observations made by remote sensing and other technics in such a way that the most important elements of the regional structure appear prominently. Often employed through efficient but limited manual procedures, this method has benefited from various correlation algorithms, such as 'Synergis' developed by BRGM. Previously based on the conventional tools of geophysics and geochemistry, it can also be applied for less conventional methods. An example is that of the anomaly in the preferred orientation of micas (Guigues, 1984), which coincides with an image lineament in the Armorican Massif (Scanvic, 1984). Both represent the surface indication of the top of a granitic axis concealed below a metamorphic cover.

3.4.2 Significance of Image Lineaments and Their Importance in Mineral Exploration

3.4.2.1 Examples of lineament pattern

From the economic point of view, image lineaments are of interest to geologists, who attempt to identify in them indicators of metallotects by comparing them with already known occurrences and assessing significant differences. In order to achieve this objective, it is necessary to define certain criteria for describing them.

- The first criterion is the length of the lineaments. The definition of this parameter is sometimes subjective but the one proposed by Nicolini (1980) is quite clear. According to him, a global lineament is an alignment of geological or metallogenic objects on a continental scale, a vincula an alignment on the scale of a mineral province and an alignment *sensu stricto* characteristic of a mineral district.
- At the three scales considered, the directional distribution of lineaments as the orientation criterion is important in mineral prospecting. Examples of such distributions are the Armorican trend in Bretagne (France) or parallel directions in the Sillon Houiller in some regions of the Massif Central. The direction, number and cumulative length of fractures, earlier determined manually, are now calculated using special software (Dutartre, 1984).

— Another method of detecting anomalous directions within a system of lineaments (Pretorius and Portridge, 1974) is based on statistical analysis of a given population. Developed by Blanchet (1957) for photofractures, this method of exploration employs the principle that frequency of angular distribution of lineaments determines a conjugate group. For a given region, each lineament can be classified into a definite conjugate set. Using the classification, it is possible to calculate a direction and a mean deviation. The values of deviations can then be transformed into maps on which the strongly anomalous zones appear distinctly. In a specific example ten sites were identified and ground checked. Five of them corresponded to intrusives visible as outcrops and four were characterised by a thick soil cover. Three of the five intrusions were found to be associated with visible indicators of mineralisation.

— Several studies in North America (Mayo, 1958; Wertz, 1963) have shown that high mineral concentrations tend to occur at the intersection of lineaments. Mapping these potentially mineralised nodes is one of the objectives of remote sensing. Applied to coal basins in France, this procedure revealed the importance of certain tectonic zones such as Bourbonnais in the north and Aveyron in the south (Grolier, 1981). A similar situation was observed in West Africa for diamond deposits (Bardet, 1963).

— The existence of a constant distance between lineaments of a directional group is considered by some geologists an important characteristic of the rhegmatic system. This was demonstrated by Kutina and his colleagues (1967) in Czechoslovakia and by Dolivo-Dobrovolsky and Strelnikov (1978) for certain regions in the USSR. In the latter case the periodicity varied according to the orientation of fractures, as can be seen in Table 3.1.

— Other geologists emphasise the importance of the criteria of fracture density and the density of lineament intersections. The principle of this approach is based on two aspects:

- Determination of significant directional intervals and retention of only the major features of a region, based on a predetermined trial using a special procedure (Sawatzky and Rames, 1980).
- Possibility of drawing the lineament groups and lineament intersections, this term being used in the sense of 'concentration of lineaments'.

Such maps have aided detection of certain important lineament structures for uranium exploration. For example, in Limousin (De Fraipont, 1982), the zones striking N-S are the most promising and all the known deposits are aligned in this direction. Interpretation of remote-sensing data has thus contributed significantly to uranium prospecting since the E-W and NW-SE structures in which uranium occurs appear on aerial photos only as rare discontinuous segments.

Table 3.1

Directions/ Regions	NW-SE	NE-SW	N-S	E-W
Baltic plate	190 km and 90 km	150 km	65-75 km and 17-35 km	145-175 km and 15-35 km
Russian platform	350-400 km and 100-150 km	300-350 km	300 km	300 km
Urals	300-350 km and 50-75 km	Indistinct	175-200 km	150-200 km and 25-30 km
Southern Kazakhstan	70-100 km	90-120 km	70-100 km	220 km, 90 km and 30 km
Tarim and Kun-Lun	?	150 km and 75 km	?	100 km 100 km

In the case of the Paris basin (Rouzeau *et al.,* 1989) multidate and multisensor thermal data was processed to enhance the information content. This resulted in filtering the data and achieving homogeneity without eliminating essential information but nevertheless enhancing the features observed on different dates and by different sensors.This method also facilitated regrouping of nearly continuous faults of variable lengths and widths but of regional extent, which are represented by high density axes. This type of representation is generally better applied to ground observation since a number of them have a deep origin and their surficial expression is relatively diffuse (Fig. 12).

Application of this technic while prospecting in Arizona (Abrams, 1983) helped to improve the perception of regional structures, minimising the subjective influence of the analyst on the interpretation. Combined with geological and geophysical maps, the lineament density map enabled the investigators to focus prospecting on highly fractured zones in which high concentrations of copper were found. In the case of aerial photographs the geomorphological differences strongly influence the perception of faulted structures but in the case of Landsat images, the synoptic effect, the high resolution and the absence of relief perception introduce a favourable preliminary smoothening.

3.4.2.2 Examples of application

— Though this approach is typically based on lineament analysis and not all metallogenists belong to this school of thought, geologists, without accepting a dogma that confines the concept of metallogeny to a narrow system, recognise its positive aspects and tend to evaluate them. According to Routhier, for example, 'this approach undoubtedly consists of part of the truth, and its dynamism is validated by remote sensing' (1980). His remark is based on the fact that the trace of a known lead-zinc deposit in the USA is visible in the 'lineament map of the USA prepared from Landsat imagery'.

However, remote-sensing data alone cannot indicate whether such a direction represents a first-order control (or metallotect) for a given metal or not, and requires a

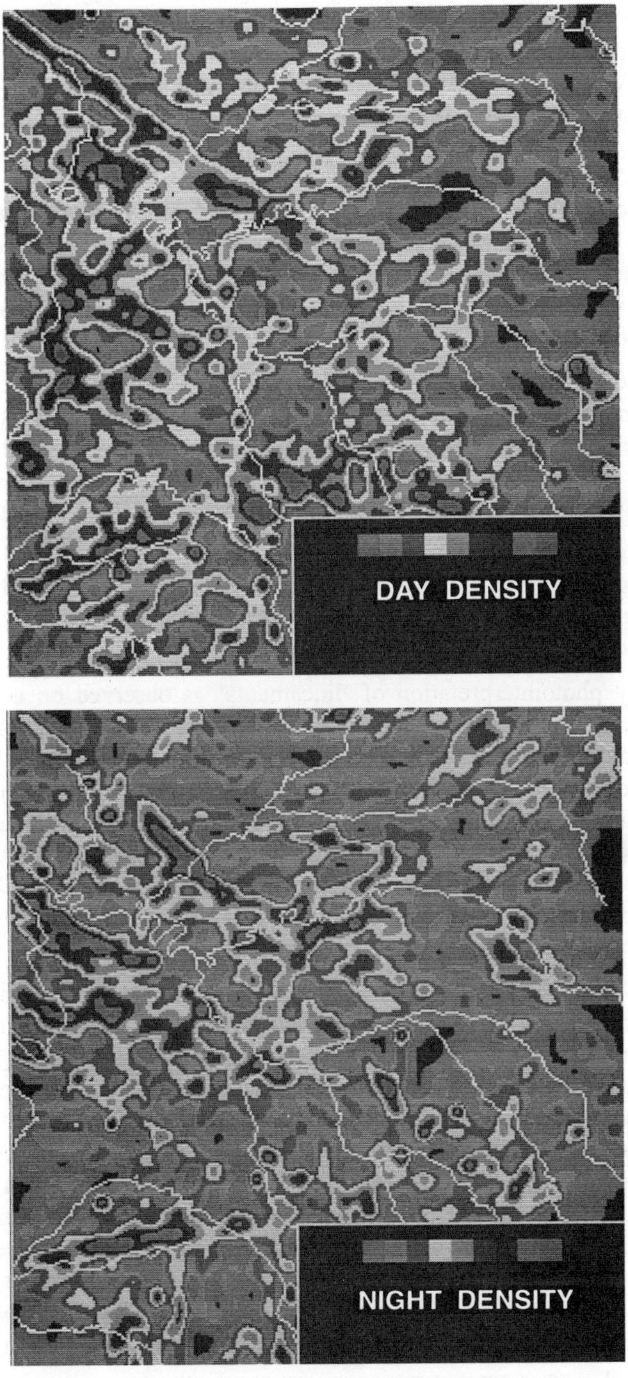

Fig. 12. Representation of the density of lineaments in thermal images of NOAA satellite (Paris basin, France) (after Rouzeau, 1989).

preliminary or simultaneous analysis of the regional metallogeny. Such an exercise was carried out for the Armorican Massif. The study (Scanvic, 1984) primarily showed that the lineaments striking 70° N quite faithfully describe the Ardennian-Vosgian core with Pb, Cu, Zn mineralisation, whose extension was indicated towards the Armorican Massif (Boyer *et al.*, 1975). Analysis of Landsat images thus led to detection of a NE-SW system of image lineaments coinciding with 'rich crests' along which, according to certain hypotheses (Aye, 1982), major mineralisations have been trapped due to ancient volcanic events that are invisible on the surface. Further, physiographic analysis of five major mineralised sites in the Armorican Massif showed that the important characteristic of their environment was localisation in a volcanosedimentary context at the intersection of lineaments striking 70° N and NE-SW.

— The concept of lineament structures is also extensively used in petroleum exploration. One can cite, for example, the existence of structural control in the Saint-Juan basin (Colorado) in which oil and gas deposits are localised by lineaments. The known oil and gas accumulations in this region occur between two series of parallel lineaments (Saunders *et al.*, 1973).

— Examples of application of remote sensing for groundwater exploration in fractured media such as basement rocks were quite rare until 1988 due to the poor resolution of data acquired by first-generation satellites. The situation has changed considerably today, however.

Priorly, only photointerpretation of 'lineaments' as observed on aerial photos of different scales had demonstrated its importance for fixing cites for groundwater wells, by identifying fissured, fractured and tectonically disturbed horizons and accurately locating them on the ground in a manner compatible with village water requirements. The procedure adopted had the additional aim of reducing the number of failures in obtaining high yields, with rare exceptions. Some hydrogeologists attempted to obtain high-yielding zones, neglecting the minor details observed on the aerial photographs in favour of a comprehensive visualisation of tectonic and geomorphological episodes.

Landsat MSS images were primarily used for this purpose, for example, in Burkina Faso (Bérard, 1982). The objective of the study was to site points for water wells with yields higher than 3 m^3/h for a water-supply scheme encompassing seven urban centres of the region. Conventional photointerpretation, which resulted in the preparation of a potential fracture zone map for areas of thick lateritic or soil cover, was combined with an analysis of Landsat images for the preparation of lineament maps. A comparison of these two maps led to a selective classification and choice of sites, made possible notably by interpretation of Landsat images, for significantly improving the yields of wells. However, this methodology has been abandoned because the possibilities of locating on the ground the phenomena observed on Landsat MSS images are not in general compatible with the precision required for locating well sites in the proximity of villages or within the areas of irrigation.

Utilisation today of multiband Spot images of high (10–20 m) resolution overcomes this limitation while still retaining the great advantage emerging from the synoptic view offered by Landsat data. The example of the Songolon basin, Guinea is illustrative (Brosse *et al.*, 1987). The project proposed was water supply to the town of Kamshar. The digital data of Spot, generated in the form of a standard colour composite and a sinusoidal enhancement was interpreted, supported by a preliminary ground survey. This study aided identification of major geological units, viz., the Telemele schists which are concealed and were only verified by drilling, the Kindia sandstones including the fluviodeltaic formations and sands and clays of the Songolon basin, which host the aquifers. Spot images also emphasised the surface effects of salt-water; interpretation revealed a fracture system which affects various formations and promotes movement of groundwater. The fracturing is very distinctly marked, emphasised by the vegetation. An excellent correlation was observed between the faults interpreted from Spot data and aerial photographs and those mapped by ground surveys. The simultaneous use of these technics indicates the possibilities of recharge of the aquifer in the Songolon basin by the fault system affecting the Kindia sandstones. In fact, the water flows from east to west and the drainage caused by the NE-SW faults ends at the major fault striking 170° N which forms the western border of the Kindia sandstones in the form of a horst. This fault acts as a major conduit because it brings the sandstones into contact with an impermeable part of the schists. Consequently, the Songolon basin is recharged by all the major faults of 170° N orientation, wherein the Kindia sandstones are covered by fluviodeltaic formations (Plate 12).

This study and its results are important because it represents success in a large part of a problematic area, validating the possibilities of remote sensing and introducing readily adaptable methodology. The application of remote sensing after a preliminary reconnaissance phase enabled the structural geologists to transform static information, viz., traces of faults, into a dynamic view of water movement in the complex tectonic system. Perfectly integrated in different stages of the study, remote sensing constituted a versatile tool in this case. The interpretation of Spot data, structural study on the ground, preparation of geophysical profiles and exploration drilling resulted in locating points for three exploitation wells in the Kamshar region, which proved productive, with an average yield of 135 m^3/h.

The data acquired by the Thematic Mapper of Landsat 5 satellite, which has a spatial resolution of 30 m, can also be used successfully, albeit with lower efficacy than that of Spot data. TM data can be used to evaluate the hydrogeological conditions of weathered formations since the spectral characteristics of the 1.6 μm band are theoretically sensitive to stresses affecting the associated vegetation and the characteristics of the thermal band indicate the water content of the soil. Because of this characteristic, if a correlation between the fractures and the existence of this stress or a particular hydrological state can be detected, it offers a means of qualifying the observed fracture systems in terms of hydrogeological conditions.

The studies and results of the Hassai site in Sudan are exemplary in this regard (Bérard *et al.*, 1990). The supply of water to an industry (gold beneficiation plant) in the arid region was very difficult in this geological terrain made up of metamorphic and igneous rocks (Plate 13), containing only disconnected aquifers. Two solutions were considered for providing water supply to the mineral-processing plant: a pipeline from Nil or a groundwater reservoir. The methodology employed was as follows:

a) Development of a model based on

- storage capacity of rocks,
- infiltration potential (fractures),
- botanical indicators of the surface from conventional water points.

b) Preparation of a 1:100,000 map of the Hassai zone using the data of remote sensing (Landsat and Spot) and lithological and structural surveys. This map (Plate 13) indicates the granitic ridges and vegetative indexes in the Wadi region.

c) Location of zones covering these three factors.

The combination of multiple criteria defined in the preceding list facilitated the selection of 14 zones favourable for groundwater occurrence. Three additional zones were selected based on the single criterion of proximity to the Hassai mine.

Six boreholes had a yield of 3 to 18 m^3/h, which is a very positive achievement in the desert environment of Hassai (42%).

Three wells had a higher yield by 1 m^3/h, which is also considerable in the given hydrological conditions and added to the success rate (64%).

Five wells were total failures.

The three wells in the proximity of the mine were total failures.

Thus the success rate of the methodology (40%), considering the objective, is the highest obtained since the first well in 1983, which was selected on the basis of aerial photographic data.

- In another field of application, geothermal exploration in the Republic of Rwanda is an interesting example of the utilisation of remote sensing in mapping the lineament density. This study was carried out by BRGM under the aegis of the French Ministry of Co-operation (Demange *et al.*, 1983) and the Natural Resources Ministry of the Rwandan Republic. The purpose of exploration was to select the most favourable zones for geothermal resources. Investigations were carried out by a multidisciplinary approach consisting of a study of aerial photographs and Landsat images, which enabled familiarisation with the topography, morphology and structural geology in particular. Especially significant contributions of this technic were exploration of the inaccessible Branca caldera as it is surrounded by a marshy zone, and investigation of the Bugamara graben, considered the best geothermal target in the region because, amongst other reasons, of the extent of fractures here.

3.4.3 Statistical Analysis of Image Lineaments

It is possible to guide subsurface exploration by means of statistical methods provided at least two sample populations such as known deposits and segments of varied length and directions of lineaments are available.

This exercise is especially adaptable to synoptic viewing of images and has been carried out in different ways.

3.4.3.1 *Study of associations by visual analysis*

This method provided significant information on the Massif Central (France) for which a detailed inventory of deposits and geochemical indicators of fluorine and baryte was done (Vairon, 1977; Boirat, 1979) and a map of linear structures prepared from standard images in the scale of 1:100,000. A tectonic boundary, the Sillon Houiller, divides the massif Central into two provinces, eastern and western. The study showed that the distribution of fluoro-baryte deposits in relation to lineaments is not random. It may be noted, for example, that the number of deposits associated with the image lineaments varies as a function of the 'tonnage' criterion. In selecting deposits with reserves of more than 50,000 tons, the correlation coefficient was found to be higher in both provinces.

These results were used for defining new exploration targets of structural and lithological controls; they demonstrated the economic importance of N-S and NW-SE trends in the western province. It may be noted that this conclusion independently supports the view of De Fraipont (1982) regarding the single block of Saint Sylvestre (Massif Central).

Based on these observations, it is possible to use the approach advocated by Nicolini (1980), which proceeds from simple operations to more complex ones as enumerated below.

— Follow the lineament known as a metallotect for finding other deposits. The application of this approach revealed the existence of a correlation between the Chaillac fluorite deposits situated in the northern part of the Massif Central and the manganese deposit with low fluorine content at Thiviers in the south. These two deposits are localised along a lineament which forms an extension of the Nantiat fault towards the south.

— Explore parallel lineaments.

— Explore lineaments that intersect a major lineament.

— Consider exploration in other scales to determine, for example, whether an investigation of a vincula is better than that of a 'lineament' at a continental scale.

— 'Also' test the hypothesis that the correlation between the deposit under consideration and the lineament is not mere chance and hence, attempt to define another metallotect or an association of metallotects.

These concepts led Nicolini to propose two approaches, viz., geometric and structural. These analyses, along with petrographic, geochemical, geomorphological and sedimentological studies, enable demarcation of the most favourable zones on provisional maps for further exploration. The author (Nicolini) employed this methodology in the Hispanolia gold province (Caribbean Isles) and proved its efficacy.

3.4.3.2 *Study of correlations by statistical methods*

This study involves measurement of the degree of correlation between a group of points, spots and lines (deposits, lithology, faults or image lineaments) to demonstrate the presence or absence of a relation between a given type of deposit and faults, or between deposits and lineaments observed on satellite images.

- A specific programme of calculation (Bonham-Carter, 1984) was developed and tested on the known gold deposits of Ontario (Canada), utilising a tectonic map and an image lineament map. The results obtained in this example are given below.
  - Local correlations exist between faults and lineaments and the dominant trends are remarkably identical in both.
  - Gold deposits correlate strongly with faults, especially with zones of high fracture density and with fractures trending 45° N and 90° N.
  - Correlation between gold deposits and lineaments is not strong except when they have orientations of 45° N and 90° N. However, the better correlation between deposits and faults compared to that of deposits and lineaments can be explained by the fact that the zones around the known gold deposits have been structurally mapped in greater detail than elsewhere, thus affecting the statistical results.
- A more recent example, the Montagne Noire (France) was described by Chen Song Lin (1987). Interpretation of satellite images (Landsat MSS, TM and Spot) led to the identification of four principal groups of lineaments in this metallogenic province. They are: NNE-SSW (20° N-30° N), NE-SW (60° N), E-W and NW-SE (140° N-150° N). All the four directions correspond to the known Late Hercynian faults and certain lineaments reflect the older geological structures.

A statistical comparison of remote-sensing results with structural, metallogenic, geochemical and geophysical data, carried out on the basis of special algorithms, showed a good correlation between the various types of observations and measurements. The lineaments are often associated with vein structures and are indicated by geochemical and geophysical anomalies. The major economic mineral deposits (Pb, Zn, As, Au, W, Sn) correlate with the 150° N vincula localised in the western part of the region. Some smaller Pb-Zn deposits are associated with the 20° N vincula and fluorite occurs in the E-W and NNE-SSW (30° N) lineaments in the Albigeois district. The lineaments, according to their orientation and geographic location, reflect specific metallogenic events and a hypothesis based on structural, sedimentological and metallogenic data is suggested. Most of the lineaments might correspond to Hercynian and Late Hercynian faults which acted in the metallogenic processes during this period. Some lineaments represent the synsedimentary structures of the geologic past associated with mineralisation during the sedimentation phase (Cambrian). Proposals for new investigations have been formulated on the basis of this study (Scanvic, 1989).

Though there is no mineral implication, still the Huelgoat formation (Armorican Massif, France) is of interest as it combines the statistical approach, the debut of modelling and application to hydrogeology in fractured basement media.

The study of lineaments in this formation (Castaing *et al.*, 1989) was undertaken in the form of *in-situ* structural analysis of fractures at the scale of outcrops (mesofracturing) and interpretation of satellite imagery, which represents one of the principal methods for understanding the fracture pattern at the scale of a massif (megafracture) in open terrains.

Based on an analysis of directions and variograms of lineaments interpreted from Spot images, studies were done on the correlation between image lineaments (assumed to represent the megafracture system) and the mesofracture pattern measured on the ground. It was observed that the image lineaments in the Spot scene over the granitic terrain present a number of similarities of behaviour with the mesofractures measured on the ground: 'families' of the same orientations, spatial distribution in 'clusters' and similarity in law of distribution of lengths.

This study established that the pattern of image lineaments represents a projection (incomplete or redundant) of the megafracture system, which in turn represents, at a different scale, the mesofracturing measured on the ground. As the validity of these correlations was accepted, the morphokinematic study of fracturing on the ground aided in classifying the image lineaments as a function of their directions and their groupings, selecting only those that had undergone or are now likely to undergo tensile strain. Thus they correspond to potential fractures with a tendency for 'opening' exhibiting significant hydrogeological implications.

Another example is presented in Chapter 7 (7.6.2.2).

3.4.4 Remote Sensing and Modelling

The contribution of remote sensing to structural mapping has been adequately demonstrated since 1972. However, other than mapping the structures and outcropping sedimentary formations, its application is limited to just the inventory phase, necessary of course, while a ground survey alone can provide a chronological and kinematic analysis. That is why structural geologists endeavour to introduce methodologies that enable them to perfect geological understanding of tectonic events through image lineaments.

These methods emerged from experience acquired in sedimentary terrains where the structural features of outcrops are well defined and have a clear expression readily transferable to scales of the imagery. The procedure adopted is quite conventional. After interpretation of image lineaments and of all other details that may present structural interest, such as the apparent direction of transverse faults, relative chronology of two events etc., transferred onto a topographic base map showing geology, the two types of information are compared and their mutual relations analysed. In the case of small-scale maps, it is thus possible to reconstruct the structural history on a regional basis. Such studies were carried out in the East African Rift (Chorowicz *et al.*, 1979), in China (Tapponnier and Molnar, 1977) and by the French Institute of Petroleum in Tunisia and Africa.

The study of the Chateaulin basin (France) is based on a more theoretical approach. The directional diagram obtained from interpretation of lineaments of the Landsat image is compared with several tectonic models (Caia, 1984). The most acceptable model was that of Riedel. It revealed five groups of image lineaments which are produced in a tangential tectonic setting with strike-slip faulting. Some groups of lineaments do not, however, fit in the proposed model. Such a situation is observed for a group oriented along the direction of 120° N, which is considered to be pre-Hercynian as it is not related to Hercynian events.

Because of significant improvements in their spatial resolution (TM, Spot) and certain capabilities emerging from stereoscopy (Spot), the data of second-generation satellites has at present enlarged the scope for tectonic investigations in two aspects, viz., measurement of dynamic phenomena and modelling.

For example, utilisation of Spot panchromatic images of 10-m resolution enables the geologist to refine his analysis and accurately estimate the rate of displacement of neotectonic faults in certain cases. Thus in the Himalayan ranges the important geomorphological features, such as glacial moraines, drainage systems, alluvial cones corresponding to the 10,000-year-old Holocene glaciation, are intersected by active strike-slip faults. These faults have a rate of displacement varying from a few millimetres to three or four centimetres per annum. Analysis of Spot images revealed several morphological features for which it was possible to accurately measure the horizontal displacements induced by sinistral movement of the major Chang Ma fault. These displacements varied from 100 to 300 m. These measurements led to determination of the rate of displacement of the strike-slip fault as two or three centimetres per annum in different cases (Peltzer *et al.*, 1987).

Development of digital models of the ground using stereoscopic remote-sensing data and launching by Spot 2 of the instrument DORIS, capable of making altimetric measurements, have considerably enlarged the field of photogrammetric applications. It is hoped that soon it will be possible to measure and follow the vertical movements affecting the surface of the globe, neotectonic and active faults, for example. Methods of tectonic analysis using digital terrain models are already being employed for Spot data (Colleau and Lenotre, 1990).

As for modelling, new developments are taking place in the direction initiated by structural geologists. Certain conventional models at the scale of the outcrop, such as shear faults, are reproduced in another scale, as in the case of the metamorphic sequence of Limousin for which such models are described (Bouchot, 1989). It appears that a true fractal dimension exists in the case of fracturing, as proven in some cases (De Fraipont, 1982) but requiring fuller demonstration. When such proofs become available, it should be possible to extract different features from the image to demonstrate reproducibility and to compare them with models.

3.4.5 Application of Thermal and Radar Remote Sensing in Structural Geology

The results presented in this chapter have been solely obtained from the interpretation of multispectral data acquired in the visible and near-infra-red bands.

Thermal or radar images are generally acquired by airborne systems as there are few or no specialised satellites in operation.[1] Case histories of these images are still quite rare.

3.4.5.1 Thermal images and structural geology

At present, no systematic data acquired by specialised satellites is available in the spectral range of 3 to 14 μm.

---

[1]ERS-1 was launched in July 1991.

The images furnished by the NOAA have a poor (1000 m) resolution and the Landsat V acquires daytime data in an inconvenient schedule. But the results already obtained under these conditions or from airborne data nevertheless demonstrate the significance of the thermal effects recorded, which often reflect geological and structural phenomena.

In the Génolhac region (France) the thermal traces observed on airborne thermal records acquired during the night under certain seasonal conditions, revealed an unexposed quartz vein. This feature is observed seasonally but is repeated (Brosse, 1975). It is related to the difference in thermal behaviour between the enclosing schists and the quartz vein which occurs below the surficial deposits.

A thermal anomaly was noticed on the airborne thermal records in the midst of a beech forest in Coat en Noz (France). It represents the surface trace of a granitoidal vein covered by three metres of surficial deposits. The mechanism that induces the thermal phenomenon is simple. The granite is highly weathered and acts as a preferential aquifer, compared to the surrounding schists. Consequently, the physiology of the leaves of the trees whose roots extend into this water-saturated medium is disturbed; their temperature increases by about three degrees compared to the surrounding forest. The phenomenon, sensitive to climatic cycles, is also reproduced during the rainy season (Lefevre, 1979).

Other recent studies (Rousselin, 1985) confirm the potential of airborne thermal data for structural investigations.

As for utilisation of satellite thermal remote sensing in earth sciences, the studies conducted using HCMM satellite data have demonstrated its importance for structural mapping. The images acquired day and night (Fig. 13) by this experimental satellite cover the whole of France with a resolution of 500 m on the ground. Interpretation of this data (Fig. 14) revealed thermal discontinuities that may or may not coincide with the faults known on the ground or visible on Landsat images; however, their signature may vary during the daily cycle (Scanvic, 1982). These discontinuities represent lineaments that are hotter or colder relative to the surroundings or linear contacts between thermally different formations. Variations in these lineaments occur in different ways during the daily cycle: from cold during the day to hot during the night (Brévenne fault), from hot in the day to cold at night (Saint-Etienne fault) and, lastly, without a distinct response during the day and hot at night (NNW-SSE faults in Bretagne). In certain cases the thermal response on the same lineament varies from one point to another Sillon Houiller. Finally, certain faults appear only on the night images across the forest whose response is partly attenuated or blurred (NNW-SSE fault in the Orléans forest).

Other studies using NOAA thermal images over the Aquitaine basin (Bienvenu, 1986) and Paris basin (Rouzeau, 1990) confirm the structural importance of the satellite-borne thermal remote sensing. This potential is also demonstrated in the results of a recent study (Rouzeau, 1988) carried out in the Condé-sur-Escaut region (France) using the thermal images of satellites Landsat V (120 m resolution), HCMM and NOAA. As a geothermal well drilled in this region a few years earlier under the *a priori* favourable conditions of a geothermal field in the east and a thermal station in the west proved a failure, reinterpretation of the results was called for. On the satellite images, thermal lineaments oriented along the two preferential directions of 110-120° N and 40-50° N appeared distinctly. The Condé-sur-Escaut well was located on one of these lineaments (40-50° N), indicated by a negative temperature anomaly relative to the surroundings. When plotted on the isohypsal map of the top of the Primaire formation, which is concealed below surficial deposits, the trace of this thermal lineament was found to be right above a sudden variation of depth (300 m), which can be interpreted as a fault or flexure. This set of observations provided a preliminary explanation for the earlier failure and aided in planning a new drilling strategy.

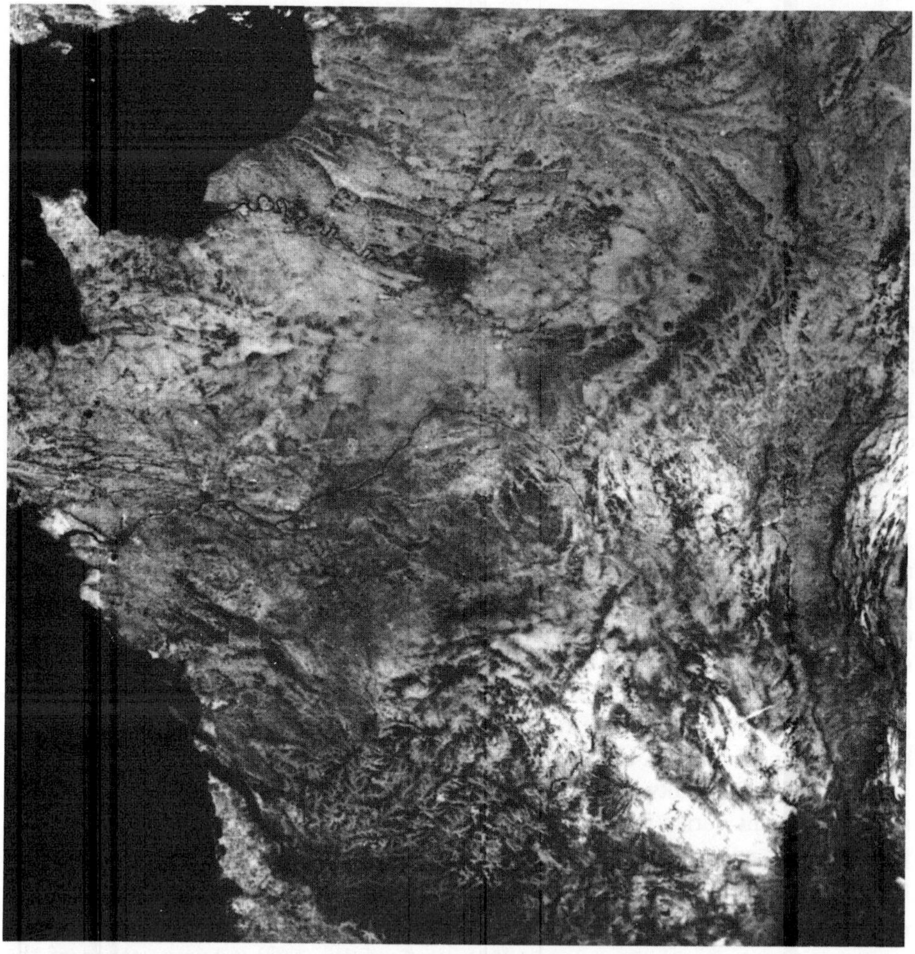

Fig. 13. HCMM. Thermal image of France, night of 17 September 1979.

Fig. 14. Interpretation of thermal image of the night (France) with differentiation of hot and cold zones (approximate scale 1:4,000,000).

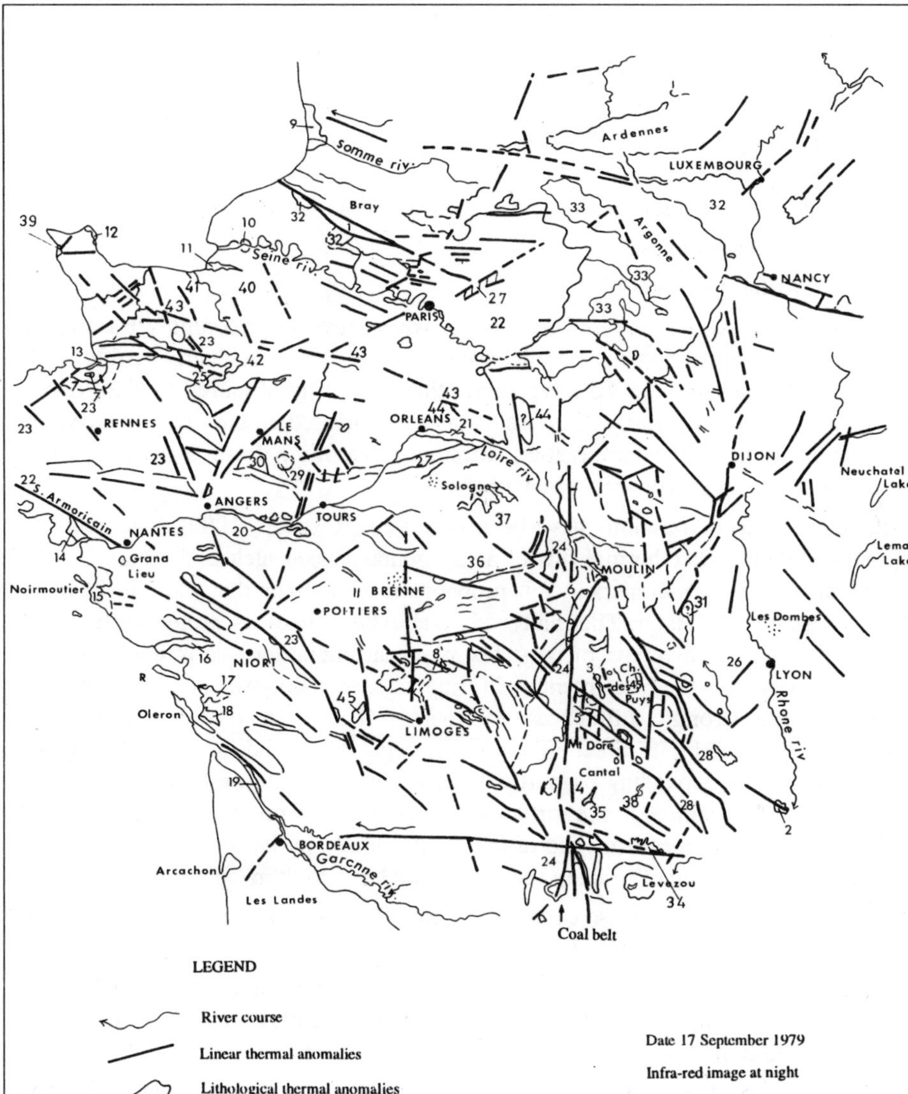

Fig. 14. (contd.).

Hot zones: Puys range (3); Mont Dore caldera (5) (in fact, this part is very complex); Basaltic flows of Cantal ragion (4); Basaltic flows of Aubrac (34); Berry Cretaceous formation (37); Lower Jurassics of Berry (36); Montmarault granite, east branch (6); Dôle and Dinan granites (7); Ambazac granite (8); Cretaceous formation of Havre region (40); Quaternary of Mont-Saint-Michel (13); Armorican trough (22); some faults of 140° N strike, a direction known as a neotectonic type (23); Sillon Houiller and Villefranche fault (24); Tertiary fault of the Paris basin (27); Avranches belt and associated granites (25); Brévenne fault (26); Margeride belt (28); fault limiting the Brioverian basement from the sedimentaries in northern Bretagne (43).

Cold zones: Bray country anticline (1); Coiron volcanic massif (2); Cantal volcanic tuffs (35); Cézallier basalt (38); Pieux granite (39) near Cherbourg; Jurassics in southern Havre (41); Bioverian (42); recent coastal deposits of: Berk (9), Havre (10), Cabourg (11), Saint-Mère-L'église (12), Saint-Nazaire (14), Challans (15), Rochefort (17), Saint-Aignon (18), Gironde (19); Sillon Houiller between Mont-Dore and Saint-Eloy; Cosson belt (27).

Thus the thermal effects of linear form observed from aircrafts or satellites represent surface expressions of different types of tectonic phenomena. These are mainly:

- Faults, which play the role of conduits favourable for movement of groundwater in the fractured media (example, the Sillon Houiller, Plate 14 and Fig. 15);
- Veins or intrusive bodies buried below a surficial covering;
- Neotectonic faults;
- Crustal lineaments.

The lineaments shown here and in Fig. 15, which reveal the complex trace of the Sillon Houiller, appearing as a parallel swarm of faults, are not clearly visible on the Landsat TM images of the 4.5.2 type since they are partly obscured by the flows of the Cantal Massif. Contrarily, they are distinctly expressed on the TM5 image where they appear due to the temperature differences resulting from a thermal effect, probably induced by hydrogeological phenomena.

Although certain correlations have been observed (Rouzeau, 1989), it is generally considered that the geothermal flux of a sedimentary basin such as the Paris basin is too small to be a thermal marker. It is probable, if not certain, that water is the major direct or indirect indicator. Thermally very stable because of its strong inertia, water by its presence and diversity of its states, plays an important role in explaining the thermal anomaly. In the free, interstitial form, water is an important component of the landscape that strongly influences the thermal behaviour of the surface. This influence is exerted by its direct effects on the vegetation (modification of vegetative growth), on soils (depending primarily on their permeability), on faults (presence or absence of faults draining to the surface or subsurface) and on the topography (which modifies the direction of flow and of drainage), as also by its secondary effects (for example, frost splitting of chalk in the case of the Metz fault in Champagne).

Thermal remote sensing is also considered a tool for studying seismic hazards, which is often based on the mapping of megafractures.

Such a study, using thermal satellite data, was undertaken by Soviet geologists (Salman and Tronin, 1990) for the first time and the results obtained by them merit special attention in view of the bright prospects they offer.

The Soviet geologists worked on NOAA data under the hypothesis that the thermal response of the Earth's surface registered from a satellite depends in part on the events occurring in the upper part of the crust. They used Central Asia as the target as it offers the most favourable conditions both meteorologically and seismically.

The systematic study of NOAA data acquired at night during the previous six years and analysis of the continuous series (100 to 200 days every year) revealed fairly extended thermal anomalies in a region that is particularly active seismically.

The anomaly at Kopet Dag has a width of 30 km and a length of 500 km. It is associated with a major fault. Another anomaly at Karatan is 50 km wide and 300 km long. It, too, is associated with an important fault. These two anomalies are constant in time and are qualified as stable. However, they include segments of variable extension in which the intensity of thermal response changes over time and thus are classified as unstable anomalies.

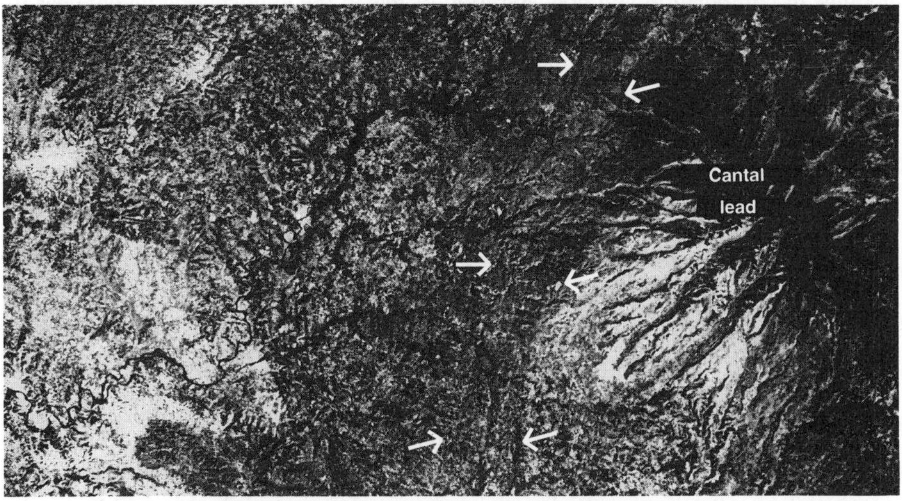

Fig. 15. Landsat TM thermal image of the Argentat-Châtaigneraie region.

For example, an unstable thermal anomaly was located in a part of the stable thermal anomaly of Karatan, at its intersection with the Tamdy-Tokrans fault which forms the epicentre of an active seismic zone (Gazli). This suggests a correlation between the two phenomena, which the Soviets attempted to demonstrate by methods of correlation analysis.

The seismic activity was represented by a logarithmic function of the number of earth tremors, divided into 13 classes of intensity for a given period of time. The thermal instability was expressed by a value of thermal radiation intensity. The study revealed a correlation over time between the seismic manifestations at Gazli and the appearance of unstable thermal anomalies. This method was later extended to the whole of Central Asia and two zones were studied in particular.

The first zone includes the Turan plate and folded zone of Tien Shan where seismic activity has a crustal origin. The second zone corresponds to the centre of Mediterranean Asia characterised by earth tremors of the Pamir-Hindu Kush region, with a rather deeper origin. In the case of the Turan-Tien Shan zones, the maximum thermal intensity of the unstable anomaly was recorded at the annual peak of seismic activity. The coefficient of correlation between the two was 0.85. In the Pamir region the correlation was poor, with a value of 0.67.

Soviet geologists have proposed some hypotheses to explain their observations but none is definitive. First of all, it is clear that statistical correlation is observed only in the case of seismic zones of shallow origin, a fact which restricts its implications. It is possible that the stable anomalies had a hydrological origin but given the rapidity of appearance and attenuation of the phenomena and the intensities attained (several degrees centigrade), it is impossible that the unstable anomalies resulted from the transformation of mechanical energy into thermal energy during the maturation phase of a shallow seismic activity. On the contrary, they may have been produced from the changes in composition and concentration of gas in the atmospheric layers near the surface above the active faults. These changes could have also been induced by an effect of compaction or by a variation of the moisture content of soils under the influence of water-table fluctuations.

Whatever the reasons be and in spite of the limitations of the method, a systematic follow-up of such unstable thermal anomalies perhaps represents a promising technic-one which the Soviets are actively exploring.

3.4.5.2 Radar images and structural geology

Radar is an acquisition system operating in the active mode in a high frequency range. Given the absence of specialised satellites until 1991, the data utilised were obtained from special aircraft flights or from experimental satellites. Their interpretation has demonstrated that they contain significant information of a structural nature.

Three experiments with radars launched on satellites or space shuttles have been undertaken: Seasat (1978), SIR A (1981) and SIR B (1983). These experiments acquired data in the 25-cm wavelength band, in analogue mode for the first two and in digital mode for the third. Geologists participated in data evaluation.

The airborne side-looking radar (SLAR) and the synthetic aperture radar (SAR) make use of the property of backscattering of high-frequency radar waves transmitted in the direction of the Earth's surface. The SLAR is characterised by analogue recording, which can be restored in the form of images of the backscattered signal. The SAR has digital recording which facilitates measurement of the amplitude of the signal and its transformation into an image.

- The first-generation radar (SLAR)[2] utilised the 3-cm wavelength (Goodyear, Motorola). It is the 'all-time acquisition' characteristic of the method which guided its choice for ensuring aerial coverage of equatorial forest regions where cloud cover is high throughout the year. The entire regions of the Amazon, Venezuela and Gabon, for which hardly any space images or aerial photographs were available, have been explored by photointerpretation of radar images.

The principle of side-looking radar is based on recording a backscatter signal which is influenced by the surface unevenness of objects and their electric permittivity. Moreover, the active mode of radar facilitates the restoration of the morphology of the regions with no special technics because of the play of shadows. The combination of textural and structural parameters makes the radar image a preferred document for tectonic analysis.

Example of Band X Airborne Radar Images

The Irian Jaya region (Indonesia), where structural analysis was carried out to investigate plate tectonics for guiding petroleum exploration here (Froidevaux, 1980), is an example clearly illustrating the capability of the radar technic in this field.

---

[2]Side-looking Airborne Radar has a natural or synthetic antenna.

The structural, geomorphological and geometric information extracted from the radar images of this region were compared with ground geological and geophysical data. This analysis helped to verify the plate tectonics model of the region. According to this model Salawati Island separated from the Irian Jaya region during the Pliocene, creating a rift zone (Sek Strait). It moved 17,500 km southwards, after a rotation of 13°. The fault associated with this movement produced a subsidence zone characterised by recent activity. This conclusion, arrived at with support from radar images, influenced the strategy of petroleum exploration in the region. It helped in understanding the dynamics of faults and in appreciating their action either as traps or as channels for migration of hydrocarbons in different epochs. This notable result was obtained due to the acquisition characteristics of radar which enabled the preparation of maps for this inaccessible region of high cloud cover and provided a synoptic view of the geomorphology with an adequate resolution, intermediate between that of aerial photographs and Landsat MSS images.

- The second-generation radars (SLAR and SAR) enlarge the scope of application. Their digital recording systems facilitate the study of the amplitude of the signal. Some simultaneously employ the frequencies of bands X, C and L; others have a facility for varying the angle of incidence of the radar waves and selecting the type of polarisation (horizontal, vertical, inclined). They have mainly been used for experiments, with the exception of Canadian airborne radar, operating in bands X and C.

Example of Airborne X, C and L Band Radar Images

In the Vans en Ardeche region (France), which served as a test zone in France for radar experiments between 1973 and 1985,[3] several flights were made with different systems (Cluzeau, 1985; Cluzeau et al., 1985).

Geological evaluation of all this data indicated the quality of perception of tectonics through a cover of surficial deposits. The analytical map prepared from the interpretation of Goodyear images of north-south flights showed three major lineaments along three directions:

— N-S direction identified on the ground constantly throughout the region studied;

— NE-SW direction, well known on the ground (e.g., Cévennes fault), represented on the radar image by long lineaments;

— NW-SE direction.

A secondary fracture system oriented in an E-W direction also exists, represented by some major lineaments traversing the whole image.

The lineament map resulting from their plotting on a 1:125,000 topographic base, when compared with the geological map, shows the magnitude of geometric deformations introduced by the radar technic. But its integrated form is useful for preparing a preliminary tectonic sketch of a region from the dissimilar maps available until now (1:80,000, 1:50,000 in different cases).

---

[3]For the Remote Sensing Club formed by the BRGM and the IFP, representing the petroleum companies of France.

The JPL images of the 1979 flight are of lesser quality with poorly expressed reliefs, large geometric distortions and radiometric differences associated with height variations. Hence the possibilities they offer for structural analysis are inferior to those of the Goodyear radar images.

However, these images taken at an incident angle of 33° are devoid of shadows and reveal interesting details along the direction of vision. Thus certain lineaments of N-S and NW-SE orientation, which were not detected by other configurations, are distinctly represented on radar images of the N-S flight, ensuring coverage perpendicular to the E-W axis of the Lozère mountains. This example underscores the significance of orthogonal coverage, especially in montane zones where the shadows can be intense.

In the Chassezac valley, between Chandolas and Lablachère, visualisation of lineaments which are already very dense was further improved by the conjugate interpretation of the axes with crossed and parallel polarisations. In particular, some of them are restored dimly on the crossed polarisation image and more clearly on the parallel polarisation image, revealing a great potential for selection albeit the phenomenon cannot actually be explained.

The inferences that can be drawn from these experimental studies are numerous. They can be enumerated as: geomorphological analysis promoted by the acquisition of X band radar image; differences in brightness recorded on L band radar image; disputable importance of various polarisations, even though cross-polarisation images appear to have a better contrast, thus enhancing perception of structural details when emphasised by differences in vegetation cover; combination of different polarisation images which indicate the response variations of certain objects; complementary nature of observation according to the look direction.

All the observations made in the Vans region reveal the structural geological importance of the information transmitted by the backscatter radar signal, in particular that concerning vegetation, which often acts as an indicator of faults and fractures because of the water content of the soil.

Seasat Experiment

While the Landsat experiment, originally conceived for agricultural applications, was successful for geological investigations as well, the Seasat experiment meant for oceanographic studies proved equally applicable to earth science studies. This is illustrated by a study (Sabins *et al.*, 1980) of the San Andreas fault (California) whose seismic effects are well known. This fault is clearly observed on Landsat MSS images throughout its known extent but in the Durmid Hills south-east of the end of the fault, an examination of Seasat radar images revealed a lineament separating a part of low brightness in the north-east from a part of high brightness in the south-west. The new observations made after this discovery have confirmed the existence of an abnormal contact between the Borrego sandstones, characterised by their high brightness, and the sands and silts indicated by their low brightness and highly irregular surface. The differences in response in this case could be attributed to the specific wavelength (23.5 cm) used in Seasat since this lineament did not appear on an airborne radar of 0.86 cm wavelength.

SIR A Experiment (November 1981)

This is one of the experiments associated with flights of the space shuttle Colombia. The radar launched on this shuttle was similar to that of Seasat. Its frequency was 1278 KHz, corresponding to a wavelength of 27 cm. An area of ten million square kilometres was covered in eight hours of flight. All the climatic zones could be observed-the main novelty of the 'Colombia' experiments.

This data was mainly used for the study of image lineaments. We shall consider three illustrative examples of the Mali region (Weecksteen, 1983). At Bougouni, in the Sahelian zone, the lineaments visible on radar images are emphasised by linear segments of river courses and by alignments or differences of vegetation. These lineaments are generally not visible on aerial photographs and the corresponding Landsat image. Contrarily, the directional diagram constructed from the interpretation of radar images correlates perfectly with that produced from ground measurements. This correlation is very distinct, particulary in the case of the gold-bearing fractures well known in the Kalama region. This example demonstrates the capacity of radar to reveal structural phenomena through differences in vegetation cover.

The geological context at Massima differs. It consists of an alluvial zone on the right bank of the Niger River where a series of parallel lineaments, oriented 70° N, can be observed on radar images. As no structural element is known from ground surveys of this alluvial field, the regional geological features were investigated to identify their structural significance. Thus fractures were detected, partly from geological maps and partly from Landsat images, in the Bandiagara sandstones. These fractures sometimes constitute the extension of lineaments visible in the alluvial zone. As the Bandiagara sandstones are concealed below the alluvium, it is clear that the fractures visible on the images result from a transport towards the surface, or from the phenomenon of 'transparency' by processes still unexplained to which the radar responds.

In the Adrar-des-Iforas, SIR A radar revealed the structural and lithological continuity of geological objects concealed below a sand cover. The structures, often very thin, were detected in certain tectonic regions of Adrar. This perception of objects below the cover poses a question regarding the capacity to penetrate through dry formations attributed to radar waves. Such a situation is observed especially in the N-S oriented fault which brings the granite basement and the Tin-Zaouatene sedimentary sequence into contact. This fault does not appear as such on the Landsat image, in which it corresponds either to a contact of sand exposure or a faint boundary of the facies. In the same region the intensity of backscatter waves helps in characterising the veins and dykes by their morphology and their linearity (Fig. 16, points 2, 1 and 3 respectively).

Thus the characteristics of radar images, especially those of band L (27 cm), make them very effective for use in structural geological investigations. First of all, the strong response of the vegetation to radar waves promotes the detection of certain lineaments. Further, the phenomenon of transparency of the cover facilitates perception of structural phenomena that are not visible on the ground surface.

Table 3.2. Existing and future SAR systems

SAR	Frequency	Polarisation	Spatial resolution	Angle of incidence	Availability
			AIRBORNE SENSORS		
SAR 580	1.28-5.3 & 9.6 GHz	HH, VV, NV, VH	5 m	13° TO 70°	since 1978
IRIS (Canada)		HH, VV, HV, VH	variable 5-19 m		1985
STAR I (Canada)	8-12.5 GHz	HH	variable 5-8 m		
JPL SAR (USA)	1.26 GHz	HH, HV	20 m		
GOODYEAR AN/APQ 102 (USA)	9.6 GHz	HH, VV, VH, HV	16 m		
VARANS (France)	9.375 GHz	HH, VV, (HV, VH)	3 m	28° to 70°	1985
C-SAR (Canada)	5.3 GHz	HH, VV, VH, HV	6 m	0° to 74°	1986
E-SAR (Germany)	9.375 GHz	HH, VV	1 m	45° to 65°	1990
			SPACE SENSORS		
SEASAT (USA)	1.28 GHz	HH	25 m	23°	1978
SIR-B (USA)	1.28 GHz	HH	25 m	15° to 60°	1984
ERS-1 (Europe)	5.3 GHz	VV	30 m	23°	1991
JERS-1 (Japan)	1.2 GHz	HH	20 m	44°	1992
RADARSAT (Canada)	5.3 GHz	VV	28 m	20° to 45°	1994
SIR-C (USA)	1.28-5.3 & 9.6 GHz	HH, VV, HV, VH	15 m	15° to 60°	1993
EOS SAR (USA)	1.28-53 & 9.6 GHz	HH, VV, HV, VH	10-20 m	10° to 55°	1994
ALMAZ I (Russia)	9.6 cm	HH	10-15 m	30° to 60°	1991
RADAR 2000 (France)			Under Study		

It is difficult to say whether this corresponds to penetration power or to variation in roughness induced on the surface by a deeper phenomenon. However, the result is obtained in both cases due to certain characteristics of the radar system.

The various studies conducted thus far in aerospace remote-sensing geology by means of radar technics in the bands X, C and L (3, 6 and 25 cm) lead in general to four basic points.

1. Radar produces for certain angles of incidence a satisfactory, albeit deformed image of the surface morphology which itself is indicative of the geological substratum. As data acquisition is possible in all cloud conditions, radar constitutes an efficient substitute for aerial photography and other satellite images, for the purpose of lithological and structural mapping in scales of up to 1,100,000. From this point of view, it can be considered as the first-priority technic for structural and geological mapping.

Fig. 16. SIR A radar (left) and Landsat MSS (right) images over the Adrar-des-Iforas region (Mauritania).

2. In arid regions the L band (25 cm) radar waves have a penetration of the order of 1 to 2 m. This effect, if confirmed in a larger number of cases, would be of great importance in geology for the study of the substratum below surficial deposits, palaeochannels and shallow structures.

3. In the same zones the L band radar waves improve the efficacy of remote sensing to differentiate rock types in spite of oxidation patina.

4. As radar waves are sensitive to surface unevenness and passive systems of the type TM and HRV to reflectance of formations, integration of the two might enhance the possibilities of discrimination, especially in surface formations.

But the effects resulting from the operating conditions of different technical parameters that can be controlled (radar is an active system), are still not known. This applies in particular to:

- Frequency, since what is observed from experiments in arid zones is not equally evident in equatorial regions in which certain experiments conducted with the L band (Champetier de Ribes and Scanvic, 1983) revealed only minor differences from data obtained with band X.
- Polarisation, whose effects on the perception of certain geological objects are not known since various experiments have given contradictory results-either demonstrating the reality (Cluzeau *et al.*, 1985) or just the opposite (Rudant and Cervelle, 1988).
- Angle of incidence, whose effects vary with relief.
- Angle of vision.
- Spatial resolution, which is still not optimised for geological applications since it strongly depends on the conditions of exposure.

Hence it is necessary to thoroughly study the implications of the above in the investigation of geological targets and in the perspective of development of new sensors. This is the objective of the space shuttle, SIR C mission and studies under way with airborne systems in France (GATT-CNES project). However, these evaluations are still qualitative since quantitative approaches are difficult to implement for natural media. Fundamental studies on modelling of a signal backscattered by geological objects as a function of surface unevenness and electric permittivity are still scarce.

3.5 CONCLUSIONS REGARDING IMAGE LINEAMENTS

Since its initial development, remote sensing has made significant contributions to the updating and enhancement of our knowledge about the tectonics that affects the Earth's crust and controls the localisation, geometry or migration of subsurface resources.

This capability is the outcome of the synoptic vision revealed by remote sensing. It shows that deep or shallow tectonics more or less determines the conditions of the landscape by introducing several components which can be observed from space.

This capability depends on the ratio of extent of ground observed to resolution of the system, which is improving with each satellite generation. Landsat MSS provides an integrated vision of structural phenomena over several square kilometres by incorporating a multitude of indicators and introducing a smoothening entirely compatible with small-scale surveys. On the other hand, second-generation satellites such as Spot or TM, offer a more accurate perception of tectonic events due to improvements in resolution. While the advantages of synoptic vision are still maintained, the images can be readily transformed into surface maps to meet the requirements of the practical geologist interested in large-scale or medium-scale mapping, with an accuracy compatible with the siting of mine openings or boreholes. Lastly, thermal and radar images are the source of new additional information, as demonstrated by the example of Villefort fault. The structural complexity of this fault was accurately mapped using various remote sensing records, viz., Landsat TM, Spot and aerial photos, while the thermal response (HCMM) clearly showed the fault in the NNW direction (see Fig. 13).

However, systematic similarity between tectonic structures and image lineaments cannot be presumed because although enhancement in resolution has eliminated the risk of confusing such with man-made features, the risk still exists for uncertain phenomena. The percentage of correlation with ground truth is generally high even though its value varies from one region to another and from one geological context to another; nevertheless certain known structural features may not be visible from space or their real effects poorly perceived. Lastly, the importance of ground truth is quite evident since structural analysis is essentially based on outcrop measurements and interpretation by modelling in remote sensing is still at the tooth-cutting stage even in the best of cases.

Interpretation of remote sensing data has thus become a necessary prerequisite for ground surveys. It constitutes an indispensable complement, facilitating spatialisation of point information gathered *in situ* for a megafracture system. The latter itself represents at a different scale of mapping the mesofracture system actually measured on the ground. Interpretation, very often static with no support from field studies, can benefit from the efforts presently under way for modelling lineament systems, which take into account the fractal theory.

Chapter 4

CIRCULAR STRUCTURES

Circular structures sometimes appear to be mythical objects and an attempt is made here to destroy this impression, especially in areas of thick pedovegetation cover. Examples of outcropping circular structures selected for their geological significance are described first for arid zones. Proof is then adduced for the existence of similar structures in other latitudes and under various environmental conditions, interpreted from landforms and validated by indirect methods such as geochemistry and geophysics.

A typology of circular structures is proposed, based on their morphology, permanence at different spatial resolutions and spectral specificities, then evaluated through multidataset analysis. The search for circular shapes is justified through an examination of circular geological structures that have a well-established mining or petroleum interest: e.g., granite intrusions, porphyry copper deposits, salt domes. Finally, examples are given of circular shapes of geological origin discovered while using various types of spatial data.

It is proposed to show in this chapter that remote sensing is a preferred tool for detecting circular structures, to establish their forms and to investigate why they can be considered targets for mineral and petroleum explorations.

The criteria for visual interpretation of circular structures are essentially morphological and spectral. The technic of digital interpretation for circular forms, which has emerged from image-processing methods, introduces certain definite measures in their mapping in which the personnel factor has hitherto played a significant role.

4.1 REMOTE SENSING, A PREFERRED TOOL FOR DETECTION OF CIRCULAR STRUCTURES

4.1.1 Aerial Photography, the First Means of Detection

Most circular structures whose geological significance has been demonstrated on the ground and whose existence has been confirmed by maps, were discovered from aerial photos until recently. These discoveries were essentially made in arid or semi-arid regions and rarely elsewhere.

In arid regions these structures are distinct because the forms are exposed and the geological phenomena they reveal are almost always explained by genetic theories postulated by geologists and mappers. Guillemot (1962) gave three examples of such structures. The first, on the northern flank of the Tindouf basin (Algeria) in the Carboniferous sedimentary formations, was produced from the emplacement of a dolerite intrusion (Lehmann and Scanvic, 1969). The second is located in Mauritania in horizontal formations of Cretaceous age. This structure is a spectacular one. The hypothesis of a meteoric impact has been proposed for it. The third circular structure is located in the eastern part of the Ahnet basin (Algeria). It is clearly represented in the geological map and exhibits the various concentric rings of a circular structure. However, the causative factor of this structure has not been explained.

In other climatic regions, in France in particular, circular structures have been reported based on interpretation of aerial photos. While the morphologies they reveal are evident on the ground, their geological significance very often remains hypothetical. Amongst the most typical structures, those of Bouenat, Mont Bessou and Bezenet (Massif Central, France) have engendered detailed investigations (Scanvic, 1983).

Bouenat is a village of Allier (France) located on a circular promontory, centred by an alluvial channel. The structure is situated in a schistose sequence. So far, petrographic and geophysical studies (Lallier, 1977) have adduced no explanation as to the origin of this structure. An aeromagnetic survey (Debeglia and Gerard, 1975) indicated an anomaly which could be related to the structure but the two are set off from each other (Note in the geological map of Gannat). This structure is in the proximity of the Echassières Massif, a typical intrusive granite, associated with an assemblage of mineralisations (tin, lithium etc.). This intrusive was the target for an experimental borehole drilled during an intensive geological programme in France. The borehole indicated the existence of a second dome and Bouenat may have been one of the two.

Mont Bessou is situated south-east of the village Bugeat in Limousin (France) and occurs in Millevaches granite. The structure has a characteristic form which can be clearly identified in the landscape and on aerial photographs. In the preliminary analysis no explanation was offered for its existence in Millevaches granite, but the presence of a micrograniic vein on its south-eastern flank indicates a hypovolcanic origin. Moreover, the uranium trace concentrations in it are statistically higher than in the surrounding rocks (personal communication).

The Bezenet structure in Allier (France) is located in the centre of a small coal basin. It is exposed by stream erosion (Sossa, 1980). Geological surveys of 1:50,000 scale (Montluçon sheet) have confirmed the concentric disposition of various outcropping geological formations. One hypothesis proposed to explain this quite spectacular phenomenon, which is visible on Landsat images, is that it represents a vast caldera. However, no sufficiently strong evidence has been adduced to date to validate this assumption.

4.1.2 Multispectral Remote Sensing and Circular Structures

Remote sensing has enabled the discovery of several circular structures on the Earth's surface because of its ability to provide a synoptic view of large areas. These landforms represent phenomena of varied nature which need not necessarily be of geological origin. While it is difficult to classify them a priori, at least a certain degree of precision can be introduced in their inventory. That is why it may be mentioned that the detection of circular structures is influenced by the resolution and the period of data acquisition on the one hand, and the degree of sophistication in processing on the other. Knowing these parameters, it is necessary to develop methodology for interpretation of data, for which the scale of the map constitutes a determining factor.

— Depending on the scale, very small structures which have no definite significance need not be considered. This principle is governed as a function of the resolution of the sensors.

— Structures of large extent should not be interpreted initially by considering independent features. Some remote sensing scientists have done so, however. Notable among them are Cazabat (1975) in France and Norman (1977) in Saudi Arabia. Norman proposes the hypothesis that these megastructures, which he calls 'Astrons', correspond to giant meteoric impacts. As they were interpreted from observations made on Landsat MSS images, it would be interesting to determine their existence by establishing their configuration on images that are more synoptic, such as those of NOAA satellites, for example.

— One should choose structures that are observable in their integrality and composed of multiple rings indicated or exhibited by variations in reflectance, temperature and brightness.

— Structures visible on images acquired in different periods, by different sensors and in different wavelength ranges should be preferred.

A study was carried out on the basis of these principles in two geological formations (Dutartre *et al.,* 1983). It resulted in the detection of nearly three hundred structures whose nature, whenever it could be determined, was found to be geological and indicated varied phenomena. An example of such phenomena is presented below.

The most frequent phenomenon associated with the circular structures in this region is that of a granite intrusion, very often of a smaller extent in outcrop relative to the size of the structure under observation. Such is the case at Montbelleux (France) where Landsat images reveal the existence of a structure of 10 km in diameter formed by three concentric rings. It is centred on a granite intrusive outcropping into Brioverian schists where it is manifest. The structure also coincides with a gravity anomaly, which indicates a larger extent of granite at depth. It can thus be reasonably considered that the structure represents the effects on the ground surface resulting from emplacement of the granite intrusive.

The Penfeunten circular structure in the Armorican Massif has been thoroughly investigated. However, no precise explanation can be given for its occurrence. The structure is found in the Saint-Renan and Kersaint granites in which it is spatially associated with uranium mineralisation. The suspected dome has no expression on the gravity map and even if it induces a low-intensity anomaly, this cannot be seen since the dome is situated in a gradient zone. Test geophysical experiments in this region would facilitate development of a methodology for detecting circular structures indicated by remote sensing.

Other circular structures are associated with volcanic phenomena. Their origin is sometimes evident if their characteristic forms are still preserved. However, this is not always so, as in the case of Puy de May where a structure was discovered in the Tertiary volcanic formations of Sancy (France). Its morphological features observed on aerial photos suggest the existence of a volcanic body and subsequent ground surveys have shown that it represents a partially eroded ancient centre of eruption (Varet *et al.,* 1980).

Even though only one example of the type has been reported in France to date, certain circular structures are of meteoric origin. Such a case has been established at Rochechouart where brecciated elements supporting such an origin have long been known. The form and extent of the impact are perfectly visible on Landsat images. This enabled Lambert (1974) to improve the mapping of the meteorite without, however, detecting any traces of rebound or other meteorites in the numerous structures identified in the surroundings. But this involved an integrated analysis which, undoubtedly, merits thorough study.

Besides these relatively obvious associations of circular structures with geological phenomena whose manifestation is observed on the ground surface, a large number of cases are more difficult to explain in the absence of characteristic surface indicators. It is sometimes possible to consolidate these observations when geophysical and/or geochemical data exists for the region concerned. Such is the case of a typical circular structure detected west of Saint-Pardoux (Scanvic, 1984) within the Sioule metamorphic sequence of the Massif Central (France). No petrographic manifestations of the structure are observable on the surface but it coincides quite accurately with a positive (29 gammas) aeromagnetic anomaly, interpreted as a concealed magmatic intrusion similar to the microgranite near Pouzol-Servant (Vennat, 1982).

A geochemical survey of the Massif Central (France) gave rise to a hypothesis based on geological origin of certain circular structures. Several occur in biotite granites and are associated with beryllium anomalies. According to Burnol (1974), beryllium is generally rare in biotite granites but is a highly mobile element. When the concentrations are high, they may indicate the shallow occurrence of leucogranitic magmas enriched with beryllium. These magmas may have been emplaced by the process of 'cauldron subsidence' and the observed circular structures could be the surface manifestation of the stresses resulting from it. The most characteristic structure of this phenomenon is that of Oradour-sur-Glane (Limousin). It is associated with a high beryllium anomaly (200 g/t). Other examples have been reported in Spain (Marconnet *et al.*, 1982), which also prove the existence of spectral transparency of leucogranite under a metamorphic cover.

Lastly, certain circular structures, such as circular grounds, prehistoric mounds, aspersion zone etc., have an anthropogenic origin. In this case the critical approach of the photointerpreter plays a decisive role in analysis.

4.2 CLASSIFICATION OF CIRCULAR STRUCTURES

The experience acquired from various remote-sensing studies using representative data of varied climatic conditions facilitates the proposition of a tentative classification of circular structures on the basis of morphological, scalar, spectral and multiparameter criteria.

4.2.1 Morphological Classification

This is based on the characteristic of outcropping which determines two major groups of circular structures:

- Structures frequently outcropping in desert and semi-desert regions (Plate 15);
- Known concealed structures, especially in regions of thick soil cover.

4.2.1.1 Outcropping structures

These are of two types: elevated and depressed.

The elevated structures appear either as homogeneous circular bodies or as a series of concentric annular bodies. Forms intermediate between the two also exist.

The depressed structures comprise a partly eroded central part and a hard peripheral ring or a zigzag of central elevated structure, eroded rim and hard periphery. This model in certain geological contexts can be explained by the phenomenon of meteoric impact on the Earth (Rondot, 1983). Images of Mars (Fig. 17) give a very clear vision of such impacts (Masson, 1986).

It is always possible to ascertain the geological origin of outcropping structures as they give testimony to indubitable phenomena.

Fig. 17. Circular structures on Mars (photomosaic from viking (NASA-JPL); supplied by the orsay planetary Photo Library).

4.2.1.2 Concealed structures

These are of two types: morphological and radiometric.

The morphological structures are generally formed by a series of concentric rings represented by curvilinear discontinuities. They may be entire or sometimes truncated by a discontinuity. It has been noticed, especially in Bretagne (France), that these structures often occur as aureoles around small granitic intrusive massifs (Gourin, Montebelleux, Scaer, Pondigny and elsewhere), identified as outcrops.

The structures resulting from the difference in reflectance are of at least two types:

- A circular form whose reflectance differs, positively or negatively, from that of its surroundings. A typical example is the Riom-Ennezat anomaly (Massif Central, France) observed above the major subsidence trough of Limagne.
- A concentric circular form in which the reflectance varies for each of the rings. This kind of structure is encountered in southern Rouez (Armorican Massif, France), where it is associated with geochemical anomalies.

The differences in reflectance are sometimes highlighted by seasonal variations which have geobotanical significance. In all cases the associated fractures, especially in the radial form, may reinforce the manifestation of the structures.

It is not always possible to determine the geological origin of concealed circular structures by conventional means of observation.

4.2.2 Scalar Classification

This is based on the persistence of the phenomenon when the scale and resolution are changed. The typical example is that of the Villefranche-de-Rouergue circular structure (Massif Central, France) detected on a 1:1,000,000 Landsat image. The interpretation of records of gradually increasing scales allows identification of the structure (Skylab record) or its composite elements (balloon photos, aerial photos) (Brosse and Scanvic, 1976).

Another interesting example is that of Bezenet (Massif Central, France), already cited, which appeared on HCMM thermal images (resolution 600 m), especially the night of 17 September 1979. This temperature anomaly corresponds to a circular structure visible on Landsat imagery of 6 May 1976 and the same is also found on aerial photographs where it is emphasised by a characteristic drainage pattern.

The structure at Thoras (Massif Central, France), analysed in detail on images and on the ground (Dutartre *et al.,* 1983), is observed on MSS Landsat, TM and aerial photos.

Other examples of structures visible on several types of data have been described in France at Mont Bessou (Sec. 4.6.1.1) and Ambazac (Scanvic, 1983).

4.2.3 Spectral Classification

This classification is essentially applicable to temperate climatic zones with little outcropping. It is based on the spectral characteristics of circular structures which can be preferentially observed in certain frequency ranges.

- The result of a comparative study of the HCMM thermal and multispectral Landsat data in the Armorican Massif (France) illustrates some aspects of this group. Three cases are presented (Scanvic, 1984).
  - — Certain circular structures are visible only on thermal images (Trevelo).
  - — Most of the circular structures visible on Landsat data do not appear on thermal data, due in part no doubt to the poorer resolution (500 m) of the HCMM system.
  - — Some circular structures are visible both on Landsat images and HCMM thermal data (Malguenac-Quilly). The thermal behaviour of these structures can be explained by a variation in the nature of alterites and their water content.
- Several circular structures were discovered in the Massif Central (France) from the different spectral bands of the Thematic Mapper (Scanvic, 1984-86). Only one of these structures appears also in the thermal image and in enhanced images of visible, near and middle infra-red bands. But its verification on the ground revealed no real geological feature. However, a quartz vein not mapped earlier was detected in the surrounding granite gneisses. This vein constitutes a new occurrence of arsenopyrite. At present the geological significance of the structure and the nature of processes which enhance its thermal response are not precisely known but several hypotheses appear probable. Its thermal characteristic for example, can be explained by gaseous emanations associated with the underlying mineralisation, and the

expansion of these gases in the air could be the cause of the temperature difference recorded. Such gaseous emanations have been studied and are known in the USSR (Karavtzov and Fridman, 1965). In France utilisation of gaseous hydrocarbons for prospecting buried lead-zinc mineralisation was successfully experimented with for the Trèves deposit (Disnar *et al.*, 1986). However, it has not been verified whether these gases affect the remote sensing signal.W9S

- In certain cases the thermal characteristics of circular structures can be determined indirectly. Such an example is described (Scanvic *et al.*, 1985) in the Armorican Massif (France) along the Armorican trough near the village of Baud. Here two closely located circular structures were identified on Landsat MSS images (point 4, Fig. 7). The separate image traces which characterise them are enhanced by vegetation. According to the geological map (Marcoux, 1980), these structures, which exhibit no spectral differences, are identified as two small granite intrusives of very close petrographic composition. Other than on a colour composite resulting from the superimposition of Landsat MSS, channels 7 and 4, and the HCMM temperature difference data, these two structures are differentiated by the amplitude of their thermal variation during the day cycle. The eastern structure is indicated by a much larger variation than the western and such a response can be explained by a difference in the thermal behaviour of the granites, emerging from the degree of alteration, thickness and water content of the altered zones.
- Similar features detected on radar images have been described (Weecksteen, 1983) for the Yambu Al Bhar region of Saudi Arabia. Close to this locality, two circular structures were observed in the SIR A radar images in the middle of granite formations. One of them is known on the ground and is very distinct on the corresponding Landsat image also. The second is masked on the Landsat image because of surficial deposits but appears on the radar image due to the active characteristics of microwaves, which accentuate the surface morphology and promote a deeper penetration of these dry soils. The latter requires verification, however.

4.2.4 Multicriteria Classification

This is based on the existence of strong spatial correlations between circular structures and geophysical or geochemical anomalies. Some examples are given below.

- A number of correlations have been observed between aeromagnetic anomalies and remote-sensing data in ancient formations. An example already described pertains to Saint-Pardoux (Sec. 4.4.1.2) in the Massif Central (France). An investigation of this type of correlation was also carried out (Scanvic, 1984) in the Saint-Brieuc region of the Armorican Massif (France), which has been covered by aeromagnetic surveys. Based on geophysical interpretation (Corpel and Debeglia, 1982), 14 anomalies were identified and transferred onto 1:500,000 Landsat images. Of the ten structures identified on the images, four coincide with aeromagnetic anomalies, which indicates their geological origin.

Other correlations of this type have been found at Monvicq, Pebrac and Cros (Massif Central).

Examples of correlation between gravity and remote-sensing data have been reported for Bresse (Dutartre and Weecksteen, 1984) in the Armorican Massif, at Montbelleux and Penguily (Dutartre *et al.*, 1983; Scanvic *et al.*, 1986), and in the Massif Central (Ziserman *et al.*, 1982) at Saint-Bonnet and Bourganeuf. They generally correspond to negative gravity anomalies. At Lafeline, in the Massif Central, the circular structure observed on Landsat MSS image is correlated with positive magnetic and gravity anomalies.

Among other, less common means of remote sensing, one can cite the photographs obtained by Spacelab with a photogrammetric camera. They indicated in the surroundings of the Bray anticline in the Paris basin (France) a number of circular structures aligned along parallel structural directions (130° N). These structures correlate with the elevated points (Weecksteen, 1985) of the biozone part of a Cretaceous formation or with the anticlinal axes often coinciding with magnetic or gravity anomalies. This result, obtained in an undifferentiated sedimentary region, shows that remote sensing reveals information emerging from the interface between surface geology and the subsurface or even deeper horizons. This is its major advantage; it also explains the difficulties encountered in evaluating these observations by conventional and economic methods.

- Examples of correlation between geochemical and remote sensing are numerous. Certain associations of circular structures with abnormal concentrations of beryllium have been described (Scanvic, 1981, 1983), especially in the Massif Central (France) at Ussel, Faux-la-Montagne, Oradour-sum-Glane (200 g/t), Cognas-le-Froid and elsewhere. Other cases of correlation have been reported in the Armorican Massif, especially in the 1:50,000 map of Nort-sur-Erdre (Scanvic, 1984). On this map a circular structure is found to coincide with a series of lead anomalies distributed along the rim and centre of the structure. Some others are associated, with spatial offset, with the known and exploited mineral deposits at Nozay (tin) and in the Pontivy Langonnet granite (tin, wolfram) situated in the Armorican Massif.

So, it is a reasonable hope that the various classes of circular structures described above can be associated with well-identifiable geological phenomena based on known models. This has actually been achieved by Russian geologists who have been studying circular landforms for a long time. Based on geological data (Dolivo-Dobrovolsky and Strelnikov, 1978), a classification of 'ring structures' is proposed and their relation with basement tectonics discussed. Five major classes are defined according to specific criteria and associated with distinct petrographic variations of different ages.

— type 1, oval granites,

— type 2, oval gneisses,

— type 3, oval structures of Proterozoic granites,

— type 4, oval structures of Middle Phanerozoic formations,

— type 5, oval structures of Phanerozoic granites.

These structures have the characteristic of large dimensions in common. They are generally situated along vinculae and constitute significant structural elements of the upper layers of the Earth's crust. They were discussed in a number of papers during the International Geological Congress of 1984 at Moscow.

4.3 GEOLOGICAL PHENOMENA OF CIRCULAR FORMS

The exploration of circular forms observed on remote sensing images is justified by the existence of geological objects presenting such morphologies. Moreover, a large number of circular or oval forms can be noticed among the morphological types associated with geological phenomena of interest in mineral or petroleum exploration.

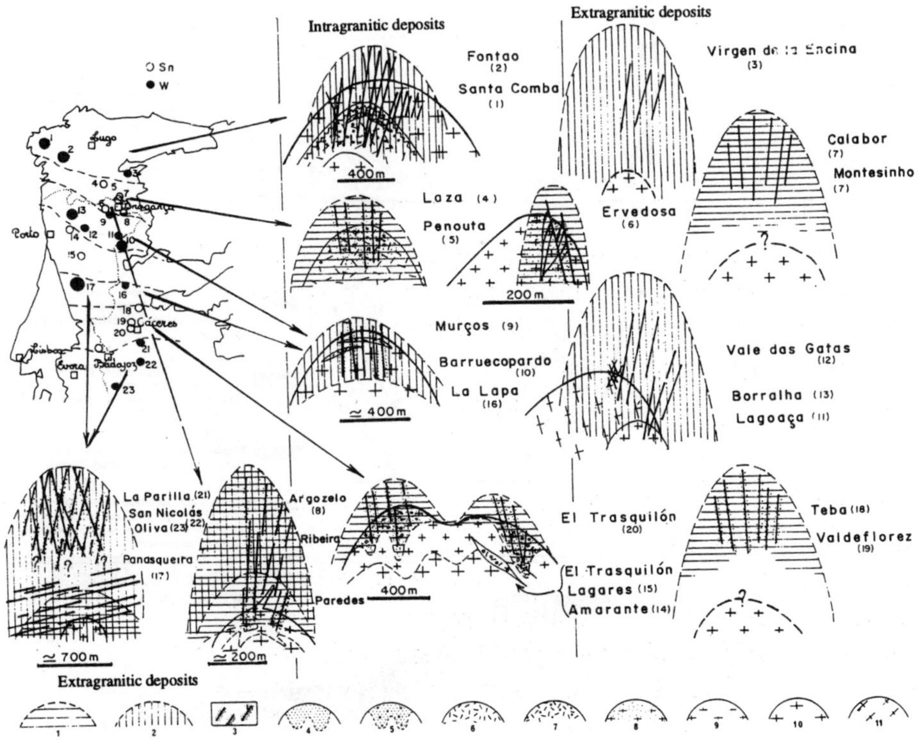

Fig. 18 . Different types of Sn-W deposits in Protugal; relation by granite domes with periptutonic zoning (after Derre and Roger, 1984)

1—zone of hydrothermal influence for dominant Sn mineralisation; 2—zone of hydrothermal influence for dominant W mineralisation; 3—veins with muscovitisation and/or tourmalisation of the contact zones; 4—pockets of greisenised granite; 5—same as 4, with disseminated cassiterite or wolframite; 6—leucogranite with albite; 7—same as 6, with disseminated cassiterite or wolframite; 8—muscovitised leucogranite; 9—albitic leucogranite with two micas; 10—granite or adamellite with two micas; 11—oriented granite.

Granite intrusions are perhaps the best known for their circular or domal form which distinguishes them from their surroundings. These are the granite or leucogranite

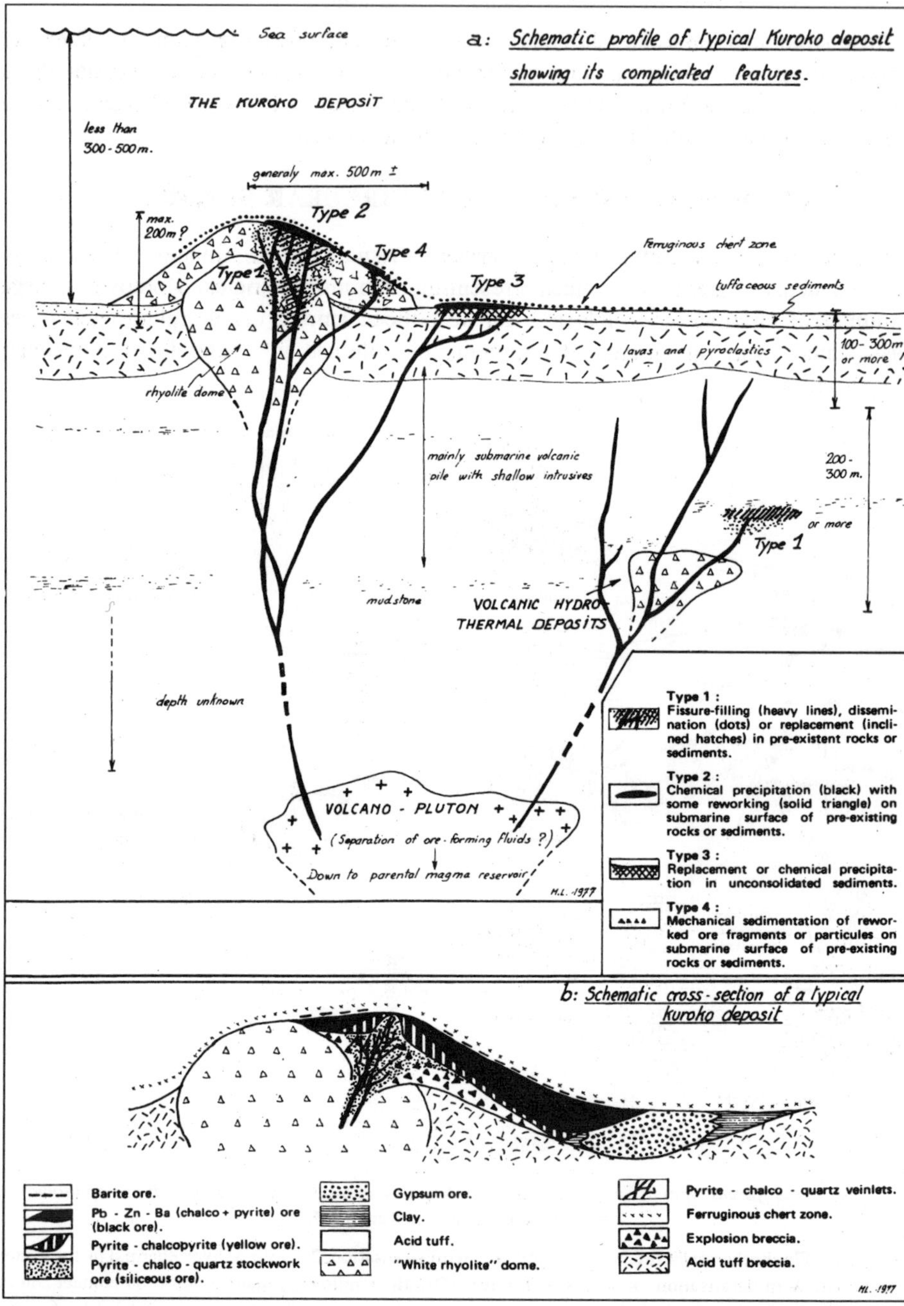

Fig. 19. Classical model of volcanosedimentary sulphide deposits (after Aye, 1982).

domes emplaced by 'diapirism', 'cauldron subsidence' or intrusion. The associated mineralisations are of pyrometasomatic type such as tin, scheelite etc. Their position relative to the structure may be intragranitic or extragranitic, as in the case of Portugal (Derre *et al.*, 1984) (Fig. 18).

The 'porphyry' type of deposits, frequently associated with quartzitic rocks within a volcanosedimentary complex, have characteristic forms often described in publications of metallogeny. Without going into details this complex phenomenon, of importance in mineral exploration, it may be recalled that such deposits are characterised by, amongst other aspects, their circular or elliptical form on the ground surface resulting from their geometry at depth. The same is the case with sulphide deposits, though to a lesser extent (Fig. 19).

Basic and alkaline magmas are also the cause of a number of circular forms. The most well known are the kimberlite pipes which constitute the major source of primary diamond deposits (Fig. 20).

Characteristic circular forms are also produced by carbonatites and nepheline-syenites with which mineralisations of niobium, tantalum, rare earths, uranium and thorium are associated. For example, the Zambesi carbonatites in Mozambique are perfectly visible on Landsat images.

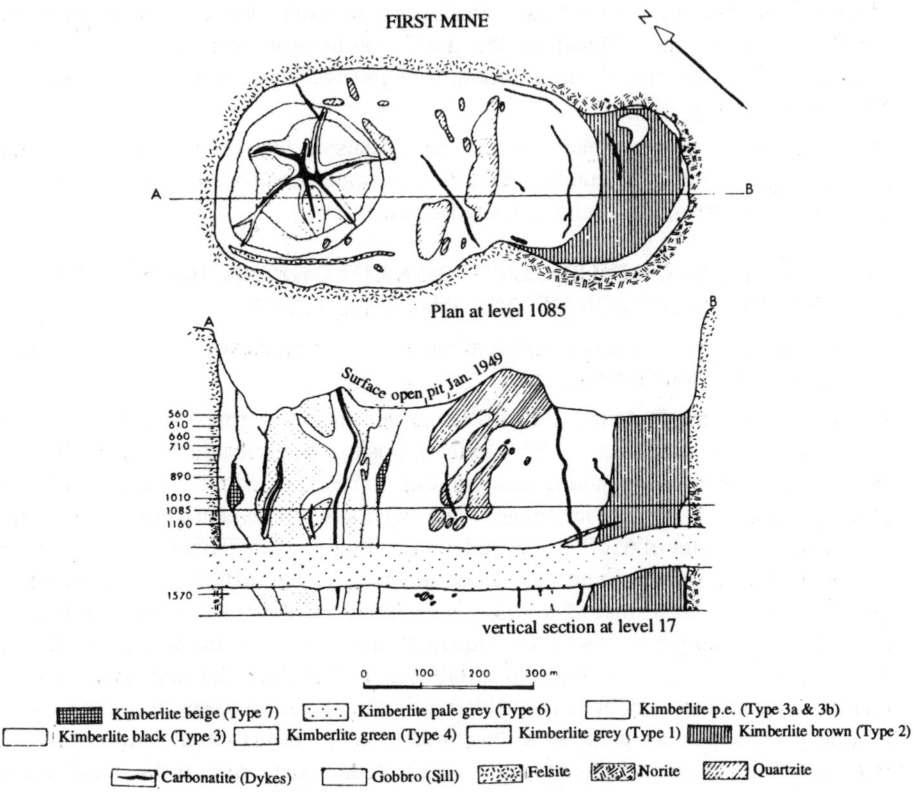

Fig. 20. Schematic plan and section of the first diamondiferous pipe at Kimberley (South Africa)
(after Bardet, 1963)

Dunite pipes, often associated with mineralisations of platinum, chromite, nickel, and the lopoliths of norites and pyroxenites constituting the hosts for chromite, nickel, copper etc., also have circular forms.

The salt diapirs or domes are explored by petroleum geologists as they constitute effective guides for petroleum occurrence. They also often have a characteristic circular form which facilitates their detection on the ground surface, on Landsat images and on radar images.

While all these manifestations of circular forms are produced by intrusions of differentiated magmas, others result from cataclysms, such as volcanic calderas, meteoric impacts and collapse structures.

Calderas, which may be gigantic, attaining a diameter of up to 50 km, are formed by ejection of large volumes of magma. Some, known as resurgent calderas, are characterised by subsequent slow uplift of their floor. These calderas are not easy to identify in the field as they are often not discernible due to their large dimensions. However, their detection is considerably facilitated by Landsat images and several have been discovered in this way in Argentina (Cerro Galon) and in Bolivia (Kari Kari). The calderas constitute important exploration targets for two reasons: they are generally situated on top of a geothermal anomaly which represents a potential source of energy and they sometimes host metalliferous deposits. One can cite in this regard the Potosi silver mine situated on the flanks of the Kari Kari caldera in Bolivia (Francis, 1983), or the Oregon uranium deposits (USA) on the boundary of McDermitt caldera.

With regard to meteoric impacts, while their effects on the surface are known, their mineral significance is disputable even though some mineral deposits appear to be associated with them, especially in Canada (gold).

4.4 EXAMPLES OF DETECTION OF GEOLOGICAL PHENOMENA OF CIRCULAR FORMS ON SPACE IMAGERY

- Role of space imagery in the metallogenic synthesis of the lead district of upper Moulouya (Morocco).

A study (Jebrak and Dutartre, 1984) was carried out on a group of deposits of this region and most particularly the Zeida deposit. Mineralisation occurs here as stratiform lenses of galena and cerusite in arkosic beds deposited in palaeovalleys of granite formation. The Landsat image (Plate 16) distinctly reveals a circular structure with a mean diameter of 6500 m, centred in the village Tyout. This structure probably corresponds to a granite dome with a composition of late differentiation (rhyolites, granophyres) similar to that of a rock outcropping farther west. Some molybdenite occurrences are known in this sector. This fact leads to the hypothesis that the circular structure might be the expression of a late granite intrusion that underwent several phases, as demonstrated by the position of mineralised arkoses of Levda on its periphery. The palaeochannels in this area diverge northwards and southwards relative to the structure, indicating the effect of the ridge zone of the structure, as suggested by the presence of a distinct bevel of Triassic and Lower Jurassic sequences. The position of the Bou Ddarne volcano in the Quaternary formations on the boundary of the structure also indicates a new igneous activity. Thus, in spite of the lack of direct evidence by drilling, the only accurate means of exploration, all the surface observations confirm the palaeogeographic nature of the circular structure inferred from the Landsat image. This result was obtained through interactive processing and thorough acquisition of surface geological information to the extent possible. It confirms the capabilities of space imagery as a tool of synthesis that can lead to a new perception of geology and consequently to the discovery of extensions of known deposits and even of new deposits.

- Study of the Magdalena Basin, Colombia

This study was carried out in 1980 by the BEICIP under the aegis of the French Institute of Petroleum and the National Centre for the Study and Exploration of Oceans (Letouzey, 1985).

The study was conducted in the Caribbean Islands. The purpose was to improve the geological and structural map of the Magdalena basin by means of interpretation of Landsat images. A number of linear faults occurring in the rock formations covered by Tertiary and Quaternary deposits influenced the sedimentation of this period and several circular structures were observed on the Landsat images. Such structures, about 5 km in diameter, were detected in the 'Serranias' of San Jacino, in which they are mainly concentrated in the eastern part of the fold zone. They are situated in a region in which the western and central Cordilleras are buried below the lower Magdalena basin and are probably the surface expression of numerous intrusions. Other circular structures of larger dimensions (about 25 to 35 km in diameter) also appear on the images of the Magdalena basin. In this case they probably represent the surface reflections of buried anticlines and diapirs similar to the mud domes detected earlier by seismic method farther north in the basin.

As the Magdalena basin is a known petroleum province, all this subsurface information, even if its nature needs to be confirmed by more direct technics of exploration, represents a significant contribution to the knowledge of the palaeogeography of this region, and should be used by all new exploration programmes.

4.5 CONCLUSIONS REGARDING CIRCULAR STRUCTURES

The geological or geomorphological features capable of producing circular forms are quite numerous. They are the manifestations of phenomena induced by external factors such as impact craters, karstic subsidences, or by internal factors such as volcanoes, intrusions etc.

Such features are revealed by the morphology, drainage pattern, vegetation, water content of the soil and spectral differences. Remote sensing is presently the most universal tool for the exploration of circular structures. But the origin of the structures identified in space images can be quite varied and their identification in covered regions is sometimes disputable due to subjectivity of interpretation. It is therefore desirable, as demonstrated in this chapter, to integrate image data with geophysical and geochemical data, as also with other remote-sensing datasets such as radar, thermal imagery etc., or to follow the more complex approach of identifying circular forms based on digital terrain analysis of the data of the Spot system (Taud and

Parrot, 1990). In fact, although visual interpretation supported by the versatile but never specific technics of digital processing is still the preferred means of mapping circular structures, digital methods of identification of forms presently under development are already giving better results.

They serve as an accurate complement to this type of inventory, especially when integrated with conventional interpretation technics, in particular stereoscopic methods.

Chapter 5

SPECTRAL SIGNATURES OF MINERALS, ROCKS AND ALTERITES: PROBLEMS OF INTERPRETATION OF REMOTELY SENSED DATA

To improve geological interpretation of remotely sensed data is a permanent research objective. As the spectral descriptor is an important parameter in integrated photointerpretation, it appeared easy to individualise in order to carry out an analytical interpretation, and the development of digital approaches was at first based on that of spectral classifications. Today the demands are much more exacting and the emphasis is on developing procedures for identifying minerals, rocks and weathering from the spectral response. The objective having been defined, the first step is to review the situation on the basis of results obtained in the laboratory; this establishes a relationship between the chemistry and the reflectance or emissivity of the minerals and the rocks, based on a set of remote-sensing spectral characteristics.

Laboratory demonstration, however, does not take into account the reality of the field. Research in this area has made it possible to understand the complexity of responses recorded for various environments in terms of vegetation, surficial formations, outcrop conditions etc. As a result, one can define the problems induced between spatial acquisition and geological objects for which the notion of spectral response is of interest. Since these objects are generally small, it is obvious that the systems to be developed must necessarily have spatial characteristics suited to the spectral potential, which is well shown through examples of detecting gossans and hydrothermal alterations. In the conclusions, which look at the potential of detailed mapping and identification of minerals and rocks from imaging spectroradiometry, one hovers between the realism of the field geologist and a less down-to-earth vision.

5.1 SPECTRAL RESPONSE OF MINERALS, ROCKS AND ALTERITES

Improving the geological interpretation of remote-sensing data is a continuing objective of research. As the spectral descriptor is an important parameter in the integrated photointerpretation, it appeared easy to identify for analytical interpretation. Application of the digital approach was primarily based on spectral classification. The need for such classifications is still high today and attempts are being made to develop procedures for identification of minerals, rocks and alterations through spectral discrimination. Starting with the concept of the spectral parameter determined in the laboratory, we shall discuss in this chapter the limitations of the spectral characteristics in field conditions as viewed from space.

5.1.1 Physical Basis

When energy in the form of electromagnetic waves encounters a solid material, it is reflected, absorbed or transmitted. The intensity of reflection, absorption or transmission depends on the physical and chemical properties of the material and the wavelength of the electromagnetic wave, the quantity of energy available decreasing with an increase in wavelength.

These phenomena result in the colour of the object when the electromagnetic wavelength lies between 0.4 and 0.75 μm, known as the visible spectrum. The same phenomena also exist beyond this range but do not indicate the colour concept. Multiband remote sensing makes their visualisation possible by means of processing according to a random colour code, known as the false colour code, usually between 0.8 and 2.6 μm.

The concept of spectral signature encompasses the whole range of frequencies used in remote sensing, including very high frequencies. It is based on a knowledge of the physical and chemical properties of the objects. It is affected by the configuration and unevenness of the surface, intensity of illumination and its angle of incidence, and transmission of the signal in the atmosphere.

The spectral signature is obtained by the acquisition of data as representative as possible of the wavelengths of interest. It is based on the assumption of the existence of a singular relation, after correction for interfering effects, between the physical and chemical nature of the object (minerals, rocks and geological alterations) and the signature recorded from space.

This assumption is tested by experiments in the laboratory, on the ground and from space to verify its validity in different contexts. Such experiments facilitate the definition and measurement of physical properties of minerals and rocks, based on which the spectral signature can be identified and its characteristics in these three conditions of measurement evaluated.

The important physical properties in this context are:

- Reflectance of rocks between 0.4 and 2.2 μm: outside this frequency range, especially in the thermal infra-red band, reflectance is combined with emissivity.
- Emissivity, a parameter related to reflectivity by Kirchhoff's law: It may be deduced from the measurement of luminescence (Becker *et al.*, 1985).

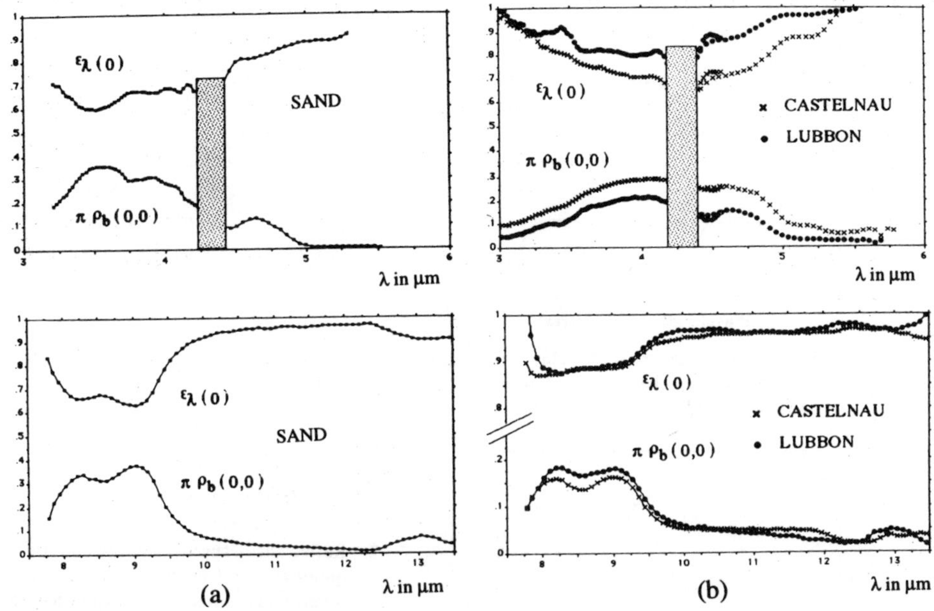

Spectral signatures ε (0) and πb (0, 0) for a quartz sand (φ 100–200 μm)

Spectral signatures for two cultivated soils: HAPEX sites—Castelnau-Riv. Basse; Lubbon, central site.

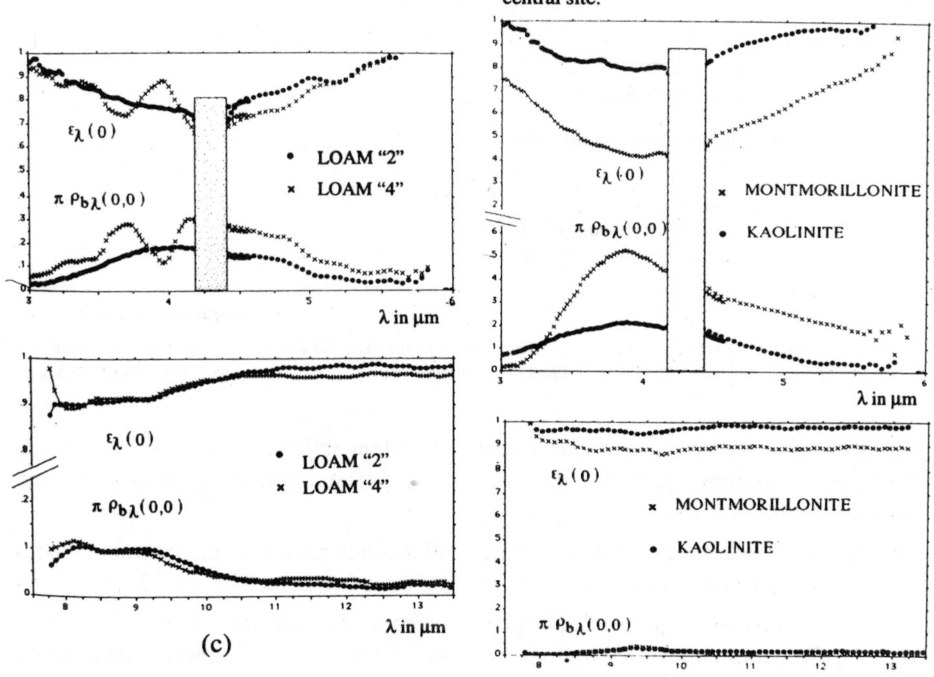

Spectral signatures for two varieties of loess; Loam '2' and Loam '4' (decarbonised); cultivated soils, Strasbourg region

Spectral signatures for two varieties of pure clays; montmorillonite and kaolinite

Fig. 21. Emissivity spectral signatures of bare soils (after Becker *et al.*, 1985).

(a) 1

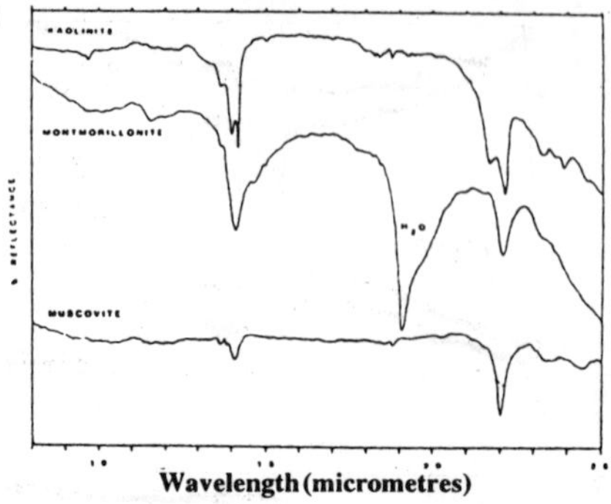

1. Reflectance spectra of clay minerals showing several characteristics between 2.2 and 2.4 μm.

(b) 2

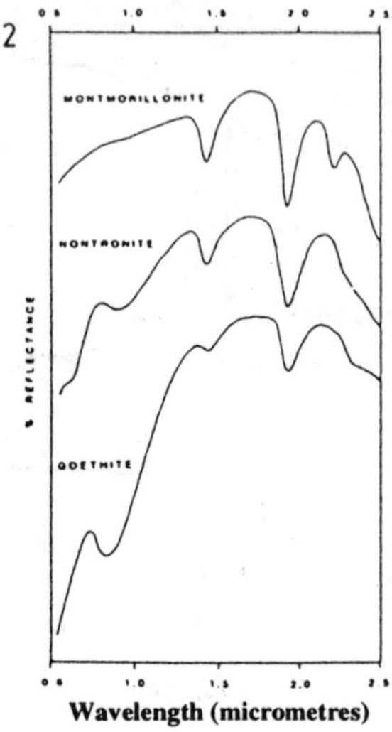

2. Reflectance spectra of montmorillonite, nontronite and goethite. The spectrum of nontronite is similar to that of iron oxide.

(c)
3

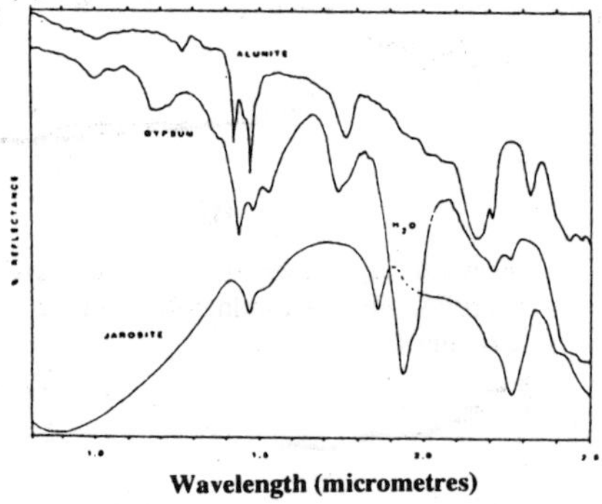

3. Reflectance spectra of some sulphates.

(d)
4

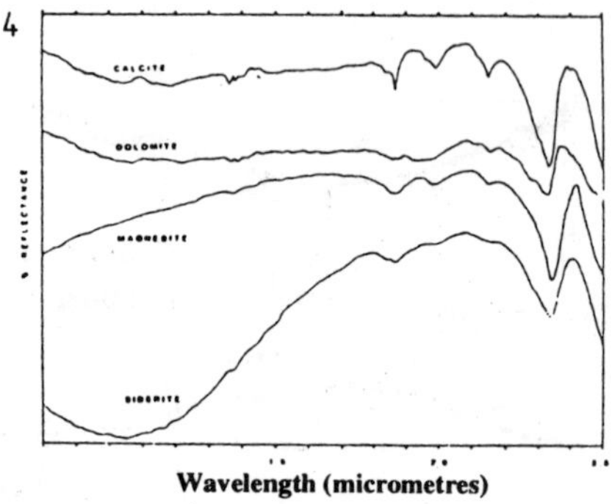

4. Reflectance spectra of some carbonates

Fig. 22. Examples of reflectance curves for minerals measured in the laboratory (reproduced from G. Whitney, M.J. Abrams and A.F.H. Goetz. *Economic Geology*, vol. 78, pp. 689, 690, 691 (1983)).

The spectral signature is given with respect to either reflection or emissivity. Some examples of spectral signatures of emissivity for different elements, sands, clays and soils are depicted in Fig. 21.

Examples of spectral signatures of minerals and rocks in reflection and emission modes in the middle infra-red and thermal infra-red regions are presented in Fig. 22.

— Surface temperature and brightness are the physical properties accessible in the thermal infra-red to high-frequency range. These parameters are affected by the chemical nature of rocks and their porosity. They vary with water content and albedo. Thermal inertia, the conventional term used in remote sensing for the heat capacity of rocks, is another physical property which facilitates the spatial interpretation of spectral signatures. Only certain satellites have the facility for its measurement.

(a)

(b)

(c)

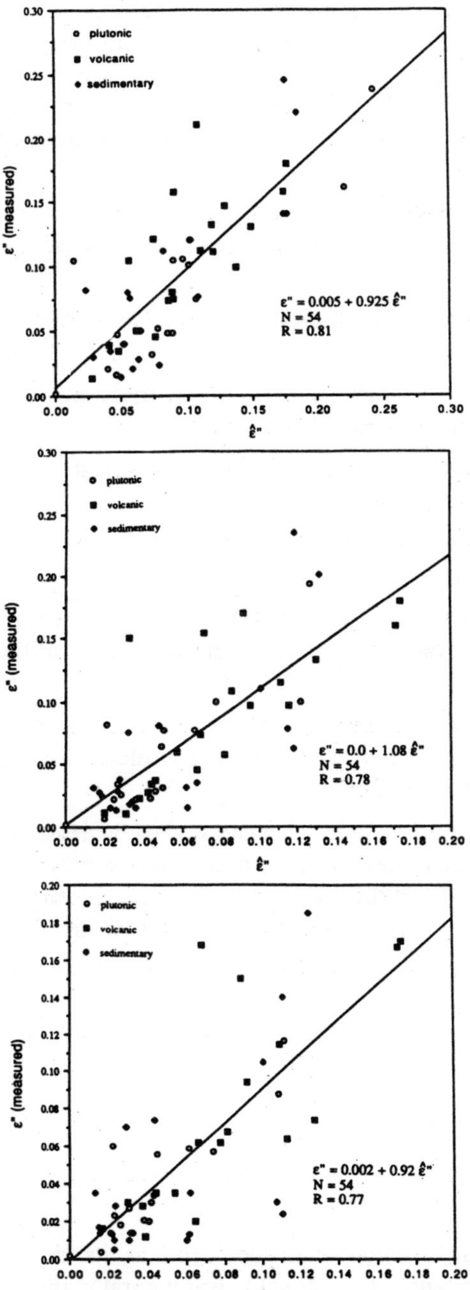

Relationships of the measured dielectric loss to the estimated loss using bulk chemistry at: (a) 1.6 GHz; (b) 7.8 GHz; and (c) 16 GHz.

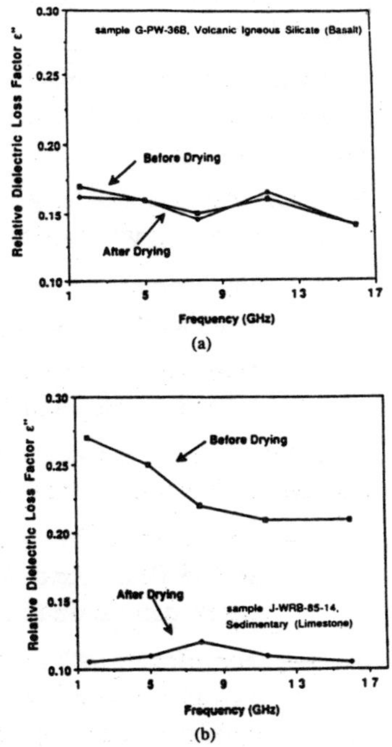

Effect of drying surface water on ε". (a) Basalt. (b) Siltstone.

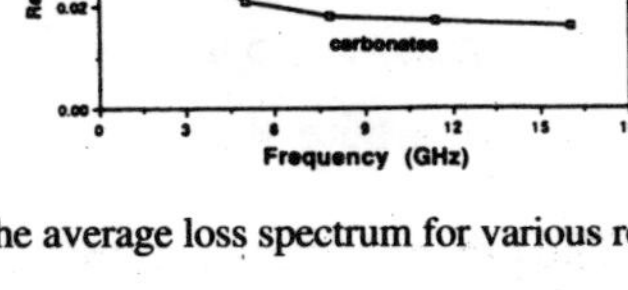

Variation in the average loss spectrum for various rock classes.

Fig. 23. Dielectric constants of some rocks (after Ulaby *et al.*, 1990).

The coefficient of backscattering recorded in the high-frequency range is not a spectral characteristic in the strict sense because of its active mode of acquisition. The laws that govern it are quite complex depending on the type of polarisation, frequency, angle of view etc. It is in particular a function of the surface geometry, its unevenness and the dielectric constant which can be measured in the laboratory or inferred (Fig. 23).

The article by Ulaby *et al.* (1990) makes a reference to the effect of dielectric constant which, together with surface unevenness, strongly influences the value of backscatter radar signal.

Summarising this article, it can be stated that the dielectric constant is a physical property of rocks, which is modified by:

— the *water-content* of rocks, which plays a major role, irrespective of whether the water is in a free state or is bound in minerals;

— *temperature* of rocks which affects the constant especially for silicates. The latter increases with increase in temperature.

In general, the mineralogical composition and texture of rocks are the major parameters that control the dielectric constant:

— For any rock, the 'true' permittivity is independent of frequency between 0.5 and 18 GHz (bands K to L). The variations observed, at least in 50% of the cases, depend on their density.

— For silicate rocks, when they have the same genetic relationship, such as volcanic, plutonic or sedimentary, 78% of the variation is explained by the combined effect of density and nature of oxides.

— For most rocks the 'apparent' permittivity decreases with increase in frequency and there is no evidence for the influence of density, except in the case of silicate rocks for which 60% of the variation in dielectric constant is explained by this parameter if they have the same genetic relationship.

According to Olhoeff (1986), the dielectric constant of some rocks is as follows:

basalt	9.4	peridotite	7.8
granite	5.6	limestone	8.7
quartzite	4.9	micaschist	6.4

Thus theoretically the radar signal can be exploited for petrographic investigations. But, is it possible in reality?

Radar data has long been used in the original analogue form, as its digital transformation appeared to be of no importance. Thus its study was naturally limited to visual interpretation with or without stereoscopic analysis. Until now the appearance of digital aerospatial radars has hardly changed the practice, except in enabling, and this is important, optimisation of image restoration in terms of geometry, contrast, filtering, combination of image sets etc. One may be surprised with this shift in laboratory results which once appeared to open up new approaches to digital classification. There are several reasons for this:

— The mixture of responses that constitute the signal: dielectric constant and surface unevenness. Numerous publications show that the second parameter strongly influences the response of rock massifs by changes in texture that it introduces; however, its real contribution is not quantified. Both these parameters are being investigated by several authors (Moon, 1991; Won and Moon, 1992; Chen and Fung, 1992; Dawson *et al.*, 1992);

— The conditions of outcropping which limit, as in the case of optical data, the extension of a laboratory model to the field.

Often a response is obtained corresponding to a 'mixture' which we do not yet know how to exploit in digital form, though research studies are undertaken in test sites that are preliminarily favourable to the use of some parameters (wavelength, angle of incidence, polarisation etc.)

— The effect of operational characteristics of radar such as frequency, angle of incidence, angle of viewing etc.

Given these facts, the examples of geological applications presented in this manual (see 2.4.6) are the result of visual interpretation like that of photointerpretation. They are chosen as they represent different geological and climatic environments.

Laser luminescence, which enables recording the specific response of certain minerals after irradiation in certain wavelength ranges (visible, thermal), can also be considered a means for understanding the spectral signatures of geological objects (Gorobets *et al.*, 1989).

5.1.2 Spectral Signatures of Minerals, Rocks and Alterites

Laws governing the spectral signatures of minerals, rocks and alterites have gradually been established for the visible to infra-red bands by laboratory measurements of samples and powders. Under these conditions, the reflection spectral response is a function of reflectance, a parameter that measures the luminescence or radiance re-emitted by the surface of the body depending on the incident intensity.

Spectral responses obtained from field measurements made on targets of very small size by means of portable radiometers result from the integration of spectral signatures of definite objects such as rocks and minerals, that are generally impure, and their surroundings.

Radiometers mounted on satellites measure from space the luminescence integrated over relatively large areas, often highly heterogeneous, for which the local values of a parameter would be variable.

That is why it is proposed to discuss the reality of the spectral signature concept for each of the three conditions of measurement separately, viz., laboratory, field and space, with the objective of evaluating the potential of remote sensing for digitally classifying geological objects.

5.1.2.1 Laboratory studies

Investigations have been carried out for several years on the use of spectral attributes of remote sensing data for improving interpretation possibilities, such as discrimination, classification, identification or qualification, quantification etc.

The aim of laboratory experiments is to establish the existence of clear and singular correlations between the physicochemical properties of minerals, rocks and alterites and the reflectance, usually known as microreflectance, under conditions of laboratory experimentation (Longshaw, 1976).

Luminescence, or radiance, is defined as the energy radiated by a unit area of the medium. It is a function of the flux received by the unit area, defined as irradiation. Both these parameters vary with wavelength.

Physicists have demonstrated under certain conditions of experimentation that reflectance, especially in the range of 0.4 and 2.5 µm, is effectively a constant for minerals, which can facilitate their identification if all the characteristics of the spectral curve are taken into consideration. Iron oxides, carbonates, clays etc. generally obey this principle (Galley, 1986; Hunt and Salisbury, 1971, 1979). Similar studies were subsequently undertaken on emissivity in the middle infra-red range, between 2.5 and 13.5 µm (Becker and Raffy, 1987; Kayle, 1987; Salisbury *et al.*, 1988). Silicates, sulphates and carbonates present specific spectral characteristics in this wavelength range that can provide their identification.

However, in the case of rocks there is generally no direct relation between reflectance and petrographic nature since they are, without exception, aggregates of varied minerals. In recent years studies have concentrated mainly on the spectral range between 2.5 and 13.5 µm (Salisbury *et al.*, 1988). These studies show that carbonaceous or argillaceous rocks, which contain one dominant mineral, have quite distinct spectral characteristics of reflectance and emissivity, while igneous rocks, whose composition is often very complex, do not follow the same laws. Nevertheless, all these rocks have significant spectral characteristics: hydroxyl (O-H) bands at 2.79 and 2.9 µm, hydrocarbon (CH) bands at 3.35 and 3.40 µm, carbonate bands at 4.0 and 5.6 µm besides a doublet at 3.5 µm, and silicate bands between 8 and 14 µm. These characteristics facilitate their identification. It is generally accepted that the success of identifying igneous rocks by their spectral characteristics mainly depends on defining a digital parameter. In this context a specific index, known as the SCFM index, was developed (Walter and Salisbury, 1988). It is defined as the ratio $SiO_2 / (SiO_2 + CaFe, Mg)$. Special tables permit its conversion, under certain limits, into reflectance and enable identification of igneous rocks (Fig. 24).

Even though a problem exists at present regarding calibration in the absence of known standards with respect to apparatus, measurements and analysis of reference samples, a certain consensus has emerged on the principles of preparing the spectral catalogues. This consensus concerns the following aspects:

a) Characteristics of the spectral curve of reflectance

Some of these are useful for distinguishing minerals and certain rocks of homogeneous composition. These characteristics include:

- Absolute intensity of reflectance which varies with wavelength and gives rise to variations in amplitude of signal;
- Relative intensity, viz., the ratio between absolute intensity of an absorption band or a transmission peak and that of another band or another peak;

(a)

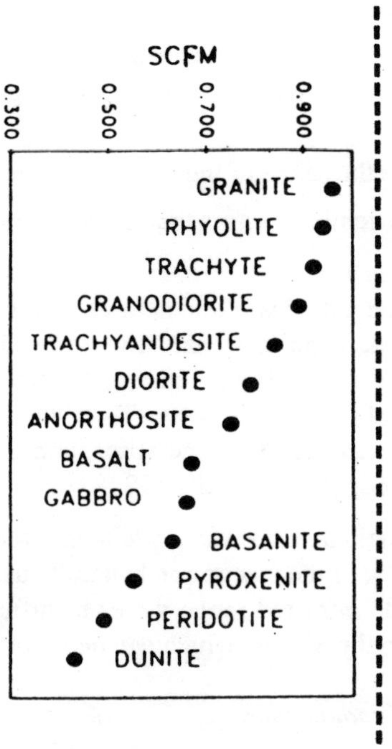

a—SCFM values for some important igneous rocks presented in decreasing order.

(b)

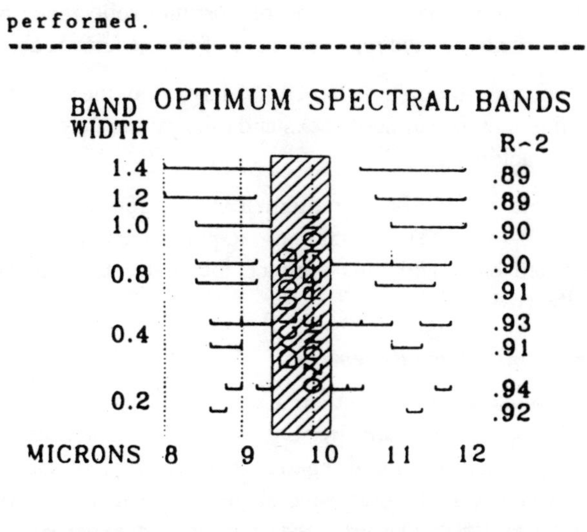

Fig. 24. Selection of the most appropriate spectral bands for determination of SCFM index between 0.2 and 1.4 μm (after Walter and Salisbury, 1988).

- Number, width, position and absolute or relative intensity of absorption bands or transmission peaks.

A combination of all these characteristics is necessary for laboratory identification of minerals and certain rocks on the basis of spectral signatures but is not always sufficient.

b) Factors which influence the shape of the spectral curve

This refers to the physicochemical factors that induce variations in the type of curve. These factors are:

- Surface state of the samples, their natural or artificial roughness (powder, polished sections etc.) and their size. These physical parameters result from grinding and considerably affect the absolute intensity of reflectance and the general amplitude of the spectral curve. Thus when the particle size is equal to or lower than the wavelength of the absorption bands, the latter are shifted towards the red bands (Morris *et al.*, 1982).
- Mineralogical variations: the water content and iron oxide affect the position, width and amplitude of the absorption bands. Thus the reflectance of igneous rocks in the middle infra-red range varies significantly as a function of the chemical indicator, the SCFM, which can be used for their identification.

Implications for design of radiometers

Using the above principles, which are not fully validated due to the lack of standardisation of analyses, calibration, measurements and apparatus, the following remarks can be made:

- Present-day radiometers-portable, airborne or spaceborne-do not have the specifications required for identifying rocks or minerals because solely designed for laboratory acquisition of absolute reflectance and for spectral ranges that are not always representative of all the known absorption bands.
- Imaging spectroradiometers generally have the theoretical specifications which enable identification of rocks and minerals as they acquire a number of spectral characteristics.
- Programmable narrow-band radiometers developed by the Japanese for satellite ERS-1 (1992), which can acquire spectral characteristics in a predetermined band and thus produce less data, are a useful compromise between the two.

5.1.2.2 Field radiometric measurements

Laboratory experiments, whatever be their accuracy, do not take real field conditions into account. The need for carrying out *in-situ* measurements with portable radiometers was imposed to guard against certain impracticabilities of remote sensing. This approach was adopted several years ago in France (Bouchet, 1985; King, 1985; Louhala, 1988; Deroin, 1990), given the experience resulting from preparatory studies on the utilisation of Spot data, mainly in the visible and near-infra-red spectral ranges.

It was necessary to establish that a clear correlation exists between the spectral signatures obtained from laboratory measurements and the responses registered in the field or, in other words, it was required to establish the relation between microreflectance and macroreflectance (Longshaw, 1976). This research has great importance at several levels of decision-making, such as:

- Preparation of database required for the functioning of expert systems;
- Designing digital data-processing technics;
- Comparison of image observations with ground truth in the middle infra-red region, in which the notion of false colour is more important than that of natural colour.

Major studies were carried out between 1985 and 1987 in the visible and near-infra-red spectral ranges. They have been continued in particular in RFA by experiments in the middle infra-red band (1.5 to 2.6 μm). They have provided a knowledge of the complexity of responses recorded in various environmental conditions-abundance of vegetation, occurrence of rock outcrops etc.

Spectral data on argillaceous minerals under field conditions is presently being collected by the Colorado School of Mines (USA) using a very lightweight (3 kg) portable spectroradiometer (Puma U), operating between 1.3 and 2.5 μm. This experiment has already been carried out in the Paris basin (Hauff *et al.*, 1990).

The factors affecting the reflectance recorded by a portable radiometer are quite varied. The major ones are:

- Mineralogical characteristics of the rock and mineral: spectral signatures are more often representative in the case of isolated and pure minerals than in rocks containing them;
- Texture of the rock: its isotropy, anisotropy, grain size, degree of alteration;
- Surface conditions: unevenness, patina and exposed area of outcrops;
- Hydrological state;
- Nature of the medium from which the information is obtained: unweathered rock, weathered rock etc.;
- Environmental conditions: state of vegetation, presence of lichens and micro-organisms etc.

Some of these factors may constitute reliable elements for identification. However, any study of the signatures of minerals and rocks must take these factors into account. Compared to laboratory measurements, it is here that the importance of field radiometric measurements lies. Further, this approach is not without errors since the objects investigated, in view of the available technology, are of small size and their representation is questionable. To overcome this, autodriven assemblies with telescopic systems or balloon surveys are sometimes employed.

In spite of its limitations, this approach is a source for better understanding of the representativeness of spectral measurements made in field conditions. This has become particularly evident from recent studies (Louhala, 1988) carried out by means of HRS radiometer developed by the CNES. Measurements of *in-situ* reflectance were made in Provence (France), a region in which the rocks are mainly covered by soils and vegetation.

Fifteen hundred spectra of reflectance between 470 and 900 nm were acquired in varied conditions over 30 rock types such as limestones, marls, mudstones, sandstones, orthogneisses, granites etc. The measurements were made on large outcrops and their surroundings, consisting of rock debris, soil and variable degrees of low vegetation cover. The area investigated in this measurement was 40 cm^2, which is rather small. The measurements were made around midday in a time interval within plus or minus three hours.

The types of spectra recorded can be classified in three categories:

1) Iron type of spectra for which maximum contrasts are observed below 600 nm;

2) Chlorophyll type of spectra, characterised by low absorption around 670 nm;

3) Monotonous spectra between 670 and 900 nm, strongly induced by mineral and organic specifications with no direct relation with lithology. They generally represent the characteristics of the patina and colour of the surfaces. Maximum contrasts are obtained for fine-grained rocks such as argillites or weathered rocks since lichen development is lower in this case. As for rock debris, 'roughness' of the surface increases and the level of reflectance decreases by about 10%, without affecting the shape of the spectrum. But repeated measurements on outcrops of the same lithology have shown variations of the order of 10%.

All these factors introduce an uncertainty of spectral characteristics between 0.4 and 0.9 µm. This uncertainty must be known in order to select useful bands for measurement. Simulation tests were conducted by Louhala by taking into account the position, width and number of certain absorption bands characteristic of common minerals. These tests showed that spectra of maximum contrast can be obtained with a high degree of correlation when eight spectral bands are used. On the other hand, use of just six bands, centred on 480, 540, 600, 680, 740 and 870 nm, reduces redundancies and satisfactorily reproduces the spectral characteristics and their variations.

Measurements made in Burundi with a portable radiometer operating between 0.4 and 2.5 µm (Lavreau *et al.*, 1991) confirm the feasibility of spectral discrimination of rocks and their alterites under natural conditions, as also the effects of weathering, which complicate their characterisation. These measurements have especially demonstrated the influence of water content, strongly expressed at 1.5 µm, and of phytosilicates, characteristic of granites at 2.35 µm. These studies confirm the need for taking into account all the spectral characteristics, especially the amplitude of reflectance.

Field radiometric studies have led to enunciation of the following remarks and recommendations which must be taken into consideration for detecting the spectral response of minerals and rocks.

- Interfering effects on the spectral signatures of minerals and rocks are not negligible between 0.4 and 1.1 µm. This signature is generally similar in overall shape and intensity to the one obtained from laboratory measurements. But some important features are partially suppressed or skewed since the surface layer of the medium (of the order of 1 to 3 mm in thickness) from which the information is received, is often weathered. Iron oxides become dominant and a surface characterised by unevenness is smoothened.

- Weathering effects persist beyond 1.1 μm but *in-situ* radiometric studies in this spectral range are very rare and inadequate for generalising the results. A study carried out at two stages, viz., laboratory measurements of powders and samples of gossan and field investigations (Bothorel, 1983), showed that the effect of weathered desert patina in the middle infra-red region between 1.5 and 2.6 μm, can be eliminated by employing the ratio type of digital processing. The depth of penetration in this spectral range is small, of the order of 50 μm (Buckingham and Sheldon, 1983), but sufficient to penetrate the patina.

- A recent study (Wood *et al.*, 1990) indicates that most spectral characteristics which facilitate identification of granites are apparent in the range of 8.2 to 13.6 μm, even though the study was limited to outcrops covered by lichens for 71% of the surface area.

- The above notwithstanding, to study the *in-situ* spectral signatures of minerals and rocks in all conditions as perfectly as can be done in the laboratory, remains difficult. It is therefore more realistic in space geology to prefer the concept of spectral response over the concept of spectral signature and to divert efforts to the development of narrow-band radiometers and spectroradiometers that are capable of facilitating the identification of minerals and rocks.

Concomitantly, it does appear possible to improve the capability of remote sensing for identification of rocks. This possibility is already high for arid and semi-arid zones and imaginable for temperate climate zones for the purpose of evaluating certain effects, namely:

— Existence of a correlation between different strata in the Earth's crust, vegetation, surficial deposits and their hydrological state and geological substratum.

— Representative spectral responses of rocks, alterations and the surrounding media. From this point of view, it would be important to create a database of spectral responses recorded in the field in addition to the database of spectral signatures measured in the laboratory. This approach appears necessary but poses the problem of how representative measurements made by means of portable radiometers would be, since at present they are inadequate.

— Ratio between the mean area of outcrops and resolution of the acquisition system. The number of pixels which represent outcrops is small and no progress can be made in this respect until spectral resolution and spatial resolution are simultaneously improved. The major problem lies in finding an optimal ratio between these two parameters because the transitions between different units on images of high spatial and spectral resolution naturally is gradual, thereby introducing the concept of 'hazy' boundaries or a continuum (Durand, 1986).

5.1.2.3 Spatial measurements

The Problem

The problem of data acquisition from space consists of identifying the phenomena affecting perception of ground objects from space, understanding their mechanisms, evaluating their effects on the data and ultimately making corrections accordingly.

It has been established that the macroreflectance measured in the field does not represent the microreflectance determined in the laboratory. Furthermore, the megareflectance obtained from space, largely influenced by transfer functions, is not equal to macroreflectance and is less than microreflectance.

For space measurements, theoretical and empirical studies are made side by side. Such investigations are conducted on representative data of various climatic environments, degrees of occurrence of rock outcrops and lithological media as varied as possible and for which factors influencing space spectral signatures are of two kinds: physical and environmental.

The physical factors pertain to clearance of objects relative to the source of energy and the relief, the two factors being closely interdependent, as well as the processes of transmission through the atmosphere, governed by the variability of its state. It is known, for example, that the presence of water vapour in the atmosphere is the cause of narrow absorption bands which disturb the shape of the spectral curve of an object.

Environmental parameters affecting the spectral signature recorded from space are similar to those phenomena influencing macroreflectance. These effects are either amplified or, contrarily, partially attenuated in the case of space observations which integrate heterogeneous areas with variable values of their characteristic parameters.

The environment can be represented as consisting of two parts-the object itself and its surroundings. The response of an object viewed from space is influenced by its surrounding features and the object per se is thus never or rarely 'pure', especially in temperate climatic regions; alterations, surficial deposits, micro-organisms, vegetation etc. distort the response. This feature implies that only pixels that are the most appropriate relative to the object under investigation should be processed. This poses the problem of effective resolution and simultaneous variation of spatial and spectral resolutions. Studies in this direction are under way.

In practice, the introduction of correction models, application of digital processing optimised on the basis of laboratory spectral data and improvements in the performance of radiometers have produced results confirming the theoretical capability of remote sensing for at least discriminating the geological objects and characterising them, if not totally identifying them from spectral responses.

Geological Objects Relevant to Spectral Response

Operationally, the presently available radiometers can record significant spectral responses of rocks and alterations in the visible infra-red spectral range (up to 2.5 µm). These responses have a regional value which can be verified over known segments in the field marked on the image. They can be enhanced by the application of specially chosen digital processing technics during the interactive session. The responses can be classified as lithological, mineralogical, metallogenic, hydrological, geobotanical etc.

Lithological responses have been obtained either with MSS and TM data of Landsat or HRV data of Spot. These radiometers facilitated identification of volcanic flows of different ages, sands and sandstones in Saudi Arabia (Chevgel, 1984-1986) and stratiform and secondary dolomites in Yemen (Christman and Scanvic, 1987).

Mineralogical responses have been obtained with MSS Landsat data for iron oxides, determined by the ratio of responses in bands 5 and 4, and with TM data for gypsum identified by the ratio of responses of 1.6 and 2.2 μm. Clays were characterised by the response of band 7 at 2.2 μm. A thermal radiometer was used for quartz, for which the response was very strong in the infra-red range between 8 and 14 μm.

Metallogenic responses have been obtained by digitally calibrating them over known reference zones on the ground, for gossans in Zimbabwe, nickel oxides in New Caledonia, bauxites in Guinea and hydrothermal alterations in the USA.

Hydrological responses are often detected using TM, Spot or thermal radiometric data, in arid and semi-arid zones, where they are readily identifiable because enhanced by those of the vegetation.

Responses that are more complex to decipher have also been observed in some cases. They are of the geobotanical type (shifts) or result from a combination of different parameters, whose association determines a unit of landscape known as the 'orthotype'. Their importance will be discussed in Chapter 6.

Under certain conditions the thermal response obtained as thermal inertia or differential temperature can identify lithological or metallogenic phenomena. This aspect of methodology is mentioned in Sections 4.2.3 and 7.6.2.

5.2 CASE STUDIES

Depending on the technic of data acquisition, two approaches exist for detecting the spectral response of a geological object. The first, the most widely used in view of the presently available data, consists of analysis by different types of processings chosen for ascertaining the spectral characteristics of minerals and rocks.

The spectral imaging technic, presently under development, forms the basis of the second approach. It consists of comparing the totality of the spectral curve recorded from space with a reference database. This approach, still in the experimental stage, is the objective of an International Geological Correlation Programme (IGCP 264), which deals with all the problems mentioned in the preceding section.

5.2.1 Spectral Responses of Rocks

In some cases the rocks identified from ground surveys constitute the source of specific spectral responses, which can be used for distinguishing them from their surroundings and to extrapolate the result to a regional scale. This possibility is closely associated with the developments in digital processing and enlargement of spectra recorded by new sensors, either spaceborne or airborne. Thus significant progress has been reported from arid or semi-arid regions, as demonstrated by some recent examples described below.

- The possibility of identifying serpentines by their spectral signature, of course in optimal conditions of occurrence of rock outcrops, has been demonstrated in Bou Azzer (Morocco) (Botherel, 1983). As this rock type is known for its metallogenic characteristics with several occurrences of cobalt, chromite and nickel, this possibility has great potential application, provided the method can be extended to other regions.

— An analysis was carried out in Saudi Arabia to study the differences in spectral behaviour of volcanic flows as a function of their chemical composition (Chevrel, 1986). This study revealed certain anomalies in the response. While the reflectance of flows in general decreased from green (band 4) to the near-infra-red (band 7), some did not obey this law. Their reflectance did not vary between channels 4 and 5 and significantly increased in band 7. This abnormal behaviour was found to be related to the high concentration of ferric ions, whose reflectance is stronger in the infra-red band, since the chemical analysis carried out independently showed that their Fe_2HO_3 content increases from 4 to 7%.

However, in the morphoclimatic conditions generally encountered on the Earth, which are unfavourable, the spectral signature alone is most often insufficient to identify or to classify rocks (Carrere, 1987).

5.2.2 Spectral Responses of Alterites and Alterations

Three reasons justify this specific research. The first concerns the importance of alterites in modifying the spectral signatures of rocks. The second, that a number of alterations have a metallogenic significance. The third, the existence of a relation between alterites and the bedrock.

Several geological and metallogenic studies give descriptions of alterites, metallotects and certain petrochemical indicators of the surface. However, integrated reports on the importance of alterations in prospecting and their classification are scarce (Meunier *et al.*, 1984). Yet a knowledge of these aspects is essential for their study by remote sensing.

Alterations can be separated into three groups: iron hats and ferruginous alterations, hydrothermal alterations and lateritic crusts. Some represent surface indicators of concealed mineral deposits. Our exhaustive search of the literature established that no specific phenomena of alteration of oil and gas deposits have been reported.

5.2.2.1 Gossans

a) *Spectrometry of Iron*

A knowledge of the spectral characteristics of iron, at least summarily, is necessary for understanding the phenomena that constitute the basis for its detection by remote sensing and for selecting digital processing technics to be employed for data analysis.

The spectral characteristics of iron are well known. The absorption bands used for spectral differentiation correspond to ferric iron (Fig. 25). They are situated (Hunt and Salisbury, 1971) in the ultraviolet range at 0.35 μm and in the visible band centred on 0.40, 0.45, 0.49 and 0.65 μm or on a wider band between 0.85 and 0.95 μm. These absorption bands characterise limonite and the melange of geothite-haematite-jarosite-lepidocrosite, which enters the composition of iron hats. Beyond the visible and near-infra-red range, it is possible to identify (Hunt, 1979) jarosite from an absorption band around 1.88 μm and goethite from a band centred on 0.90 μm. Geothite can be identified with certainty from the absorption peak at 0.90 μm in the absence of peaks at 1.88 and 0.45 μm. Some of these characteristics were also found in field conditions over a gossan at Rouez (France), which represents a small sulphide deposit. Field measurements in this zone using a portable radiometer revealed variation of reflectance with iron content (Scanvic *et al.*, 1985) for values of 23%, 17% and 8%. These *in-situ* measurements showed the existence of two absorption peaks at 0.90 and 1.15 μm and a mean reflectance value lower than that of the surroundings for all the wavelengths. However, in the case of rocks with smaller variations of iron content, detection of these absorption peaks is not easy even though the total effect of iron is manifest in all the spectral bands (Girard and Girard, 1977) since reflectance decreases with an increase in iron content.

b) *Gossans: Problem of Identification by Remote Sensing*

Iron hats are the outcrops of different residual products resulting from meteoric alteration of the underlying parent rocks enriched in oxides and hydroxides of iron. They are often the surface indicators of mineralised accumulations concealed at depth. Two terms, viz., 'iron hat' and 'gossan' (Kosakevitch, 1979) are used for these alterations but the latter is limited to oxidised zones of only sulphides. In remote sensing, however, the term gossan is used in all cases due to its extensive use in American research in this field.

It is a fact of experience that certain types of digital processing of Landsat MSS data facilitate identification of iron hats or rocks with a high concentration of iron oxides. This is achieved by ratio analysis in which the ratio of responses in channels 4 and 5 (4/5 ratio) is often utilised (Rowan *et al.*, 1977). However, its magnitude is not always related to the percentage of iron content and the results are sometimes contradictory. Thus a low ratio value is observed for certain highly ferruginous soils while for others, contrarily, this ratio increases with an increment in iron content (Lyon, 1975). For discrimination of gossans, Lyon preferentially utilised a complex colour composite prepared from the ratios 7/4, 6/5 and 6/4. At present, the spectral characteristics of the Thematic Mapper radiometer are better adapted and have enhanced capabilities.

- While the experiments and their results demonstrate the feasibility of identifying iron hats or gossans from MSS data, they also indicate the limitations. The first, if not the most important limitation is related to the resolution (79 m) of Landsat MSS data. As a matter of fact the size of the targets is too small to be perfectly compatible with such a resolution since gossans 20 m or even less in length are common. Also, attempts have been made to improve this parameter by employing certain digital processing technics such as resampling due to which the initial resolution can be replaced by a pseudoresolution of 50 m. The onboard radiometers of second-generation satellites, the Thematic Mapper of Landsat 5 and the High Resolution Visible of Spot have brought this limit down to 30 m and 20 m respectively, and thus are capable of identifying smaller iron hats.
- A second limitation is due to the density of vegetation cover or the percentage of soil cover. When this percentage is variable or the quantity of vegetation

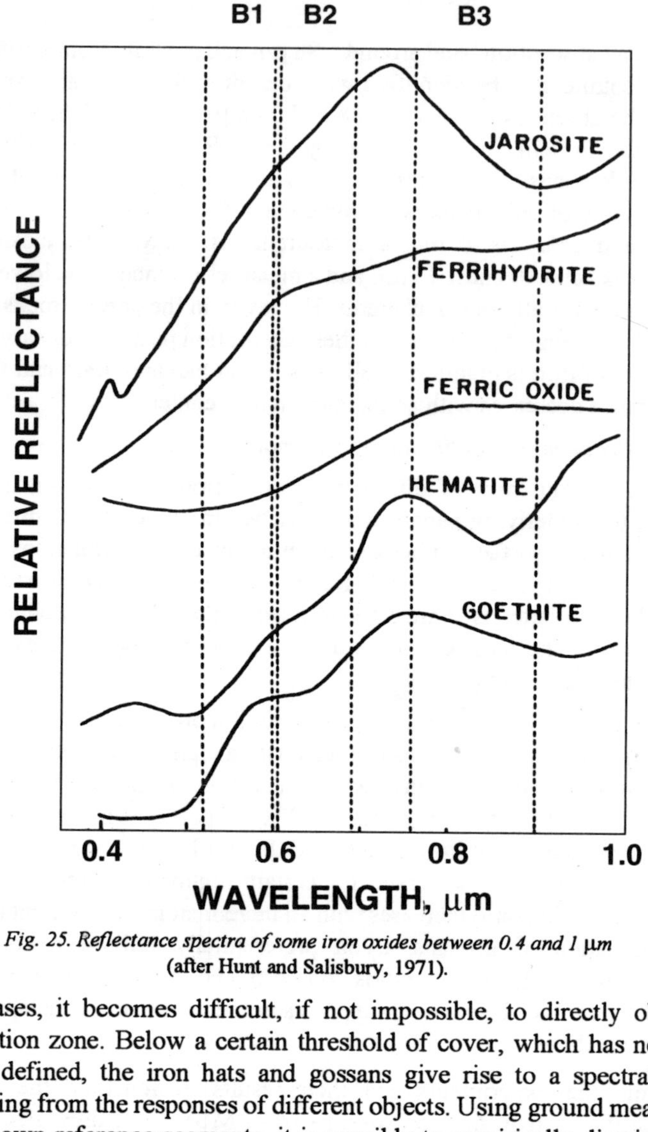

Fig. 25. Reflectance spectra of some iron oxides between 0.4 and 1 μm
(after Hunt and Salisbury, 1971).

increases, it becomes difficult, if not impossible, to directly observe the oxidation zone. Below a certain threshold of cover, which has never really been defined, the iron hats and gossans give rise to a spectral response resulting from the responses of different objects. Using ground measurements on known reference segments, it is possible to empirically discriminate and identify them based on the technics of ratios (Siegal, 1983), especially the ratios 4/5 and 6/7.

- While the problem of vegetation, a severe constraint in several regions, disappears in arid and semi-arid zones, a third limitation is encountered in these climates, emerging from the common existence of a ferruginous alteration layer on the rock known as desert patina. This problem, which prompted the famous statement 'nothing other than iron is seen there' (Cervelle, 1989), can be partially resolved with the help of radiometers operating in the middle infra-red region between 1.6 and 2.2 μm, since these waves penetrate the patina.

c) *Examples of Detection of Gossans*

- The application of certain special processing technics on the data, acquired by means of a Daedalus type of airborne radiometer, led to the detection of iron hats in Greenland in the Werner-Bjerge region (Favard *et al.*, 1981). The different spectral bands used were 0.50-0.55, 0.55-0.60, 0.60-0.65 and 0.65-0.70 μm, and the data was regrouped in the form of ratios: 4/5, 4/6 and 4/7. The spatial resolution was 10 m.

A colour composite combining any two of these ratios with a neutral colour associated with the ratio 7/10 (red) depicts the oxidised zones in a reddish-brown tone. Thus 365 anomalies could be identified in an area of 1500 km^2 having a good occurrence of rock outcrops. Their average size was 250 m. Ground verification of representative samples showed that the anomalies correspond to oxidised zones and all the oxidised zones known from earlier surveys were clearly distinguished. The few countable exceptions corresponded to zones of high relief where conditions of observation were poor because of associated shadows. This was the only limitation in the well-exposed conditions of Greenland.

- The need for high spatial resolution is confirmed by a more recent example of Spot XS (multispectral) data. The study was carried out in the Lomagoundi basin in Zimbabwe, region of Copper Queen, characterised by dominant soil and vegetation cover (Scanvic, 1986). Earlier efforts with special processing of Landsat MSS digital data had yielded no results. The size of the gossans, on average 50 × 40 m, was too small to differentiate them using their spectral response. Other experiments were conducted on the high-resolution (20 m) digital data of Spot, choosing the processing programs on the basis of data of known iron hats that were accurately located on the images because of their form. Some of the spectral ratios described in the literature, such as 1/2 (red), 1/3 (green) and 2/3 (blue), which had facilitated the mapping of gossans under other conditions, did not give a positive result. On the contrary, the colour composite of the ratios 2/3 (red), 1/3 (green) and 1/2 (blue), optimised using known targets, aided in regrouping all the known gossans into a single class of reflectance (yellow) and in discovering new sites (Plate 17). However, the final selection of potential sites was made by combining the criterion of spectral response with that of shape suggestive of the existence of granite gneiss domes, considered one of the metallotects.

- The sensors presently available are amenable for optimal choice of spectral bands specific to iron but in some cases perceptible improvements have been obtained by 'playing' alternatively with MSS and TM data (Abrams, 1983). In Arizona, over a chosen zone of experimentation use of MSS data optimised by specific ratios aided in separating weathered rocks rich in iron oxides from unweathered rocks on top of copper porphyry. This mapping was subsequently completed by the processing of airborne Thematic Mapper data generated in the form of a colour composite of the ratios of bands 0.83/1.65 μm, 0.66/0.56 μm and 1.65/2.2 μm, following the blue-green-red code. This composite depicts zones rich in iron oxides, which are characterised by an iron absorption peak in the ultraviolet and also extending to higher wavelengths. Clay zones are indicated on these ratio images as red since the minerals in them produce an absorption band around 2.2 μm. When iron and clay are mixed, they are restituted as yellow or orange depending on which predominates.

Similar experiments in arid zones (Podwysocki, 1984; Bailey and Dwyer, 1985) led to an identical conclusion that discrimination of gossans is possible in regions of rock outcrops if the choice of spectral bands, their centres, widths etc. is optimised. These results justify the ongoing technological developments for producing programmable narrow-band radiometers.

5.2.2.2 Other examples of ferruginous alterations

Various ferruginous alterations of gossans considered surface indicators of mineral deposits, have been described in certain geological situations (Peters, 1983). Such is the case of certain mineralised sandstones in which specific ferruginous alterations are developed around zones of uranium concentrations. These ferruginous zones were studied on the ground by means of a distribution model for demarcating the exploration targets. This study showed that the alteration zone had resulted from the conditions of deposition of uranium. Its location relative to the mineral deposit follows precise laws. That is why the possibility of identifying these alterations, of red to dark red coloration, was evaluated using the multiband spectral data in regions of fair occurrence of rock outcrops where the laws of association have been established.

To some extent the principle of identification of these alterations is identical to that used in mapping gossans. In fact, sandstones, with or without mineralisation, always contain goethite but the latter is more abundant in mineralised samples than in normal ones. Contrarily, haematite is rare in unweathered sandstones and abundant in mineralised sandstones. Hence its response can be used as a discriminating factor. Accurate measurements of reflectance made on laboratory samples showed that haematite has an optimum response at 0.57 and 0.85 μm and goethite at 0.9 μm and 0.5 μm. These responses can be theoretically used for their discrimination in the spectral range accessible with Landsat MSS radiometer. In practice, however, the results obtained by the processing of this data using the ratio technic are not satisfactory. The ratio 4/5, often presented as a model (0.5-0.6 μm over 0.6-0.7 μm) does not distinguish goethite from haematite. The ratio 6/7 (0.7-0.8 μm over 0.8-1.1 μm) also does not overcome this inadequacy since band 7 of MSS is too wide to discriminate haematite from its absorption peak at 0.85 μm.

Subsequently, Peters evaluated the utility of the data acquired with an airborne radiometer in the spectral bands 0.4-0.5 μm (A), 0.5-0.6 μm (B), 0.6-0.7 μm (C), 0.7-0.8 μm (D) and 0.8-1.1 μm (E), which are equivalent to those of MSS with an additional band (0.4-0.5 μm). Based on the reflectances measured in the laboratory, he computed the ratios A/B, B/C and D/E, coded as red, blue and green respectively on the colour composite. When goethite is present (unweathered sandstone), the ratio A/B is low and red is rare on the image. If the ratios B/C and D/E are high, magenta-blue dominates the image. When the haematite content is high (mineralised sandstone), the ratio B/C is low and is represented by green, facilitating its detection.

However, if the ratio A/B is high or medium, yellow to greenish-yellow predominates; discrimination weakens and cannot be improved (as in the case of MSS) due to the large spectral width of band E.

The possibilities of discrimination can generally be improved using spectral band A but without eliminating the inherent limitations of band E. The result is nevertheless significant since unweathered sandstones and altered sandstones can be correctly differentiated in 65% cases and can be improved by optimisation of processing. Further, while altered zones were sometimes classified as unaltered, the reverse was never observed. Certain altered zones which were not mapped in ground surveys could be detected by this digital processing technic.

This example shows that the available design of onboard radiometers makes the digital processing required for discrimination of different components of alterations cumbersome without advantage. This justifies present attempts to develop a narrow-band radiometry.

5.2.2.3 Hydrothermal alterations

Solutions rich in sulphides are often injected in the host rocks during the final phase of volcanism. The process of injection and the supergene alteration which results from it modify the mineral content of the host rocks in the vicinity of sulphide deposits. The most intensive alterations are produced at the contact of the ore body and the degree of alteration varies as a function of distance from the contact. On the ground surface these alterations manifest as whitish zones and the affected rock is frequently coated with iron oxide resulting from the transformation of pyrite. They are also accompanied by silicification, sericitisation etc.

The minerals of hydrothermal alterations have been studied in the laboratory to determine their spectral characteristics and to evaluate the possibility of their detection by remote sensing (Hunt and Ashley, 1979; Lowell and Guilbert, 1970; Whitney *et al.*, 1983). They are listed in Table 5.1 and minerals with a characteristic spectral signature are marked with an asterisk.

Some of these minerals have specific spectral characteristics. In particular, two absorption peaks occur at 2.2 µm for kaolinite, shifted for montmorillonite and muscovite. Characteristic peaks for magnesian clays are observed between 2.3 and 2.35 µm; the absorption peak of alunite occurs at 2.17 µm; absorption peaks of carbonates are recorded between 2.35 and 2.53 µm.

The results of laboratory measurements have shown that it is theoretically possible to differentiate clay minerals, sulphates and, to a lesser extent, carbonates, using their spectral characteristics between 2 and 2.5 µm. For this, however, radiometers of high spectral and spatial resolution are required as the targets explored are of small size and the generation of photographic images needs to be supported by the ratio type of digital processing to enhance differences.

Table 5.1

Zone	Minerology
Propylitic	Albite plagioclase
	Chlorite *
	Epidote *
	Carbonate *
	Montmorillonite *
Argillitic	Quartz
	Kaolinite *
	Chlorite *
	Montmorillonite *
Phyllitic	Quartz
	Sericite *
	Pyrite
Advanced argillation	Quartz
	Alunite *
	Pyrophyllite *
	Kaolinite-dickite-opal *
	K-mica *
Silicified	Quartz
	Alunite *
	Pyrophyllite *
	Kaolinite-dickite *
	K-mica *

Following these measurements, the possibility of *in-situ* discrimination of some of these alteration minerals was experimentally explored in the field by means of a portable Barringer radiometer operating between 0.4 and 2.5 μm (Gladwell, 1984). The experiment was conducted on the Kerr Adison gold deposit (Ontario, USA). The important results are as follows. The degree of sericitisation is characterised by the ratio 2.1/2.2 μm. The abundance of carbonates is revealed by the ratio 2.15/2.33 μm. Low values of the ratios 2.14/2.33 μm, 2.15/1.95 μm and 2.1 / 2.2 μm indicate a lower degree of alteration of siliceous material compared to that of the tuffaceous host rocks, thus facilitating detection of cherts with gold mineralisation.

Experiments with airborne measurements are still rare. We can cite the example of the Eocene copper porphyry deposit of Stafford (Arizona, USA) whose alterations, known to be of the propylite (epidote) and quartz-sericite type, are emphasised by a whitening and chloritisation of the rocks (Abrams, 1983). Digital processing by the 'canonical transformation' technic makes these alterations appear in three different colours on a simulated Thematic Mapper image, exhibiting an excellent degree of correlation with ground observations.The example of Silver Bell (USA), presented by the 'Joint NASA/Geosat project', demonstrates the utility of certain spectral bands. The airborne data collected by means of the radiometer NS001, having seven channels between 0.46 and 2.36 μm and a thermal channel, was digitally processed using different algorithms such as ratios, canonical transformation, principal component analysis and supervised classifications (Abrams, 1985). An excellent result was obtained for the discrimination of rocks and alterations on images of the following ratios: 1.65/2.2 μm in red, 0.66/0.56 μm in green and 0.83/1.15 μm or 0.83 / 1.65 μm in blue. Twenty-four types of alterations could be differentiated on this image versus only 6 on a colour photograph and 10 on the images of M2S radiometer, which has 10 channels between 0.5 and 1.1 μm. These ratios have also indicated in the same region zones rich in iron oxides as green (absorption in ultraviolet band) and clay zones as red (absorption at 2.2 μm). Hydrothermal alteration zones consisting of clay, sericite and limonite appear as orange and yellow, without which it is not possible to discriminate the potassic zone from the phyllitic zone. Finally, the peripheral zone consisting of propylitic and argillitic minerals was characterised by greenish-yellow to brownish-yellow.

In this procedure, essentially based on the investigation of reference segments defined on the ground, there was no identification of minerals in the strict sense of the term, since the radiometers utilised lack this capability. However, other experiments are in progress using spectroradiometers coupled with databases (Abrams, 1985). They may be used in certain cases to obtain spectral signatures. In particular, analysis of data acquired by means of the imaging spectroradiometer AVIRIS during an airborne mission over the Goldfield mining site (USA) demonstrated the true capability of this system to map and identify the minerals of hydrothermal alterations in this specific case (Carrere, 1990).

5.2.2.4 Laterites sensu lato

Lateritic crusts are produced from the alteration of vast areas. Depending on their composition, their economic significance can be high. A simplified but clear terminology is adopted here (Grubbs *et al.*, 1984), grouping the laterites into four types: laterites proper, altered rock containing secondary iron, nickel, aluminium and manganese; nickeliferous laterites, rich in nickel and derived from ultrabasic rocks; aluminous laterites or bauxites; and terra-rosa bauxites, closely associated with limestones or dolomites. Figure 26 presents the reflectance spectra of major lateritic minerals.

In general, lateritic formations have an areal extent compatible with the resolution of remote-sensing sensors. This feature justifies the study of the spectral characteristics of their constituents in order to determine the potential of spectral radiometers to discriminate them.

Several case studies are presented in the literature. They are concerned with ferruginous crusts, bauxites and nickel oxides. A study of ferruginous crusts was carried out in the Kéniéba region in Mali (Roquin *et al.*, 1987). Spot images over an area of 25

km^2 were compared with multielemental geochemical maps prepared from 1000 surface samples collected in a 100 × 200 m grid. The most significant correlations observed between the geochemical and mineralogical concentrations on the one hand, and the alteration and vegetative indices on the other, were represented as diagrams for all the samples together and also separately for each facies. The diagram of correlation coefficients calculated for the whole group shows that the samples fall into three classes according to their affinity with the two radiometric indices. These classes correspond to three principal poles of geochemical differentiation:

— Quartz, silica and trace elements (Y, Zr, Ce, Ti), associated with heavy minerals that accumulate in valley soils on which vegetation is more abundant, correlate positively with the two indices.

— The opposite pole consists of elements enriched in oxyhydroxides of iron in laterite crust (V, P, Nb, As, Mo, Cr).

— The intermediate position is occupied by elements accompanying the aluminium phase with kaolinite predominant (Cu, Ni, Ba, Sr, Zn, K, Mg).

On the other hand, when each of the four facies observed on the ground is considered separately, two different types of responses, essentially related to vegetative indices, can be noticed:

— Among the three facies of ferruginous crusts (aluminous, siliceous and gravelly, the vegetation is most abundant in zones rich in aluminium and kaolinite.

— For alluvial soils the index of vegetation is related to the elements Mn, Ba, Ce, Sr and Co. This relation represents the effects of coprecipitation along the drainage axes where vegetation is best developed.

One of the aims of this study was to demonstrate the existence of direct and indirect correlations between the composition of laterites and their spectral responses.

At Boké in Guinea (Brosse *et al.*, 1987), at the request of the Bauxite Company of Guinea, digital data of Spot XS was used to study the extension of a deposit under exploitation. The Spot data was restituted photographically after sinusoidal display type processing, which proved perfectly adaptable for spectral identification of bauxites. On this image the bauxite basin is clearly distinguished relative to its lateritic surroundings due to the reflectance differences resulting from the iron oxide content. Even within the zone of bauxites, the reflectance differences correlate with the variations in composition determined from chemical analyses. Thus pure bauxites and ferruginous bauxites are characterised by their spectral reflectance, which varies with the concentration of iron oxide (Plate 18).

In the region of Thio, New Caledonia, Spot images restituted as colour composites facilitated spectral discrimination of nickel oxides and their surroundings. On the sea coast and west and east of Thio, a peridotite sill forms a continuous band differentiated by the alteration products associated with it. It is least affected in this region by fracture tectonics and the major faults observed on the Spot image have a NW-SE (110⊃ N) orientation typical of New Caledonia. These are in general strike-slip faults (Scanvic, 1988; Brosse *et al.*, 1987). Using the spectral analysis of

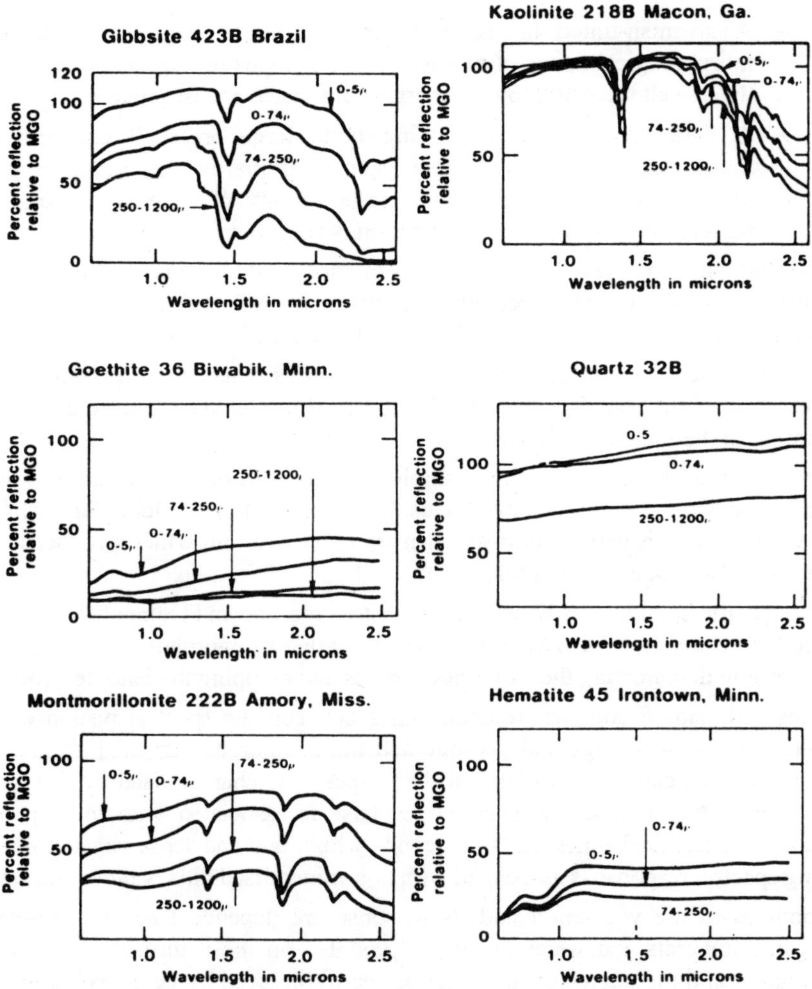

Fig. 26. Reflectance spectra of some common minerals found in lateritic formations (after Grubbs *et al.*, 1984).

Spot imagery, it was possible to map this group of faults in detail from the mineral exploration point of view. By extrapolation of ground observations and simulated data of Spot (Lajoinie and Weecksteen, 1985), the following subgroups could be identified (Plate 19):

— A white facies with which all the known deposits of nickel oxide are associated. Even though the site was disturbed by mining activities, the response was significant.

— A bright red facies over unweathered or slightly weathered peridotites that are generally characterised by a well-developed vegetation cover. This facies, which can be demarcated when it represents homogeneous zones of large extent, is of least importance for mineral exploration.

— A greenish-tinted facies, occupying a smaller area. It corresponds to morphological zones of relatively gentle slopes or large crests with which diverse alterites of known mineral significance are associated.

— A dark green facies occupying quite large areas. It corresponds to morphological zones of steep slopes, primarily the rocky slopes of unweathered peridotites where the possibility of coverage of significant mineral targets by soils or vegetation is very low.

Depending on the context, the spectral information could be classified digitally. The correlation between nickel content and value of reflectance, wherever it was observed, was high, as shown by Table 5.2 (after Lajoinie and Weecksteen, 1985).

It may be noted that the different spectral responses detected are produced on the eastern part of the island, where they can be used to identify the different levels of alteration in the scale considered.

However, in a broader sense the examples of detection of laterites are not always based on spectral criteria. In New Guinea the discrimination of nickeliferous laterites is mainly based on variations in the density of vegetation, which produces strong responses (Taranik and Reynolds, 1978).

In the specific context of a Venezuelan forest (Brojanigo and Delpont, 1989) studies carried out using multiband Spot images suggest that geomorphology and vegetation are useful in demarcating the exploration zones and mapping the bauxite deposits.

In Los Pijiguaos a complex relation exists between the spectral response of the vegetation, its physiology and geology and the geomorphology of the substratum. These relations can be classified into two types: one characterised by low-density vegetation cover whose low spectral response is associated with the top of the tabulated mineralised zones, and the second by luxuriant and dense vegetation whose strong spectral response is associated with non-mineralised hills, valleys and ravines.

In both cases the vegetation and its response are dependent on the presence of inorganic minerals and water in the soil, as also on the nature of the geological substratum and the degree of its alteration, partly induced by geomorphology. This methodology aided in detecting an important zone 40 km east of the Los Pijiguaos deposit, in which preliminary ground studies indicated the existence of a bauxite deposit. Subsequent exploratory pits and trenches revealed bauxite reserves of 150 million tons with zones containing 50% Al_2O_3.

5.3 PERSPECTIVES FOR UTILISATION OF SPECTRAL SIGNATURES IN REMOTE SENSING

The data available on spectral responses and spectral signatures of rocks and minerals can be analysed to evaluate present and future possibilities of identification offered by digital processing and technics of acquisition of remote-sensing data.

Table 5.2. Description of sites used for a study of correlation between radiometric response and nickel content (after Lejonie and Weecksteen, 1985)

Curve no.	Description of minig sites	Extreme radiometric values C_1	Extreme radiometric values C_2	Mean Ni content (%) (minerals exploited or stocked)	Observations
1.	Old mine MOULINET	30–40	60–80	0.8–1.0	'Dry' rocky slopes; numerous talus zones that reduce the radiometric response
2.	St. Martin outlet	50–60	100–110	1.2–1.4	Laterite:homogenous response
3.	WELLINGTON outlet	60–70	100–120	1.3	2 levels⁺ of radiometric response: lowest (50-60 for C_2) corresponds to talus
4.	Mine platform MOULINET	30–50 55–65	60–80 100–120	2.0	Low values of spectral response correspond to non-mineralised rock
5.	La Recherche		100–125	2.45	
6.	Gironde		90–100	2.5	
7.	BORNET 1	70–90	140–180	2.6	
8.	Ore 'block' N of St. Martin outlet	60–80	120–140	2.8	
9.	L'ESPERANCE	80–100	170–190	3.0	
10.	Garnierite stock (Nakala River)	30–40	110–120	3.0	
11.	Elise		130–160	3.1	
12.	Etoile		120–150	3.2	
13.	Aurore		120–150	3.2	
14.	Mariette	90–100	160–180	3.3	
15.	EIFFIE MICKLE	80–100	170–190	3.3	

⁺Second level not described in original report—JYS.

5.3.1 Inferences from Laboratory Studies

5.3.1.1 According of laboratory measurements, a correlation exists between chemical composition and reflectance or emissivity of minerals and, to a lesser extent, of rocks

However, this correlation is observed only when all the spectral characteristics of the objects are obtained in the whole frequency range of remote sensing. Nevertheless, laboratory studies help in selecting the most significant spectral bands.

- In the visible and near-infra-red range the spectral behaviour of vegetation determines the response. In particular, a strong absorption band, indicative of chlorophyll, exists between 0.5 and 0.65 μm, with a maximum reflectance at 0.7 μm. Being important for the detection of stress in vegetation, these characteristics may also constitute secondary indicators of lithological and mineralogical variations.
- In the zones of rock outcrops or least vegetation cover, the existence of an absorption band at 0.87 μm indicates the presence of iron oxides.
- It is possible to characterise and differentiate most carbonates, sulphates, nitrates and clays from responses between 1.6 and 2.5 μm.
- The spectrum between 3 and 15 μm facilitates identification of different rocks such as carbonates, nitrates and sulphates from the data between 3 and 5 μm and silicate and carbonate rocks from the response beyond 8 μm. However, it is necessary to compute complex digital parameters such as the SCFM index for identifying composite media such as igneous rocks, or to determine elaborate physical constants such as thermal inertia or day/night temperature differences.
- The utility of the coefficient of backscattering of high frequency waves in lithological discrimination has not really been proven to date. However, it reveals indirect guides that enhance the responses. These guides include macro and microrelief, drainage network, vegetative and soil conditions, often induced by lithological variations. The combination of radar and visible spectra may give excellent results in such cases.

5.3.1.2 Correlation between chemical composition and reflectance

This correlation is less evident in the case of radiation recorded by onboard radiometers. The radiation is modified as a function of transfer from the ground to space and is altered by integration of the signature of the rock outcrop with the response of its surroundings. *In-situ* measurements made with portable systems have shown the complexity of parameters affecting the response. These parameters are represented by the outcrop and its surroundings, affected by alterations and surficial deposits, their water content, the micro-organisms which cover the rocks in their natural surroundings, the vegetation and the micro and macrorelief. Pixels corresponding to a single parameter are rarely observed.

Modelling and filtering procedures provide corrections for the effects of illumination of relief and transmission in the atmosphere. It is sometimes possible, mainly in regions of good rock outcrops, to quite faithfully correlate the recorded signal with definite geological objects in the reference segments. But in general the signal results from the integration of several types of responses which cannot readily be separated. Hence it is more desirable to study the response in its totality.

5.3.2 Some Remarks

The various functions preliminarily attributed to remote sensing, viz., discriminating classifying or characterising, identifying and quantifying etc., are not effectively or operationally in much use today.

Visual methods of discrimination of rocks are presently superior to digital processing methods because of the experience of specialists. They are benefited, however, by the developments in digital processing of data, which covers a wide spectral range. Their limitations can be further overcome given the latest innovations in acquisition technology.

Methods of supervised classification make use of reference training sites which are standardised in all aspects. The efficiency of these methods varies depending on morphoclimatic conditions and must be significantly improved to become comparable to that of the photointerpretation method.

Methods of identification are in a highly experimental state, which is closely related to technological improvements and the creation of regional reference databases.

5.3.3 Foreseeable Developments in Spectral Remote Sensing

The major studies and observations regarding spectral responses follow two approaches:

— An analytical approach which facilitates the study of processes of interaction between radiation and the geological objects according to deterministic models; the reconstruction of spectral response of objects with well-designed parameters and determination of the characteristics of the object under study using a signal.

— An empirical approach to establish a correlation between the parameter of interest and the radiometric response.

The empirical approach is a potential tool for improving the efficiency of geological remote sensing in regions of no rock outcrops. In addition to establishing such relations, this method also provides the definition of laws governing the characteristics of geological features, the objects which they consist of, their relation, and demonstrates how and why they are indicated and the extent of their representation. Some aspects of this problem will be discussed in Chapter 6.

Nevertheless, efforts should also be directed towards improving radiometer design. Several solutions are presently under way, which include:

— Radiometers with narrow and continuous bands whose centres and widths can be programmed;

— Spectroradiometers that can simultaneously acquire the total spectral curve in a frequency domain as well as the more conventional data in well-defined bands (Figs. 27, 28, 29).

Nowadays an airborne spectroradiometer, the PMI (Programmable Multispectral Imager), is commercially available. It operates between 430 and 800 nm with a resolution of 2.6 nm in a pure spectral mode or restitutes 8 bands in the conventional mode. This Canadian instrument was developed from the Fluorescence Line Imager (FLI).

NASA (USA) has developed a prototype instrument, AVIRIS, which operates in the range between 0.4 and 2.4 µm and can record 224 spectral bands. JPL is presently developing a SISEX instrument, with 192 spectral bands, which is expected to be mounted on an orbital platform in 1995.

CNRS and INSU (France) have developed a spectroradiometer, ISM, for interplanetary exploration, which operates in the visible and infra-red ranges. This spectroradiometer will be used in the airborne mission with ARAT[1] aircraft, preparatory to the launching of an orbital platform.

The GER (Geophysical Environmental Research) developed in the USA is a radiometer with 63 channels operating between 0.442 µm and 3.478 µm. This spectroradiometer can also acquire data in 500 channels in the range of 300 nm and 3 µm within 10 seconds and can produce the complete spectrum in 40 seconds. It is presently offered on a semi-commercial basis. An example of application of this unit in Cévennes (France) is presented in Plate 20.

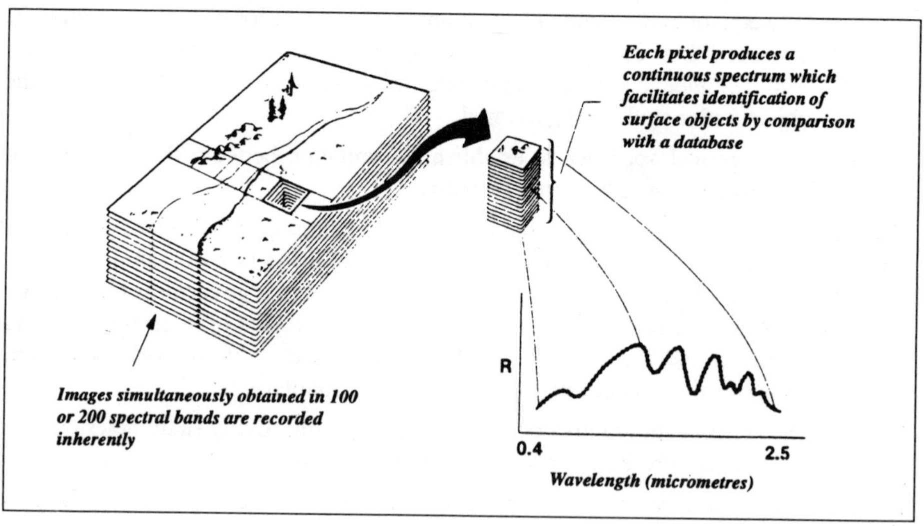

Fig. 27. Principle of imaging spectrometer (after Abrams *et al.*, 1985).

[1]Atmospheric Research and Remote Sensing Aircraft.

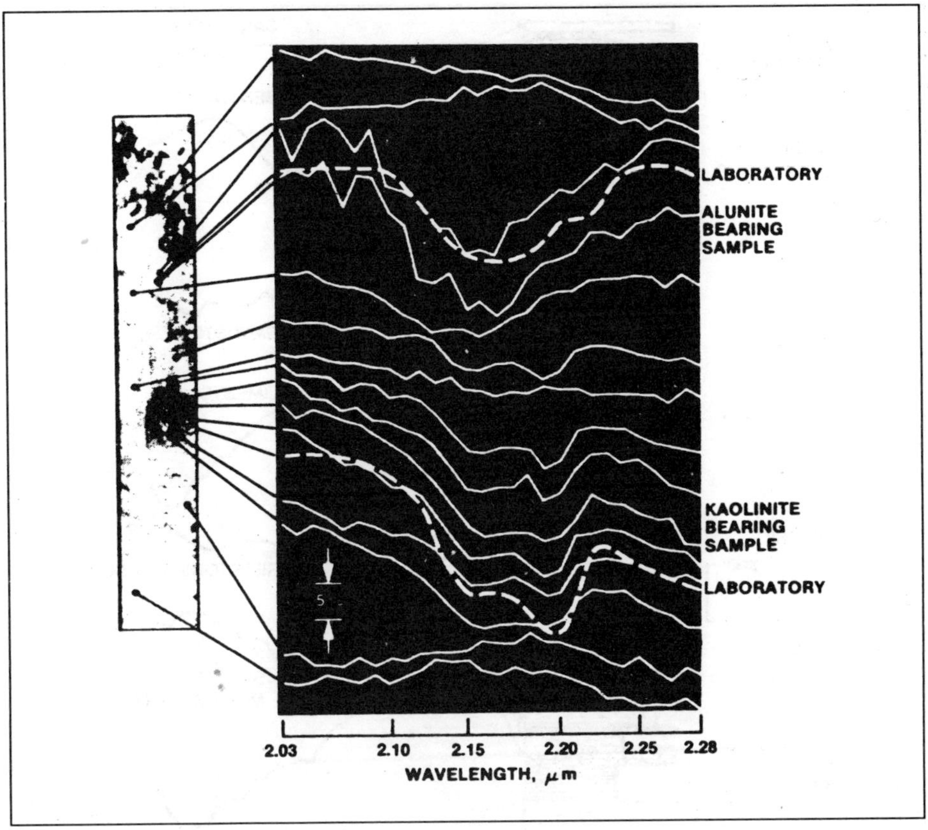

Fig. 28. Correspondence between laboratory spectra obtained with the AIS over a mineral district in Nevada (USA) (after Abrams *et al.*, 1985).

DLR (Federal Germany) has developed a prototype instrument 'ROSIS', which operates between 400 and 900 nm, with 256 bands, changeable to 128 channels. Extensions to other wavelengths, viz., 1.4 to 2.5 μm initially, then 3.5 to 5 μm and eventually 8 to 15 μm are under study.

A large inflow of data is expected from these technological improvements. Investigations are also under way to reduce the quantum of data to the necessary levels by supervised filtering or digital preprocessing of preliminary data before transmitting it to the receiving stations on the ground. Such technics will be used with the Japanese satellite JERS-1 for identifying clay formations.

However, according to the morphoclimatic conditions of the medium studied, the spectroradiometer might quite rapidly attain the limits available to a field geologist but which eliminate 'exposure by exposure' observation. Remote sensing can and must take advantage of the numerous technological innovations that are emerging. Even though the ambitious plan of being able to identify rocks and minerals in all contexts from space may not be very realistic, such aspirations may lead to improvement in technics of digital classification.

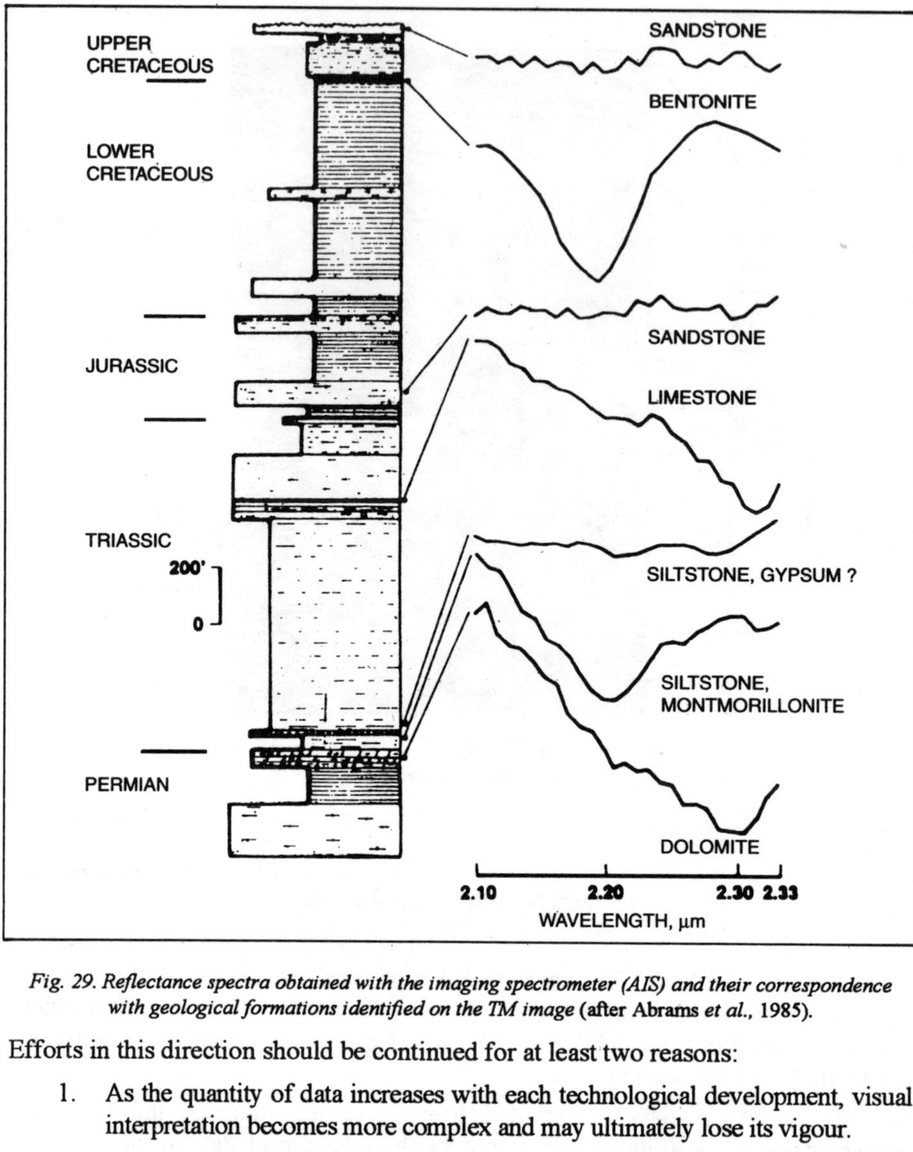

Fig. 29. Reflectance spectra obtained with the imaging spectrometer (AIS) and their correspondence with geological formations identified on the TM image (after Abrams *et al.*, 1985).

Efforts in this direction should be continued for at least two reasons:

1. As the quantity of data increases with each technological development, visual interpretation becomes more complex and may ultimately lose its vigour.
2. The integrated and geographic information systems which respond to the need for synergistic information for aiding real-time decision-making, necessitate an elaborate digital form of remote sensing which is a common factor for several disciplines.

If it is admitted that in a number of natural contexts encountered on the Earth the concept of spectral signature will not suffice to satisfactorily improve the digital classifications of geological objects, other solutions need to be explored. The concepts and investigations emerging in this context are inspired by the visual procedure of photointerpretation based on simultaneous analysis of spectral, morphological and textural criteria. The most important of the solutions concerns the analysis of texture which brings in complex mathematical principles used in certain domains of geology, for example petrography.

However, in the case of remote sensing the definition of texture given in photointerpretation may not accord with the definition formulated mathematically, wherein it is described as a class of equivalence of an element, such as the pixel. It is therefore necessary to take care that this definition does not lead once again to reducing analysis to the sole criterion of reflectance while new developments trend towards the criteria of proximity and location. This is only one part of the problem.

Chapter 6

REMOTE SENSING, GEOBOTANY AND GEOLOGICAL LANDSCAPE UNITS

Since most of the land on our planet is covered by vegetation, soils and surficial formations, through which some outcrops emerge, the recorded spectral responses are only partly characteristic of the lithological formations one is trying to map. This is why methodologies are now being developed that take into account the overall perception of the natural environment. This logical approach to spatial observation makes it necessary to push back the natural barriers separating the different disciplines of earth and environmental sciences.

Two complementary concepts are considered: the geobotanical concept and that of landscape units in their relationship with geology. For the former a review is given of the different changes caused to vegetation by rock and soil chemistry, indicating those that are a priori detectable by remote sensing. Specific reference is made to stress metal in order to highlight the present limits of satellite detection.

For the concept of geological landscape units, reference is made to the few studies that have been done on the interface between geological substratum, alterites, surficial formations and vegetation. These studies, and the results emerging from them, are encouraging for continuing along this path, which is the source for a global surface mapping best suited to the present challenges for a balanced management of the environment.

A major part of the surface of our planet is generally dominated by vegetation, soils and surficial deposits through which some rock outcrops emerge. Hence the spectral responses recorded by remote-sensing sensors are only partially characteristic of the lithological formations to be mapped. That is why methodologies are now being developed that take into account the overall perception of the natural environment. This logical approach to spatial observations makes it necessary to overcome the natural barriers between different disciplines of earth and environmental sciences. Remote sensing presently acts as a driving force for multidisciplinary and interdisciplinary exchange of views whose objective is to demonstrate the validity of a hypothesis that provides clues for classifying the geological units.

Two complementary concepts, viz., the geobotanical and the concept of landscape units, are analysed in this chapter.

6.1 GEOBOTANICAL CONCEPT

6.1.1 General

The geobotanical concept is based on the principle of the existence of a close relation between the characteristics of vegetation and geological conditions. The first research in this field was undertaken in the USSR almost a century ago and is presently being developed under the impetus of remote sensing. These studies show that the nature of the rocks and the presence of metal concentrations in the subsoil exert a strong influence on the anatomy and chemistry of plants, the quantity and distribution of vegetation species since they induce variations in the chemical composition, hydric state and toxicity of soils and surficial deposits.

This chapter discusses the status of research in geobotany and the present and future role of remote sensing in this field. Several publications are cited, viz., Cole and Brown (1976), Hawkes (1957), Cannon (1955, 1960), Milton *et al.* (1983) in a special issue of *Economic Geology*, Barbier (1984) and the proceedings of the UNESCO IUGS symposium at Orléans (Teleki and Weber, 1984) which, together with the proceedings of the ERIM symposium at Calgari (1989) may be referred to for further information on this subject.

6.1.2 Review of Vegetative Modifications Induced by the Chemical Composition of Rocks and Soils

6.1.2.1 Almost total absence of vegetation

This is the most extreme case, long known in the form of the celebrated 'copper clearings' of Zimbabwe, produced due to soil toxicity. Trees are generally absent in such clearings because they rarely survive. However, the soil in such zones does support some kind of vegetation despite high metal concentrations-14% copper in the Congo (herbs) or 10% cobalt in Shaba (specific plants, e.g. cobaltophiles).

Zones of intensive chemical alterations are sometimes indicated by water seepages and springs which induce specific hydromorphic and even marshy environments. These phenomena have developed especially in the Scandinavian regions where the glacial cover hinders direct observation of outcrops. Such zones also exist in France in which the numerous springs have a significant influence on the geochemistry of certain soils or stream sediments. In this case the concentrations of Fe and/or Mn in soils near the springs are high and the cobalt content increases up to 50 ppm.

In general, these phenomena are highly localised and may not be detected by remote sensing even if the spatial resolution of the radiometers employed is compatible with the size of the targets.

6.1.2.2 Vegetative modifications

Cannon (1955, 1960) and Yost (1971) identified different types of vegetative modifications concerning the anatomy of plants, their state of health and the nature of vegetation associations.

- *Modifications in anatomy* are manifested by different phenomena:
  - Gigantism or nanism (the most common phenomenon).
  - Modifications in carriage or in some organs of the plant. Certain modifications of plant organs were observed in Cévennes in the Cham du Pont region, France (Deroin *et al.*, 1990). In this site of beech trees the deficiency of manganese in the soils associated with an arkosic substratum is manifest by variations in volume of foliar tissues and in chlorophyll concentration. These modifications lead to an increase in reflectance of the leaves, which relates to the signature of the arkoses, thus facilitating their detection on Spot images.
  - Anomalies in vegetative cycle.
- *Modifications in the state of health* of the plant are represented by chlorosis, senescence and other manifestations that induce a change in foliar reflectance. Chlorosis develops due to the inhibiting action of the transition metals of the iron group (Ni, Co, Cu, Cr, Zn) when they are excessive in soils (Cannon, 1960). It increases only when the equilibrium is disturbed by human activities (Duvigneaud, 1958). However, this is not always the case since the toxicities described in Norway (Bolviken *et al.*, 1977, 1978) produce no trace of chlorosis in the clearings or adjoining forests. A probable explanation for this could be as follows: If the human action is not significant, the chlorotic and sick plants are replaced by toxic-resistant species. Finally, according to Hornler *et al.* (1980), chlorosis results from varied causes, such as poisoning by metals, salinity of soils, dryness, insolubility, diseases and senescence. In these cases there would be no specific spectral response.
- *Modifications in the foliar reflectance* have been studied in recent years, both in the laboratory as well as in the field, since they correspond to the phenomena that can be detected by remote sensing beforehand. Laboratory studies have been carried out by comparing the foliar reflectances of diverse vegetation cultivated on soils of known metal concentrations. Press and Norman (1972) showed that the intensity of reflectance of bean leaves decreases with an increment in lead concentration in the soil, especially around 550 nm, which corresponds to the absorption band of chlorophyll. According to Collins and colleagues (1983), any toxicity modifies the relation between chlorophyll and the cellular tissue. This results in poorer absorption of light, particularly in the infra-red band, and the reflectance is altered.

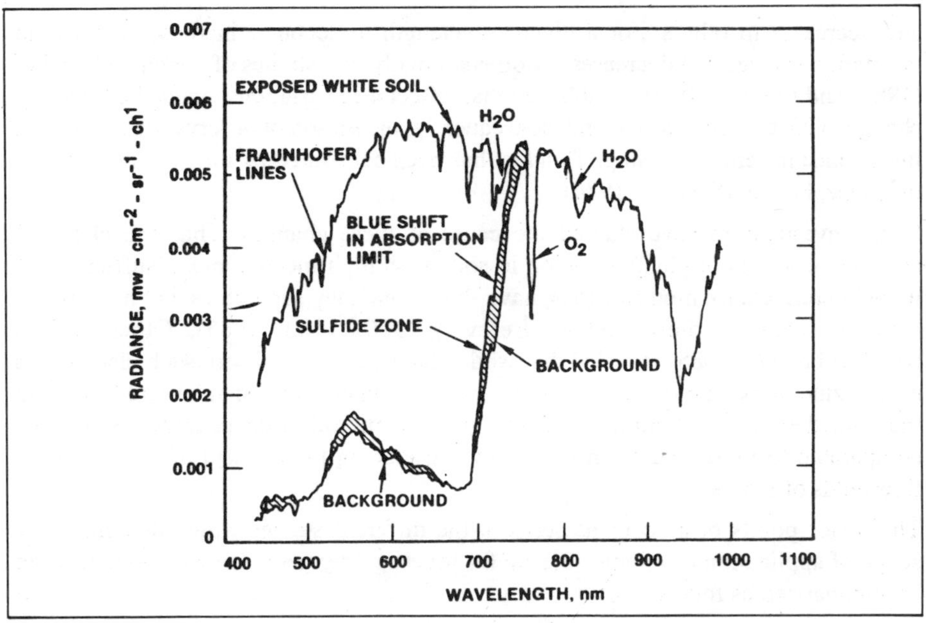

Fig. 30. Spectral variation (shift) induced by mineral stress: example of conifers over a sulphide mineralisation (reproduced from Collins *et al., Economic Geology*, vol. 78, p. 727 (1983)).

> Chang and Collins (1983) found that soil with concentrations of copper, zinc and nickel greater than 100 ppm retarded the development of chlorophyll in plants cultivated in the laboratory. They also studied the effect of toxicity on the spectral characteristics and demonstrated the existence of a slight shift of certain characteristic points of the reflectance curve towards shorter wavelengths, such as the blue and red shifts of the chlorophyll band (Fig. 30). The effect of the metal also varies according to the element considered. In this case Cu and Zn were found to be the major toxic elements whereas Fe and Pb had a negligible effect.

Radiometric studies have also been conducted on the reflectance of vegetation growing on mineralised soils. Yost (1971) observed an enhanced intensity of reflectance of diverse species of conifers in zones of Mo and Cu mineralisation. Howard and colleagues (1971) have reported an identical phenomenon in pine trees on soils rich in copper. Even though the conclusions resulting from measurements were affected by experimental parameters, a correlation was established between the variation in reflectance of wheat observed with a portable radiometer and the known geochemical anomalies of copper and lead in the same region. Finally, in a review of articles on this subject, Hornler and colleagues (1980) confirmed the occurrence of variation in the reflectance of plants growing in mineralised soils, but emphasised that the increase results from stresses of varied nature and is not specific to toxicity alone.

An integrated study on this aspect (Labovitz *et al.*, 1983) led to a generalisation of the preceding observations, viz., that foliar reflectance increases in certain wavelengths and decreases in others. For a given wavelength, reflectance increases for certain vegetation species and decreases for others. Finally, the studies of Darch and Barber (1983) and Lefevre (1980) established that reflectance variations induced by toxicity change in the course of a year and, consequently, the season of observation has a great importance in remote sensing. This fact has been confirmed by measurements made in Rouez, France (Scanvic, 1984).

Some investigators have studied the cause of these changes. Thus the chemical analysis of leaves of plants growing in soils with high metal content, sufficient for inducing reflectance modifications, have shown that they sometimes, but not always, contain traces of these metals (Lefevre, 1980, 1982, 1983). Other studies (Greffard *et al.*, 1985) revealed that while plants generally absorb such elements as boron, zinc, manganese and, to a lesser extent, copper, molybdenum and lead from the soil, their concentrations in plants growing in soils rich in minerals may be comparable to those that are not, thereby introducing the concept of concentration thresholds of plants.

The major points of agreement between the different studies which determine the scope of application of remote sensing for investigating geobotanical anomalies can be summarised as follows.

- Stresses that modify the reflectance of a plant can be of varied nature.
- Metal stress is one. Its effects are indirect and are expressed either as enhancement of chlorophyll or as the lesion of cellular tissue. Generally, no abnormal concentrations occur in leaves of affected vegetation. The plants absorb the elements according to their requirements but the contents of toxic metals are not dependent on the degree of pollution of soils.
- Certain deficiencies of metals are the cause of anatomical lesions which induce a stress.
- Depending on the species affected and the wavelength considered, there may be enhancement or reduction in absolute reflectance.
- Modifications in the shape of the spectral curve are small. The shifts observed occur only over a few nanometres (blue shift, red shift). Their measurements should be refined and require narrow-band radiometers.
- Modifications in reflectance vary and are sometimes compensated during the vegetative cycle.

- *Modifications in distribution and type of species* in the midst of vegetation communities induced by an excess of heavy metals have been observed in some cases. These are important only if they occur in an association of plants that grow naturally in a soil. As these modifications of vegetation associations have no universal value, they have not been studied systematically. Only a few special publications deal with some of these modifications. So the experience of Cole (1982, 1984) is significant. The mapping carried out by her in Zimbabwe showed that some abnormal vegetation communities, especially those consisting of *Helichrysum leptolepis*, are related to copper mineralisations in clay horizons. This mapping was extended to the entire region with good results and two zones, viz., Witvel and Ngwaku, were investigated in detail. At Witvel the geobotanical anomaly was manifest by numerous plants of *Helichrysum leptolepis*. When this species exists, it does not cohabit with other plants found outside the anomalous zone. Thus it represents a specific marker which effectively demarcates the concealed mineralised zones. These zones are detectable on aerial photos but are not visible on Landsat MSS images as they are too small. At Ngwaku the bedrock with copper mineralisation, which is covered by 4-m thick calcrete, is also indicated by a geobotanical anomaly represented by *Ecbolium lagardae*, a specific bushy plant with blue flowers. Other species constituting the general vegetation in the surroundings are totally absent in the anomalous zone. In this specific case chemical analysis of the leaves revealed abnormal metal concentrations, thus clearly demonstrating the difference between metal stress on different types of vegetation and the adaptation of flora to the environment.

6.2 GEOBOTANY AND REMOTE SENSING

Laboratory experiments and *in-situ* measurements have shown that an excess or deficit of metals in the soil produces a considerable effect on plants. The accord between the two methods may be debatable as it masks the divergence, which is often quite high, bringing the capability of remote sensing to perceive this discrete effect also into question. Be that as it may, some illustrative examples are presented below.

6.2.1 Detection of Vegetation Clearings Due to Toxicity

Certain zones of vegetation clearings resulting from the toxic effects induced by an excess of copper salts in the soil have been identified in Landsat MSS images. Processing by the spectral ratioing technic, in particular using the ratio of signals in bands 5 and 7, led to their detection. On the images the poisoned zones differ from their surroundings by a spectral value lower than the normal by 20%. This contrast can be enhanced by additional processing of the data, in particular using discriminant analysis (Bolviken, 1978).

6.2.2 Detection of Reflectance Modifications of Vegetation

Investigation of reflectance modifications of vegetation is a means of detecting changes in anatomical and/or health conditions of plants resulting from metal stress. Two methods, one qualitative, the other quantitative, are presented in publications on this subject.

6.2.2.1 Qualitative approach

This involves observation of temporary variations in the reflectance intensity during the vegetative cycle in which they occur. The problem can be stated as follows.

- How to determine the nature of the stress which modifies the intensity of reflectance and whose source may be geological, anthropogenic or phytopathological.

- How to record the highly fluctuating modifications of reflectance, often of very low amplitude, within a given time before they are obliterated at the end of the cycle.

The first point of the problem, if not properly investigated, can lead to the appearance of too many reflectance anomalies (Lefevre, 1982). This can be checked by basing the analysis on two criteria-radiometric and geometric. The radiometric criterion forms the primary element of the state of reflectance. The geometric criterion involves the search for lineaments or circular forms which often have a geologic and metallogenic significance. Employed in the Armorican Massif (France), this method resulted in the location of nearly twenty radiometric anomalies associated with circular forms and geochemical anomalies (Scanvic, 1984).

One way of optimising the detection of reflectance modifications is to follow the multidate approach using data representative of the entire vegetative cycle for a given climatic zone (Lefevre, 1983; Darch and Barber, 1983). However, meterological obstacles may be numerous and need to be taken into account. In France the normal growth cycle for low vegetation occurs in spring between March and June and for bushy vegetation between April and September, thus representing a highly variable period depending on the climate. In practice, it is often difficult to obtain two or three datasets that are representative of a vegetative cycle and one has to be content with the data acquired during two consecutive annual cycles.

Finally, digital image processing promotes detection of low-amplitude reflectance variations. An example is that of Echassières (France), where a reflectance anomaly on the Landsat MSS image indicates the geochemical variations of an underlying mineralised granite dome. The multidate data (March-April-June) in which the anomaly appeared was digitally processed using the ratioing technic, particularly the ratios 7/5 and 7+5/7–5. This technic gave excellent results. These ratios were combined in the form of principal components. Component 1 enhances the pixels showing abnormal retardation of biomass and which are stable in time. Component 2 perfectly depicts the pixels whose reflectance varies rapidly from April to June. Component 3 characterises the pixels whose reflectance varies rapidly from March to April. The retardation phenomenon is essentially indicated in the first component of the ratio 7/5 corresponding to three dates. If a rapid phase of variation capable of compensating this retardation exists, it is observed in the second and third components of the ratios 7+5/7–5. These components characterise the pixels that present an indication of relatively weak biomass (Lefevre, 1983). The method is quite complicated but has given good results.

However, data acquisition is the major obstacle in qualitative detection of reflectance variations. In practice, it becomes necessary to enlarge the choice of images to several consecutive years, which is neither precise nor desirable. In spite of this limitation, good results have been obtained (Scanvic, 1984) in the Armorican Massif (France). Applying the ratioing technic on multidate Landsat MSS data, nearly thirty reflectance anomalies were detected in the western Bretagne region. These anomalies were indicated by radiometric differences whose magnitudes vary with time and are compensated at the end of the vegetative cycle. They occur in circular form. In areas where systematic geochemical data, such as the 1:50,000 maps in France, 18 of the 19 reflectance anomalies recorded on the Landsat images were found to coincide with geochemical anomalies (Cu, Pb, Zn, Sn) that were quite distinct from the monotonous background. These anomalies were selected for further studies. In areas outside the zone of geochemical coverage, 7 of the 12 reflectance anomalies were investigated on the ground. Metallogenic indicators such as quartz veins and rubefaction were detected in the anomalous zones in six cases.

In Cévennes (Deroin and Deroin, 1990) the spectral parameters of beech trees have been confirmed as a discriminating characteristic of arkosic substratum.

Lastly, remote sensing has been successfully employed over the vast forests of Canada to study the senescence resulting from metal stress (Singhroy, 1987, 1991).

6.2.2.2 *Quantitative approach*

The quantitative method consists of measuring the small-amplitude reflectance modifications by comparing them with a reference index or by precisely quantifying the magnitude of changes in the spectral characteristics of vegetation.

The feasibility of the method has been demonstrated in the first case by ground measurements. An increase in reflectance by about 25%, between 0.75 and 0.9 µm, was observed for conifers (Yost, 1971). The pines grow in soils having concentrations of copper salts much higher than normal. Other experiments (Birnie and Francica, 1979) on pines growing on top of mineralised zones of copper and molybdenum confirm these observations. In both cases the reflectance variations were detected on Landsat MSS images.

The second case was initially demonstrated in laboratory studies (Chang and Collins, 1983). Elaborate measurements on various plants cultivated on soils rich in various base metals, showed that the spectral characteristic at 700 nm shifts by 0.2 to 0.4 nm (Fig. 30). The present developments in spectroradiometers facilitate recording the whole spectral curve from which these characteristic phenomena known as 'shifts' can be visualised. Various airborne surveys have already demonstrated the feasibility of the method (Milton *et al.*, 1983) but its systematisation appears to be difficult (ERIM 89).

6.2.3 Study of Modifications in Vegetation Communities

Contrary to reflectance variations whose 'metal-stress' origin is neither established nor accepted, the modifications in vegetation communities induced by lithology and mineralisations are proven based on undisputable observations. In spite of the generally small size of such features, the feasibility of their detection on Landsat MSS images has been demonstrated within defined limits. An illustrative example is reported for Mont Kelly (Queensland, Australia) (Cole, 1982, 1984).

Situated in a metalliferous province with several working mines, the stratiform mineralisations of Mont Kelly are associated with specific lithostratigraphic formations of the Lower Proterozoic. Some deposits are controlled by large faults visible on Landsat MSS data. Others are indicated by abnormal vegetation associations relative to those known in the surroundings. These anomalies are identifiable on aerial photos.

A colour composite analysis of Landsat MSS data of the Mont Kelly region provided an appreciable amount of new information. In particular, it helped in distinguishing surficial deposits, rock outcrops and partially covered rock outcrops, and in visualising the reflectance differences resulting from specific associations of plants characteristic of the underlying lithological formations. Some typical examples of these indicator plants are:

— *Eucalyptus brevifolia* and *Triodia punges* over quartzites;

— *Enneapongon polyphyllus* over dolomites;

— *Acacia shirleyii* on ferruginous sandstones etc.

The extension of these geobotanical differentiations was detected on a regional scale from Landsat data in which the synoptic effect emphasised the complexity of a phenomenon which otherwise appeared simple in ground mapping. In fact, quite significant spectral variations were observed inside these geobotanical associations, mainly induced by the laterites or ferruginous alterations of the bedrock, as indicated by subsequent ground verification. Some of these variations occurring as linear structures were carefully examined on false-colour aerial photographs and on the ground. Subsequent studies showed that these anomalies reveal the existence of different ferruginous rocks and laterites and are characterised by specific vegetation communities consisting of certain indicators of copper (*Polycorpaea glaba, Jacksonia ramossissima* etc.). Boreholes drilled in these zones indicated the presence of cherts and copper-bearing dolomites.

This example of mineral discovery, to which remote sensing contributed largely, is the outcome of an integrated methodology involving structural, lithological and botanical observations and concepts. However, the mean area of the anomalies is generally insufficient to be detected on Landsat MSS images having limited resolution. The second-generation satellites, Spot in particular, remove these limitations as a more recent study in the same region has demonstrated (Cole, 1987). In this case the colour composite images of principal components and certain ratios have facilitated identification of smaller anomalies.

Examples of application of geobotany to petroleum exploration are very rare. A study was carried out with this objective on an experimental site, Lost River (USA), where an aerial survey was conducted using the Thematic Mapper type radiometer (Rock, 1984). It indicated seasonal changes in plant behaviour, with the appearance of a spectral variation in autumn. Utilisation of an imaging spectroradiometer subsequently confirmed this spectral change and certain species were identified by the supervised classification method which was found to be exact to 89%. Thus a detailed mapping of vegetation species and their distribution could be obtained and anomalies in the geometry of populations and their density were detected. Also, geochemical anomalies of methane, ethane, propane and butane were found by analysis of soils over which anomalies in the populations of maple trees were observed. The relation between these two anomalies was explained by a relative tolerance of maple trees compared to the abnormal growth of certain species of fungus under the action of different gases and specific anaerobic conditions of the soil. The existence of these anomalies and their explanation form the primary basis for establishing the capability of remote sensing for detecting them, since they depend neither on a given geographic situation, nor on specific vegetation species and the fungi responsible for these anomalies have a universal distribution.

6.3 CONCLUSIONS REGARDING GEOBOTANICAL REMOTE SENSING

Vegetation masks the surficial deposits and the geological substratum in most of the land area. This is a major obstacle to using the correlations between the two, which has been demonstrated by numerous experiments.

The geobotanical method of exploration is an established technic but the problem of geobotanical remote sensing is very complex. Nevertheless, the latter has tremendous scope and has stimulated research into methodology. Current developments in technology have notably enhanced the capability of the methodology. We can cite in particular the improvements in spatial resolution which facilitate better perception of the changes of concern, often on faint surfaces. Spectroradiometry is useful for indicating the effect of metal stress while radar is the preferred tool for studying vegetation. Combined, they help us to understand the geobotanical processes.

While the feasibility of geobotanical remote sensing in large-scale mapping is confirmed, its place vis-à-vis multielemental geochemical prospecting is yet to be examined. It cannot pretend to replace the latter since it is neither as qualitative (it cannot specify which metal is present) nor as quantitative (it does not indicate even the approximate range) as the geochemical method. Thus remote sensing appears to be only a complementary to geochemical methods in the following two stages of exploration:

- Reconnaissance phase: systematic geochemical survey involves long time durations and remote sensing can be used for guiding the sampling strategy.
- Phase of selection of anomalies: In spite of some filtration through statistical methods, the number of geochemical anomalies recorded in a survey is generally too high and selection not always easy. An integration of these two methods of measurements and observations would facilitate selection of anomalies that merit further investigation.

In any case, geobotanical remote sensing will really become efficient only when it provides a means of identifying the detected anomalies in a metallogenic or regional geological context.

6.4 CONCEPT OF A GEOLOGICAL LANDSCAPE UNIT

6.4.1 General

Geological classifications using remote-sensing data are essentially based on just spectral information. Published case histories mainly pertain to the most favourable regions, such as arid zones. Furthermore, a few examples of relative successes notwithstanding, the results have generally been uncertain.

Nevertheless, significant correlations have been detected between the soil type and the geological substratum. Such a situation is observed in the 1:50,000 scale map of Mâcon (Plate 21). It is clear once again that the scope of digital classification in the case of rocks with extensive soil and vegetation cover can be improved by following the photointerpretation approach based on the integration of spectral, morphological and textural parameters. The combination of the human eye and brain has demonstrated its advantage to the full in this domain.

Compared to aerial photography, space remote sensing provides a larger synoptic view of the Earth and an overall perception of its natural environment. It thus reveals the existence of entities or units of the landscape modelled by the morphology, vegetation cover and geology. This concept, familiar in photointerpretation but not really explicit in its mechanisms, can be useful in understanding geology in a different manner.

6.4.2 Concept of a Landscape Unit

Without going into a complex history, inclusive of Tricart's article (1978), it may be said that the concept of landscape unit (Dewolf, 1987; Deroin *et al.*, 1989) and its use for geological applications are of recent origin. This concept emerged from a remote-sensing study conducted over the Armorican Massif for the purposes of geological mapping and mineral exploration. This study, based on geobotany (Scanvic, 1984; Scanvic *et al.*, 1985), demonstrated that the selection of reflectance anomalies of vegetation indicated by temporal analysis of Landsat data could be optimised by combining the spectral criterion with the morphological criterion, i.e., the circular shape of the anomaly (see Sec. 6.2.3). This represented the initial concept of a unit of geological landscape.

The term orthotype (Dutartre and King, 1986) is defined as a group of variables determined by functional relations between the geological substratum, its fracture system and the vegetation cover. Contrary to the geobotanical concept, which is primarily ground based, the concept of orthotype is closely related to the spatial view, which alone can define it. An orthotype is distinguished and can be identified by morphology, soil, vegetation cover and geology. For this purpose, two types of images are investigated, viz., an image representing large regional variations of spectral responses and another emphasising the small localised variations. Using these images, the reference units are interpreted by identifying several descriptive criteria for each of them. One ground observation station is fixed for each unit defined and the parameters identified. The most characteristic parameters are selected on the basis of conventional measurements.

All the observations and measurements gathered at the ground stations constitute a database, comprising information of highly varied nature and complexity, both qualitative and quantitative. These parameters are analysed individually, using univariate statistics for establishing the dominant characteristics of the medium. Possible relations between all these characteristics are identified and worked out by factorial analysis of correlations. The factorial axes, which are the linear combinations of variables that constitute the cause for maximum dispersion of the whole data or for dispersion of individual 'stations' are then computed. The distribution of stations in the new reference plane defined by these axes determines the orthotypes. The characteristic parameters of each of these groups are associated according to certain laws.

Thus orthotypes are indirect rather than direct reference groups, being both descriptive as well as explanatory. They can be compared radiometrically by examining the spectral distribution of the stations or spatially by determining the correspondence between an orthotype and the type of unit being interpreted.

This method was experimentally studied in the Rhodope montane region (Greece) through a collaborative programme with the European Community. The study resulted in the identification of seven orthotypes showing significant correlations between vegetation and the fracture system (Dutartre and King, 1986). Certain orthotypes are produced by the presence or absence of carbonate in the substratum, which influences the type of surficial cover, and by the distribution of vegetation, which is independent of the fracture direction. However, the responses obtained are solely of a geological nature and do not represent any metallogenic effects, as was the expectation.

The most recent development of the concept (Deroin *et al.*, 1989) has enhanced the number of parameters to be taken into account. Deroin (1990) defined the landscape unit as an area of the image, of any extent, presenting constant relations between some of its environmental characteristics. These criteria describe different parameters of the natural environment and are verified by means of existing databases, such as topographic maps, vegetation maps, geological maps etc. Using these relations, Deroin proposed the selection criteria which facilitate digital as well as visual identification of the landscape units considered as systems consisting of different types of datasets, such as geomorphology, vegetation, substratum geology and its alterations and fractures, derived from two levels of observation-space and ground. Some of the parameters identified would be specific to space observations, some to the ground and others to both. The definition of the levels of information is considered a preliminary step indispensable to the study of landscape units since we do not seek to recognise the contextual parameters necessary for digital processing, but to take them into account in toto for classifying and discriminating them.

It is important to study the interrelations between the different types of observation in order to demonstrate the mechanisms of interaction that are basic to the concept.

6.4.3 Status of Research: Relation between Geological Substratum and Vegetation Cover

We shall not revert to the acquisition of geobotanical data which plays a significant role in understanding these interrelations, but instead restrict ourselves to a description of the relations between alterites and surficial deposits and vegetation and geological substratum.

6.4.3.1 Relation between surficial deposits and bedrock

The concept of the landscape unit is amply justified since it is already recognised by the relations of one of its components, viz., surficial deposits, with the geological bedrock. Without going into details, which the reader can find in a reference work (Campy and Macaire, 1989), we shall briefly present the notions regarding surficial deposits. It is common to distinguish the bedrock consisting of hard rocks (mainly visible in quarries) from surficial deposits, which are soft and poorly consolidated with thicknesses of a few centimetres to several tens of metres. The 'geological landscape' is developed in surficial deposits. The latter occur at the interface between the lithosphere and atmosphere and are always exposed or partly covered, supporting the alteration products and the soils. They also provide the essential nutrients to the vegetation.

When surficial deposits occur as remnants of physical or chemical weathering of the bedrock that supports them, they are referred to as alterites. An allochthonous type of surficial deposit also exists, resulting from the sedimentary, volcanic etc. processes which give rise to glacial, fluviatile, lacustrine, eolian or volcanic deposits.

The nature and distribution of surficial deposits relative to the bedrock provide the basis for an analysis of the geological landscape using remote-sensing data. This is done at two levels.

— Level of reflectance: The products of alteration have a characteristic spectral response. Such products include phytosilicates (illites, chlorites, kaolinites etc.), the oxides and hydroxides of iron.

— Level of organisation of the medium: This includes the geomorphology and the nature of the soil.

The distribution of soils is mainly controlled by the climate, topography and parent rock. The parent rock exerts considerable influence when its chemical composition is characterised by a well-developed specific feature. Such a situation is observed in the case of carbonate rocks on which calc-magnesian soils are developed. The parent rock also affects pedogenesis through its texture. Thus a clayey texture in humid climate inhibits the movement of water and produces hydromorphic soils even in topographically elevated regions.

Studies (see Sec. 7.6.1.8) concerning the interrelation between the soil and the bedrock have been carried out in Beauce (King, 1985).

6.4.3.2 Relation between alterites and bedrock: Example from Mauges (France)

Large-scale data of the middle infra-red and thermal remote sensing (Thematic Mapper) helped in studying the response characteristics of the surficial deposits and alterites in the Mauges region (Maine-et-Loire, France), and in proving the existence of their strong correlation with the bedrock (Berthiaux *et al.*, 1987). The data acquired by different technics of remote sensing, such as Landsat TM and Spot, was compared with the results obtained from other methods, viz., geological, geophysical and geochemical investigations.

The Mauges series is a lithological metallotect constituted by epizonal to mesozonal mica schists of Precambrian age. It consists of different members, viz., metapelites and metagreywackes at the base followed by predominant amphibolites in the middle, which form the Beaupréau volcanic sequence, and metapelites and metagreywackes of sedimentary origin at the top. The known mineralisations (Au) are associated with the base of the Beaupréau formation, mostly of graphite schists. Those observed in eluvial deposits in fields are associated with iron oxide blocks of variable size.

The surface of the Mauges plateau is affected by a strong meteoric alteration which penetrates deep into the bedrock of mica schists. This alteration results in argillation extending up to 10 m or more, making geological and mineral mapping very difficult. Mapping is essentially carried out by multielement geochemical analysis with the objective of reconstructing the lithostructural framework below the cover of surficial deposits and in particular to deduce the bedrock lithology using the characteristics of the alterites derived from it. The task of remote sensing thus involves detection of the surface indicators characteristic of the bedrock formations.

The interactive processing of satellite data on video screen helps in selecting the optimum combination of spectral bands that give information about geology as it is presently known. The lithological units are best differentiated using an appropriate combination of TM bands. These bands are four in number and are quite distinct. On these images (June 1985) the best discrimination of the facies was obtained by a combination of bands 4, 5 and 2, coded as red, green and blue. In this combination, the TM1 facies* represents the spectral response of the altered products of metavolcanics of the Beaupréau formation, such as those found in the regions of Le Fief-Sauvin, Beaupréau, Jallais and Cossé d'Anjou. This formation outcrops only on the slopes exposed during rains and promotes erosion of surficial deposits. The criterion used in the ground surveys for characterising the alterite-parent rock pair in blind zones is the relative rarity of quartz compared to the alteration products of other members of the sequence. This difference appears to influence the spectral response recorded from space but this criterion is not totally reliable and cannot be used alone.

The TM2 facies correlates quite well with the carbonaceous shale horizons in the base of the metavolcanics of the Beaupréau sequence and indicates its boundary in the known area of its extension. Ground observations have confirmed the correspondence of spectral facies with alteration of graphite schists that give dark soils. Their areal extent is more restricted than that which appears on the TM images, however, no doubt because remote sensing also takes into account the alterites that are not *in situ*.

In the Evre valley and in northern Saint-Pierre-Montlimart, a similar spectral facies is observed in a region in which only the lower metasediments of the Mauges sequence outcrops in the closed valleys. This spectral response of the TM2 type does not represent carbonaceous shales here but appears to indicate a relative abundance of bedrock outcrops and a shallower thickness of surficial cover. The same interpretation can be extended to most of the shores of the type TM2 facies distributed along the tributary valleys in the Layon region (Western France).

The TM2 facies on the Mauges plateau is depicted in the form of wide belts which have been verified on the ground. They correspond to the altered pelites and outcropping pelites, even though their relation with the base level of the Beaupréau formation cannot be confirmed.

The TM3 facies, white to bluish-grey, with a diffuse boundary corresponds to different types of alterites resulting from diverse bedrocks, the Tertiary and Cretaceous formations in particular.

In general, the spectral response of surficial deposits is distinct in band 5 on this image, which means that it is influenced by the hydric state of soils and vegetation. This state is normally dependent on the grain size and mineralogical characteristics of the surficial deposits, which does not suffice for affirming that spectral responses are always related to a definite horizon of the Beaupréau formation. While this spectral class invariably has a geological significance, the observed relation is not unique.

* Facies—unit defined on image based on texture, structure and radiometry—General Editor.

The image obtained from a combination of bands 6, 3 and 1 (Plate 22) presents great similarity to that of the preceding combination. The influence of band 6, the thermal band, becomes significant and thus increases the scope for identifying new facies (TM4, alterites on different bedrocks, but different from TM3) or the thermal gradients inside a facies, which reflect the behaviour of two types of alterites-those of the Beaupréau formation and the Cossé d'Anjou granodiorites. Other zones of smaller thermal gradient in other places on the plateau consist of dense spots, without an apparent order. Some of these zones have been verified on the ground. They correspond to deposits of sand or shell marls. Their extension is exaggerated in the image either due to the effect of the transfer function which takes the near environment into account, or because of the colluvia fed by the outcrops of sands or shell marls, which spread out the spectral response in a natural way.

In other regions, such as the Evre valley, the thermal responses recorded are influenced by variations in the thicknesses of alteration zones which represent the metasedimentary sequence.

An investigation of the correlation between spectral responses and multielemental geochemical maps indicated a variable relation for all the members of the sequence together but an excellent correlation for the Beaupréau formation alone. A very distinct correspondence is observed between the distribution of chromium and vanadium concentrations (Plate 22), and between extensions of spectral facies and formation shape wherever it has actually been mapped. Moreover, in the extension of the Cossé d'Anjou structure, beyond the zones mapped on the ground, the anomalies of chromium and vanadium correlate well with the spectral facies similar to that of the Beaupréau formation. Besides, certain image lineaments which are the potential fractures are correlatable with geochemical anomalies, thus demonstrating the metallogenic nature of these structures.

The spectral responses and conductivity anomalies detected from electromagnetic prospecting also show an excellent correlation in the case of graphite schist which forms the main marker formation of the Beaupréau sequence. The mineralogical characteristics of the graphite horizons also make them an excellent geophysical marker and the associated soils have a distinct spectral response. The correlation for other members of the sequence is less evident.

Thus in the context of the Mauges plateau, the spectral data of middle infra-red and thermal bands carries significant information for mapping various formations. Ground observations have confirmed the responses associated with the alterites of the Beaupréau formation. However, these responses are not totally unique, even though distinct when the sequence is exposed or when a specific type of alteration dominates. Their geographic distribution confirms the map prepared from geochemical analyses. Hence exploration should not be based solely on remote sensing.

6.4.3.3 Relation between surficial deposits and vegetation

The study of part of a schistose region in central Bretagne (France) is an example of methodological investigation of image processing of remote-sensing data based on preliminary analysis of correlations between surficial deposits and their vegetation cover. The application of different types of analyses during this study (Seyler, 1986) led to two thematic results which have an important bearing on identifying the interface between the two. One involves identification of the differentiations and transformations of surficial soil horizons that are dependent on the alterites. The other is based on determination of the correlation between the distribution of vegetation groups and surficial deposits, the former being marked by hydromorphic characteristics.

This study also demonstrated the role of digital processing of data for its interpretation. The image obtained after spectral and textural processing revealed the boundaries of the structural depth of surficial deposits, which are not recognisable on the initial images in which only colour changes were observed. The processing methods chosen facilitated giving the same weightage to data pertaining to reflectance and that related to texture, which contributed to a more objective classification.

Other examples of geological landscape are illustrated in Plate 23.

6.4.3.4 Relation between forest cover and bedrock

A remote-sensing study was carried out (Fauconnier, 1988) to investigate the relation between the forest cover and the geological substratum in the regional forests of Vouzeron-Allogny-Saint-Palais, included in the 1:50,000 map of Saint-Martin d'Auxigny (France). The vegetation cover in the vast forest region was mapped using Thematic Mapper data. On the image produced from digital data, one can distinguish the major forest units, young, old and resinous foliage, the essences predominant in the foliage and the peduncle oak and red oak. The mutual correlations between the nature of vegetation cover, the 1:50,000 geological map and the soil map showed that the oak populations are associated with flint clays, most often covered by brown or sandy-clay soils, and that the resinous plants are found in alluvial and podzolic soils. Judiciously chosen processing methods can thus facilitate forest inventories, the first layer seen from space, which can serve as a guide for geological and soil surveys.

6.4.4 Perspectives

As knowledge in the spectral domain is gradually increasing, the need for taking other parameters into account is also being felt more and more. It has now become evident that identification of geological objects should be based on multiparametric classifications or combinations that are valid on a regional scale. Investigations are required to understand the interactions within the natural system consisting of geomorphology, vegetation and the bedrock, and such studies necessitate the development of mathematical tools to facilitate modelling.

The investigations carried out in Cévennes (France) (Deroin *et al.*, 1989; Deroin, 1990) have resulted in the formulation of integrated procedures for discriminating geological formations or sequences. Analysis of an image of the test zone situated at the Velay border led to the preparation of a map of landscape units which revealed different ground truths. Some of these units are essentially determined by the vegetation (colour, shape and nature of the cover over the Cévennes crystalline formations), others by the geomorphology (Urgonian* mountains of the Cévenole border) or by the surficial processes of erosion and alteration (karst plateau). The method of study in any case provided a necessary reference to the bedrock. However, the analysis of ground data of the targets, though validating the representation of defined landscape units, showed a large heterogeneity, prompting consideration of more parameters for better interpretation and understanding of landscape units.

At this stage of development of investigations, the limitations of applying this method to conventional geological mapping are already recognised. Further, it is more imperative in the near future that the objective of such investigations be to map the whole environment from the point of view of management. Such maps, increasingly in demand, do not exist at present. Individual thematic maps are the order of the day whereas managers demand multidisciplinary guidance maps. Being a common factor for a number of earth and environmental sciences, remote sensing could and should play a co-ordinating role by proceeding along this line of research.

* Lower Cretaceous facies in France—Translator.

Chapter 7

GEOLOGY AND DIGITAL IMAGE PROCESSING

A brief description of the technics of digital image processing is given in order to show how an optimised approach can benefit from the use of satellite imagery for geological purposes. This is again based on representative case histories of different types of processing for different environmental conditions and varied thematic objectives.

An example is given of processing various data obtained from the same place, with a view to photointerpretation and classification, through superimposing the various remotely sensed data on themselves, or on geophysical and geochemical data from a multidataset analysis. The use of GIS is discussed.

In conclusion, two themes are developed. The first concerns present developments in 3-D photointerpretation carried out on a screen, whereby the resultant map is directly in the digital mode. The second concerns visualisation of the synergy that results from a digital combination of information obtained through different methods of geological observation; on this theme, although statistical analysis makes it possible to define zones of interest, imaging methods still lack maturity, at least in thematic exploitation.

DIGITAL PROCESSING IN REMOTE SENSING

We do not intend to detail the digital image processing technic in this chapter but only to show how its use enhances the value of the information acquired by means of space remote sensing for geological applications. This chapter largely draws from the experience acquired by the remote-sensing group of the BRGM and in particular from specific reports (Malon, 1984; Dutartre, 1984). Two distinct ideas have emerged from these experiences:

— Significance of ground-image interaction in geology;

— Role of information technology in geological remote sensing, primarily considered as a supplementary tool for geological mapping, which is still mainly based on photointerpretation.

7.1 GENERAL

Among the major factors that have given aerospatial remote sensing its present status, the most important is the fact that it is the source of an information coded and stored in digital format. The wide scope for its application in varied domains is principally due to the possibilities it offers to process digital data by means of information technology. These means facilitate the exploitation and management of information for specific purposes (Fig. 31).

The computer algorithms used in image processing can generally be grouped into four major types:

— Linear programs based on signal processing: Fourier transforms, high-pass and low-pass filters are of this type.

— Syntactic programs that allow identification of shapes and extraction of original forms.

— Morphological algorithms.

— Statistical analysis.

Looking from a thematic point of view and without attempting to make an exhaustive review of the digital processing technics available or employed, we propose to classify them into five major groups:

— Enhancement processing

— Geometric processing

— Thematic processing

— Specific processing

— Multicriteria analysis

This classification presents the advantage of being consistent with the sequence generally chosen for the application of digital processing programs.

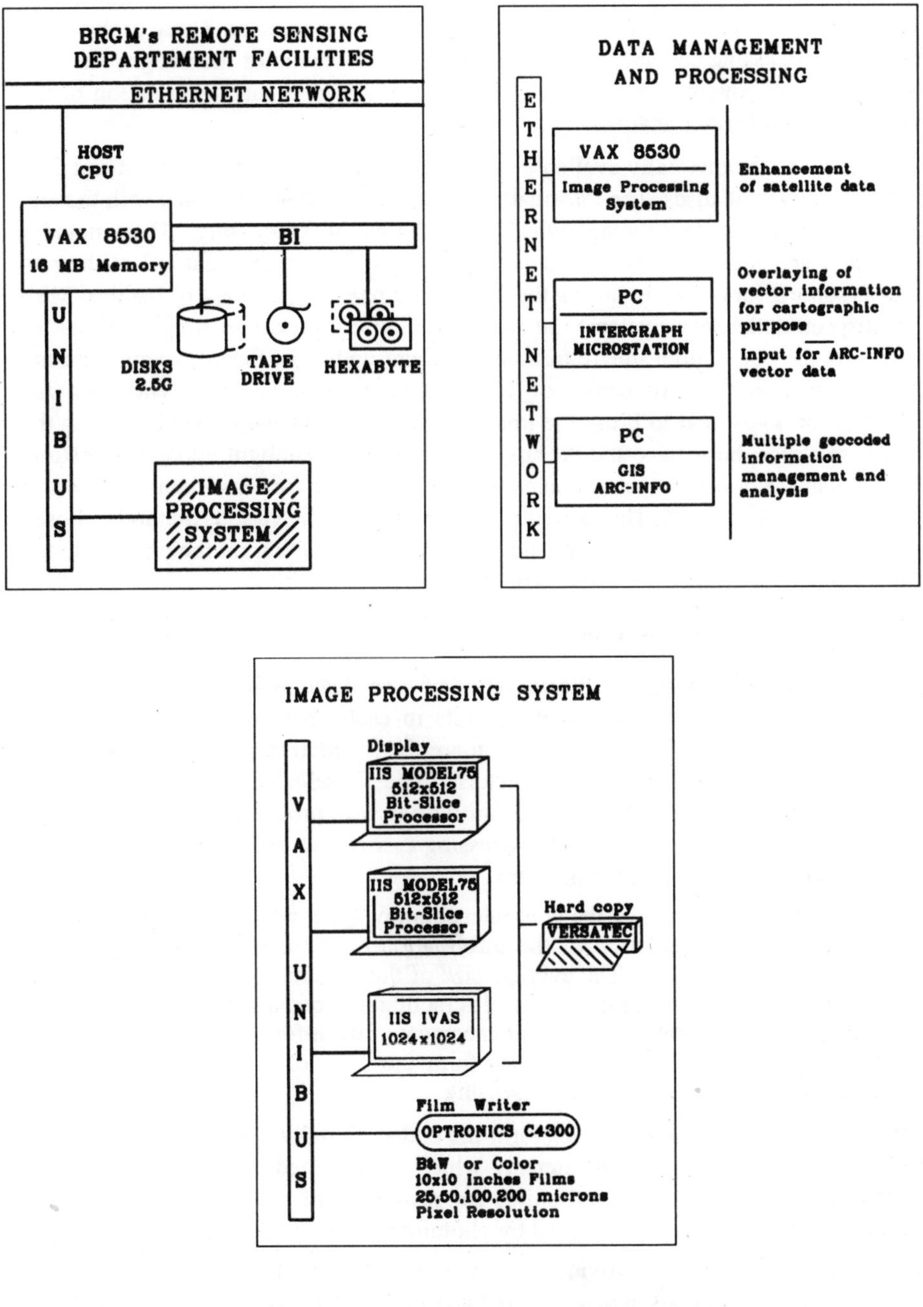

Fig. 31. Example of a digital image processing system: BRGM configuration.

7.1.1 Preliminary Stages of Processing

Before detailing the various processing technics, it may be mentioned that the thematic users of space imagery have an uphill task of going through a large amount of processing programs which should be kept to a strict minimum. These preliminary stages of processing concern reading, storage and restoration of the image. They are executed in a conventional sequence, resulting in a preliminary formulation of the information for further analysis.

Satellite data, commercially available in the form of magnetic tapes, is read and then stored on magnetic discs. The information structure is then standardised with the aim of optimising and facilitating subsequent phases of processing. This includes geometric and radiometric corrections (at least to a first approximation) of the raw data as some errors creep in during the acquisition phase. In the case of Landsat MSS (Multispectral Scanner System) the basic geometric corrections simply consist of rendering the pixel into a square form and accounting for the Earth's rotation during data acquisition. The objective of radiometric corrections is to eliminate the 'information gaps' and to bring the images to a certain standard level by taking the effects of atmospheric haze and the angle of incidence of sunlight into consideration. The final phase of image restoration consists of correcting for sensor noise (effects of pixel and scan-line drop). This correction can be effected by various methods, chosen according to the nature of the data and its mode of acquisition on the one hand, and the desired level of quality of results on the other.

7.1.2 Modes of Visualisation

In order to control and manage the operation of specialised processing programs, it is necessary to be able to visualise the results in each phase of computations. The presently available modes of visualisation are the interactive console, photographic copy and ink-jet printer output. The pictorial mode is selected according to various criteria. Some important ones are the definitive nature of processing, the need to work on an interactive basis, the type of processing carried out and lastly, the feasibility consideration vis-à-vis quality/cost ratio.

Visualisation consoles are of the computer monitor type, which provides rapid and simple pictorial representation of the information on a screen due to a system of storage on a magnetic disc. The average size of the screens is 512 × 512 or 1024 × 1024 points. The structure of these consoles facilitates a quasi-real time visualisation of the results of computation and an enhancement of the information according to the thematic objective. Their characteristics form the basis of the interactive mode, a methodological innovation in remote sensing.

Photographic restoration units transform the coded digital value of each elementary cell (the pixel) into an optical intensity which is imprinted on black-and-white or colour film. This technic is well suited for restitution of the geological landscape which is to be subsequently analysed by photointerpretation technics.

The ink-jet printer outputs provide more rudimentary restitutions in black-and-white or in colour. But they are less expensive and perfectly suitable for certain types of processing, such as classification. Graphic restorations can be made as crude displays, as line drawings or according to a ternary code of blank, point and arrow or by dot matrix.

7.2 ENHANCEMENT PROCESSING

Enhancement processing forms the basis for utilisation of remote sensing in various domains of application. In effect, after the restoration phase, just described above, the images are unsuitable for use in a given thematic problem. They have to be subjected to a number of specific processing programs designed to meet a particular need, which are essayed in real time on an interactive video system.

In general, digital images have a large primary dynamic range which does not allow thematic specialists to use them directly. Hence the data is systematically processed by a contrast enhancement technic, generally referred to as 'stretching' (the terminology of digital processing is predominantly American in origin). This technic consists of isolating a geographic or digital subgroup of the total image and suppressing the remaining part of the image by threshold masking. Different types of stretching exist, such as linear, average, variance, Gaussian, sinusoidal (Plate 24), uniform and logarithmic. These technics specifically enhance only those radiometric levels which carry the information sought, since the values recorded in any image are distributed in 256 levels, this number being the full dynamic range of data restoration.

A second type of enhancement processing consists of colouring the primary black-and-white images to simplify analysis. Even a trained human eye cannot effectively distinguish more than about 15 different levels of grey but this discriminatory power is considerably improved by the use of a colour code. The images in this type of processing are generated in the form of false colour composites, in the absence of data corresponding to the colour blue in most satellite recordings. The colours conventionally used are blue, green and red, associated with channels 4, 5 and 7 of Landsat MSS, channels 1, 2 and 3 of Spot HRV and channels 2, 3 and 4 of Landsat TM.

7.3 GEOMETRIC PROCESSING

The importance of remote sensing for geological mapping increases to different degrees according to the level of geometric correction achieved depending on the types of application. This level is dependent on the spatial resolution of the data, the photogrammetric precision of the reference maps and the number of control points on which the calculations are based. The geometric processings, highly specialised, enable us to prepare base maps of required quality for a given application. The IV Commission of the International Society of Photogrammetry and Remote Sensing (ISPR) (Baudoin, 1988) proposed a classification of maps derived from space images based on two criteria-quality and content.

Two classes of quality-map quality and image quality-are generally considered. Map quality refers to precise geometric specifications irrespective of whether it is represented in the form of tracings or as an image base. In the latter case the map is known as an 'image-based map' or a 'spatio map' (Paul, 1988). The image is a document which does not respond to any precise geometric specifications.

Two classes of content are defined. The first type corresponds to a tracing map in which the initial image no longer appears. The second refers to the image-based map on which the image pixels still appear, though in a transformed form, along with other datasets facilitating definition of subclasses. Examples of such maps are the orthoimage, corrected geometrically relative to a digital terrain model, which may be superposed on a reference grid and may include several toposheets, and the corrected image coupled with a geological, planimetric, elevation or topographic map.

Transforming the image in conformity with the corresponding topographic map is also an essential operation of processing. It is based on the selection of reference points common to different maps depending on which a deformation model can be calculated. The precision obtained is related to the number of control points taken into account. The ISPR has formulated the following recommendations:

— For maps in the 1:250,000 or 1:200,000 scales, corrected Landsat MSS images have to be employed using 40 control points per scene and additional points have to be used for preparing the mosaics.

— For maps in the 1:125,000 or 1:100,000 scales, TM or Spot XS images have to be used with accurately located (15 m) reference points, numbering 20 to 40 for TM and 6 to 10 for Spot. Use of digital terrain models is necessary in the case of montane regions.

— For 1:50,000 maps, a Spot panchromatic stereo pair may be used for obtaining altimetric data or a vertical XS image for preparing an image base.

— For 1:25,000 maps, a Spot panchromatic stereo pair or a vertical XSP image resampled at 5-m intervals may be employed.

In the last two cases the locations of the reference points should be known to an accuracy of a few metres. This indicates the importance of spatial resolution of the data used for analysis. Depending on the quality of corrections, the geometric transformation of remote-sensing data facilitates the preparation of documents of geological interest, which are often used to generate derived products that are of greater utility in different geological applications.

According to the norms of the ISPR, these products are also known as poster images, digital mosaics, thematic spatio maps and geospots, if they include a specific thematic component of the Spot satellite. The poster image does not qualify the precise mapping specifications but it forms a communication support to certain intermediate aspects. The choice of the subject depends on the visual criteria designed to meet the geographic requirements of the user such as characteristic coastal features, islands, towns, mountain chains or administrative units. Fulfilment of these criteria requires no complex processing to prepare the mosaics.

The spatio (satellite) geological map strictly follows the rules governing the preparation of topographic maps. The preparatory programs such as digital mosaicing, geometric corrections, smoothening etc. require computations. This type of representation is well suited to initiate the study of geological maps because the spatio map enables us to associate the landscape in its totality with the graphic, planimetric and topographic forms. Spatio maps can be incorporated in digital form into the geographic information systems, wherein they are associated with other variables. They serve as base material for specific thematic maps such as geological maps, hydrogeological maps etc.

However, the optimal products are the orthoimage and the orthospatio map which are based on a digital terrain model. Until now, digital terrain models were prepared using topographic maps produced by photogrammetry. Today, even though the mode of acquisition is not optimal, the Spot system allows development of spatial digital terrain models within certain limits of accuracy. The Spot satellite has the capability for acquiring data in a vertical mode and in an inclined mode. Thus the same terrestrial zone can be observed from two different orbits making a stereoscopic vision possible and photogrammetric technics applicable.

The digital elevation model (DEM) of the terrain is generated using a stereoscopic pair of Spot images by the photomaps and spatio map services of the National Geographic Institute (NGI) and other service companies in France based on two methods.

The first method consists of restituting the topographic relief map on an analytical restoration apparatus (such as the Traster de Matra) using two Spot images on a film base. The operator follows the elevation contours.

The second method is based on automatic correlation and appears more promising since it employs digital images. The procedure adopted by NGI is as follows:

— An elementary image cell (window) is displaced on the first image (for example, by every 80 m). For each cell of the first image, a cell of the second image which is similar to the first one is searched. In order to avoid the search in a vast zone of the second image, the window is moved on both images along the same lines which are the curves that correlate with each other in the geometry of the Spot images.

— The central pixels of the image cell are considered as similar when the two image cells present a maximum coefficient of correlation. However, the measure of correlation is deteriorated by the distortion introduced in the image cell by the local relief. This effect is minimised by carrying out several iterations and correcting one of the images of its distortion in each iteration.

Thus two similar grids are obtained whose image co-ordinates (line, column) of each point are known. A simple calculation gives the elevations of the nodes of this grid.

In France the NGI and ISTAR have developed a number of operational programs for obtaining digital elevation models using this method. The accuracy of estimating the elevation by these programs is better than 10 m depending on the quality of the images and the angle between the axis of view by the sensors. Research teams are presently studying means for improving this accuracy. The mean contour interval in this case is of the order of 20 m.

The importance of preparing digital elevation models of the terrain using remote-sensing data lies in two levels-geographic and thematic. The geographic importance is due to the fact that such models can be prepared for almost any part of the globe without national constraints. Thematically, the technic facilitates preparation of derived maps useful for several applications. Thus a direct perception of the topography is obtained by integrating elevation data of the digital models with space imagery. This integration is achieved by simple orthogonal projection of the original image onto a reference plane. The merit of this type of representation is due in particular to the free choice of angle of observation that makes the ratio between the scales of the length and elevation variable (superstereoscopy). It almost makes it possible to obtain oblique views required for certain applications, such as detection of lineaments, ground movements and intrusions etc. It also forms an important document of communication between the technologists, elected representatives and administrators concerned with urban planning applications (Plate 25).

The preparation of products derived from digital elevation models allows formulation of remote-sensing data in a useful mode compatible with the geographic information systems. Maps of elevation, exposition, slopes and curvature form the base for digital analysis of the landscape, leading to certain possibilities of identification of land-forms, mapping the drainage network, detection of linear and circular structures, measurement of dips and thickness of sedimentary sequences and horizontal and vertical movements of tectonic disturbances.

Lastly, the orthoimage obtained in a vertical direction or in a virtual angle provides an ideal support for stereoscopic photointerpretation.

The contributions of remote sensing are significant and will become more and more so if it is optimally integrated with the present developments of quantitative geomorphology to solve the problems of resource management of our planet.

7.4 THEMATIC PROCESSING

Thematic processing programs are more specific than enhancement processing. The boundary between the two is debatable except for the fact that the former is more application oriented. The important technics of thematic processing are spectral ratioing, principal component analysis, hue saturation, filtering, resampling and classification.

Spectral ratioing is useful to overcome certain problems associated with illumination variations (topographic shadows) and to enhance the low contrasts between some objects. The technic consists of calculating a new image based on pixel-by-pixel division of the digital values of two spectral bands. The choice of spectral bands for ratioing depends on a knowledge of the spectral characteristics of the objects to be differentiated. In general, the extent of the terrain studied is so large that systematic use of the ratioing technic is often inapplicable for investigating the same objects in different environments. Nevertheless, systematic use of ratioing is recommended in specialised publications. Such ratios include the vegetation index and the biomass index in agronomy and the iron ratio in geology for determination of gossans.

Principal component analysis aims at obtaining totally uncorrelated images, thus eliminating redundancy, which is generally a characteristic of remote-sensing data. Principal component analysis is also a tool for mixing data, which facilitates integration of geophysical or geochemical data with remote-sensing information.

Computation of an IHS (Intensity Hue Saturation) image is based on a description of colour according to the three parameters of brightness, colour in the strict sense and purity. This technic is especially useful for combining ratio images, which are generally noisy, with other images that offer a better visualisation of the topography. An example is provided by the following combination:

Band 5 for controlling intensity of colour (brightness),

Ratio 5/7 for controlling the colour,

Ratio 2/3 for controlling the purity of colour.

Detection of textural boundaries facilitates determination of contacts between different objects and consists of enhancing the low contrasts between different grey levels. The operation is carried out along three preferred directions of the image. Three new images are thus computed, which individually amplify the horizontal, vertical and diagonal differentiations. As the method also takes man-made phenomena into consideration, its utilisation for the study of natural objects needs to be corrected by visual interpretation. This spatial filtering is achieved by means of high- and low-pass filters which enhance the features of variable extensions, from very fine textural variations to large regional zonations. These filters also eliminate certain noises, enhance certain structural features and improve the image.

Resampling is used for improving the readability of the image by overcoming enlargement limitations associated with the spatial resolution of a given sensor. Resampling, for example, allows use of a Landsat image of 80 m resolution up to scales of 1:100,000. This processing also facilitates the superposition of multisource images by making their different resolutions compatible. It consists of carrying out a bicubic interpolation along the directions of lines and columns or applying a method known as 'near neighbours'. The image obtained is thus totally computed and is of a superior quality compared to the original one.

Classification processing is used for searching and classifying the objects into zones that present common spectral properties. Its efficacy depends on the number and quality of reference training sets. The procedure of supervised classification involves an experimental phase during which the thematician selects the reference zones judiciously to represent pure classes and a phase of determining the number of classes and boundaries between them. The most common applications of this method concern soil type, vegetation and geology. It represents an essential supporting tool for topographic studies and in particular for geographic databases. However, improvement in the efficacy of the technic is possible only by an integrated usage of other methods, such as analysis of texture or identification of shapes.

7.5 SPECIAL PROCESSING PROGRAMS

Certain statistical processing methods are used for specific representations of fracture data. They are applied either on structural data interpreted from remote-sensing images or directly on digital data.

Digital analysis of maps of image lineaments or fractures can be carried out by three different procedures: optical (laser), electronic (image analyser) and digital processing. Today the digital technic is the most widely used. It requires a preliminary digitisation carried out by means of a digitising table or scanner. Digitisation is generally done in two modes:

— Vectorial mode, in which each peak of the grid is designated by a pair XY;

— Raster mode: In this case the image is represented by elements or pixels of square or rectangular shape. Each point of the image is represented by a brightness intensity.

The programs utilised are based on determination of frequency histograms of azimuth, number and/or cumulative length per each directional group. These programs lead to the preparation of fracture-density maps either for all the groups of fractures together, or for specific groups, or for points of intersection of fractures. The programs 'Lineam' and 'Dense' (Dutartre, 1984) developed by the BRGM are of this type.

Extraction algorithms can be directly used on remote sensing data. The only limitation of this approach results from the interference of man-made linear features which introduce strong deviations and distort results in some regions.

The final stage of these procedures includes analysis of digital elevation models obtained from Spot data.

7.6 MULTICRITERIA ANALYSIS

Through the integration of information networks (Fig. 32), multicriteria analysis has developed as a tool of decision-making because it aids in evaluating the thematic importance of the original data acquired through different means of exploration. It is the intelligent and dynamic part of the geographic information systems. In this domain remote sensing information is integrated with the data of other tools of geological investigation (geophysics, geochemistry etc.). As far as remote sensing is concerned, the existing programs (described in earlier chapters) offer different approaches:

— Multidate: facilitates the combined interpretation of spatial data acquired in different periods to infer the differences (variability, thermal inertia, temperature etc.).

— Multisensor: enlarges the spectral perception of surface features and enhances the possibilities of differentiation. Datasets of visible and thermal regions, TM and HRV sensors and visible and microwave systems, which yield information of complementary geological nature, can be superimposed on one another.

— Multivariable: developed for correlating remote-sensing information with other types of data, such as geochemical and geophysical results. Programs such as Synergis (BRGM) belong to this category. The technics of integrated analysis are well developed but nevertheless certain problems are encountered with regard to the quality of the data to be processed and the variations in the sampling interval, which differs considerably from one variable to another. In the case of remote sensing in particular, the choice of form of data, such as noisy data, enlarged data, classified data etc., to be processed, poses difficulty. A technical solution is essential because it determines the place of remote sensing in the geographic information systems and decision-making.

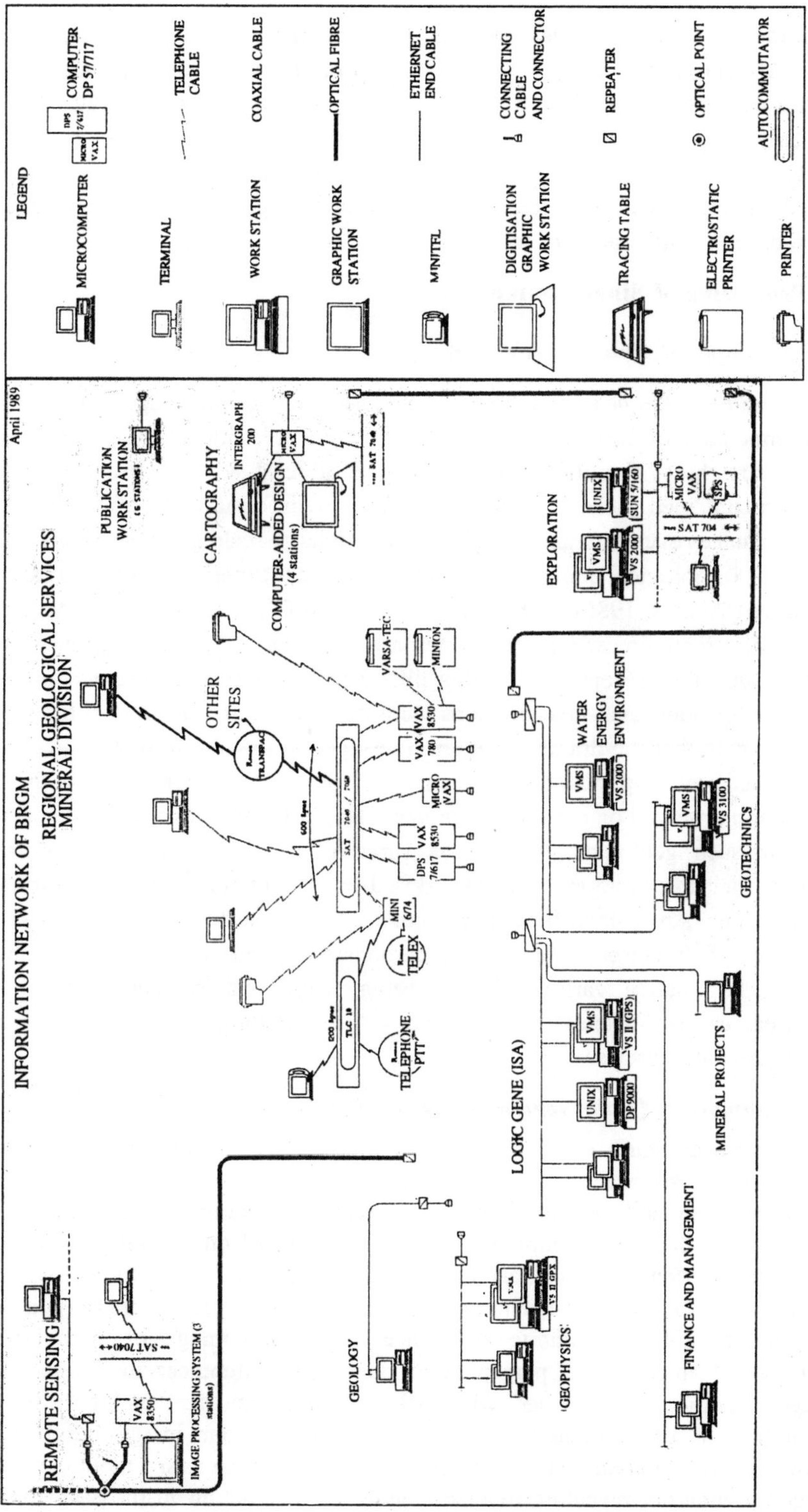

Fig. 32. Example of integrated information network: BRGM

To conclude this brief note, it is clear that utilisation of aerospace and multispectral data for different geological applications depends on the supervised and interactive digital processing technics which produce images to be analysed or interpreted maps. The examples presented below illustrate their importance. These examples demonstrate the specific contribution of each of the digital processing programs employed, the efficacy of the interactive methods and the potential of the synergistic approach based on integration of multisource data.

7.6.1 Processing of Single-source Data

7.6.1.1 Example of Challans (France)

As demonstrated in a large number of cases, the colour composite images obtained by the superimposition of the data of three bands after dynamic enhancement processing (Gaussian stretching) facilitate interpretation of at least 80% of the geological information contained in the original space data. New information is often obtained from these images even though it is compared with large-scale geological data such as the 1:50,000 geological map of France. In this regard the map of Challans (France) is illustrative (Minoux, 1986). The Landsat TM image of this region reveals a textural and colour differentiation within the Quaternary formations (Plate 26) which consist of the Bourgneuf bog deposits. As this differentiation was not observed on the geological map, additional investigations were undertaken. It was noticed from the topographic map that the drainage pattern developed in the bog formation is much denser in the western part than in the eastern, both of them coinciding with the differentiation observed on the Landsat image. Such differentiations have often been induced by such geological features as variation in grain size of Quaternary formations, their thickness and porosity. Based on this interpretation, new surveys were undertaken. Analysis of the samples collected showed that a correlation exists between the photointerpretation facies, carbonate content and grain size, thus indicating a large lateral variability in the sedimentation pattern. These geological aspects were earlier difficult to understand with conventional methods and were missed in previous surveys.

7.6.1.2 Example of Qasim region (Saudi Arabia)

We have already emphasised the importance of interactive algorithms in the choice of appropriate processing technics for solving the various problems encountered in geological mapping and exploration in a given region. Based on such methodology, several programs have been simulated and five visualised on photographic paper using Landsat MSS data of the Qasim region (Saudi Arabia). Their study (Chevrel, 1984) has largely facilitated the mapping of sedimentary formations from the point of view of identifying their economic importance. In this study images of optimal grey levels provided an enhanced perception of the small differences in the spectral responses in homogeneous zones, while false colour composites enabled a better understanding of the saturated shades. On the true colour image lithological distinctions were apparent due to the reconstitution of the real colours, which introduced a zonation revealing diverse degrees of rubefaction in the sandstones.

Rock outcrop zones could also be traced on this image due to their specific colours. The colour composite ratio image precisely demarcated eolian deposits, dunes and surficial deposits that indicate the most oxidised horizons in sands, and certain lithological boundaries. Lastly, the image produced by sinusoidal stretching of the dynamic range aided in discriminating different lithological units with similar spectral responses and in identifying a zone within a homogeneous patch of the image.

All these observations incorporated in a single map depict the distribution of materials of economic importance in this region.

7.6.1.3 Example of Nafud (Saudi Arabia)

The example of Nafud At Thuwayrat (Saudi Arabia) is also highly instructive. The MSS image (Plate 27) in this case corresponds to a key zone in the regional geology because an unconformity appears here between the Cambrian-Ordovician sedimentary formations (Saqi sandstones) and various formations of the volcanosedimentary and gneissic basement. All the rock formations are affected by a N-S fault, known as the Al Amar fault, whose effects are also found in the Triassic horizons (Jilh formation). Its perceptible recurrence is the cause for the Al Khoshm al Juar anticlinal folding which affects the monoclinal sandstone layers. All this information was extracted from the true colour image which helped in making finer lithological distinctions, understanding the geomorphology and the structure, and in placing the different units in a regional geological context. The other processing programs provided additional information. Thus the colour composite of the ratios 5/4, 6/5 and 7/6 indicated subtle differentiations in the volcanosedimentary sequence, which were not initially observed in the basement, but at the cost of geomorphological information. The colour composite of ratios 7/4, 7/5 and 5/4 provided a remarkable visualisation of structural phenomena (folds in the basement) which were not observed in any other image.

7.6.1.4 Example of Maqna (Saudi Arabia)

Significant results have been obtained for the Maqna-Aynunah region of Saudi Arabia. A study was carried out here with Landsat MSS data (Motti *et al.*, 1980) prepared in the form of different processed images through an interactive program, using a reference image and a colour composite of stretched bands. In particular, the colour composite of the three principal components revealed interesting details concerning the structural and lithological information. Some of the semi-circular forms correspond to the diapiric structures known or inferred from ground surveys. The distinct linear forms are associated with the fractures observed locally on the ground. The variations observed within the marl-gypsum belt correspond on the ground to facies differences in the gypsum beds, surficial detrital deposits and small outcrops of interstratified conglomerates, whose origin and morphological evolution are related to tectonic activity.

7.6.1.5 Example of application to hydrogeology in arid regions

The study by Fourniguet *et al.* (1984) in Saudi Arabia is an important example of the application of remote sensing to hydrogeological prospecting in arid regions. The efficacy of two types of processing was examined in a region of the Arabian Shield, selected because preliminary ground truth data was available. The study was carried out on black-and-white derived images used for mapping fracture density and on a colour composite obtained from a contrast enhancement and filtering of MSS band 7 (near infra-red) oriented towards exploration of groundwater occurrences.

In this particular case interpretation of three derived images (N-S, E-W and diagonal) enabled reconstruction of the total fracture pattern, while the colour composite revealed various detrital formations characterised by their low reflectance and their association with soil moisture content. In general, these formations have a distinct spectral signature in the red band (from white to clear red), contrasting over the colour background of the blue band. However, these images have to be analysed with great care because, while the intensity of the signatures (tending towards clear shades) is related to water content, it still depends very much on the lithology. Moreover, the variations observed in general represent extensions of the zones with temporary or permanent moisture content. That is why the authors of this study recommend cross-checking of this information with the results of geomorphological analysis for reasoning in terms of hydrogeology. Lastly, it may be noted that this type of analysis could be further improved as it is conditioned by the resolution of the radiometers and the period of data acquisition.

7.6.1.6 Example of Puys volcanic range (France) in a temperate region

In an earlier chapter (4.3.4) we showed the importance of the resampling type of processing in the structural analysis of the Pontgibaud area (Massif Central, France), wherein the accuracy of interpretation obtained by such processing was confirmed as compatible with that of 1:50,000 maps. Analysis of this data (Scanvic, 1984) also provided an answer to the question often raised by geologists regarding the volcanism of the Manzat region. A body of volcanic tuffs occurs in south-western Clermont-Ferrand whose eruptive centre was not known. Investigations by Venat (1982) indicated its location farther north in the Pouzol-Servant granitic body, but his findings were not totally convincing. However, the 1:100,000 colour composite MSS image (Plate 2) obtained from the resampling processing revealed a circular form at the southern end of the volcanic massif truncated by a lineament. According to the topographic map, this structure is associated with a concentric drainage network and a characteristic geomorphological feature which, interpreted in the volcanic context, can be explained by a caldera type of eruptive origin. Recent geological surveys in the 1:50,000 scale (map of Saint-Gervais d'Auvergne) carried out independent of the interpretation of Landsat images neither contradict nor confirm this hypothesis; they simply reveal the existence of volcanic tuffs in the centre of the structure, which differ from their surroundings. This finding justifies additional petrographic studies.

7.6.1.7 Example of geological classification

Due to a general lack of reliability, the classification type of processing methods is seldom used in geology. However, its future significance cannot be underestimated because modern improvements in resolution could lead to the development of interpretation methods that are less dependent on the analysis of landscape. The synoptic view would lose part of its importance and the perception of details become more important. Landforms interpreted from stereoscopic analysis will increasingly be correlated with radiometric forms. But though stereoscopic interpretation of geomorphology can be mastered making use of the experience of photogeology, this is not so for radiometric interpretations, especially when we look for details. In this context digital classification technics might provide a better geological understanding. For example, in Beauce (France) a classification technic (King, 1985) based on a sequence of radiometric measurements on the ground and correlation with geological and soil maps was developed using simulated Spot data. Certain classification errors notwithstanding, the results regarding geological formations were encouraging. Silts were clearly identified, as were marls and sandstones of the Burdigalian sequence in particular, because they are associated with characteristic soils. Since the soils are not specific to the lithology of the 'Beauce' limestones and the 'Blamont' marls, the results for these are more heterogeneous.

Another approach, based on the principal component analysis of simulated Spot data, was experimentally essayed in the same region. This processing minimised the differences induced by the surface inhomogeneities and enhanced the contrasts resulting from the bedrock. The homogeneity of responses of the lithological formations increased, especially for the two primary components of the silts, sandstones and limestones. Thus according to King (1985): 'In spite of the arbitrariness of comparison between Spot data (analytical and digital) and the geological map (integrated map), an excellent correlation is observed between the two forms of data, indicating that the soils reflected the lithology due to the processing techniques employed, which increases the significance of radiometric data.

An experimental application of the classification technic in arid regions was carried out on ophiolitic formations during the 1:100,000 geological mapping of Oman, which was supported by the processing and interpretation of Spot images (Chevrel *et al.*, 1989). The Spot data of this region was classified into 11 groups and the results compared with the ground truth and with a colour composite obtained from stretching. This study provided a high-quality interpretation. A statistical computation indicated the lacunae of this classification, partly disturbed by the topography. However, a significant improvement was achieved by applying the classification technic to the ratios established by dividing the signal of each Spot band by the sum of the signals of three bands. This new classification represented the major lithological units.

7.6.1.8 Mapping of basaltic flows

An experiment for discrimination and identification of the Harrat Lunayyir basaltic flows in Saudi Arabia was conducted on the basis of classification processing (Chevrel, 1986). The MSS image under reference, consisting of a standard colour composite, facilitates an excellent visualisation of geological information but the expression of the different basaltic flows is totally lost. The contrast stretched image enhances perception of lithostratigraphic details, revealing the differences resulting from spectral response variations between flows of different ages. Lastly, the image obtained by thresholding, taking the part of the histogram corresponding to only basalts, exhibits a closer correlation albeit some differences do exist. This result indicates the possibility of classifying flows using their spectral signatures, since the correlation between age and response is established in this case. This classification, which presents a strong similarity with the geological map, demonstrates the efficacy of the programs developed for arid regions.

All these examples simply demonstrate the potential usage of thematic processing technics. Other applications have been described elsewhere in this manual. They include the vegetative index in geobotany and the ratios for detection of ferruginous alterations (see 5.3.2.2 and 6.2.3).

7.6.2 Processing of Multicriteria Data

This type of processing has the objective of optimising the interpretation of measurements and observations made by means of various tools of investigation, viz., geological, remote sensing, geophysical and geochemical methods, by simultaneously correlating them (method of cross-correlation) under the constant supervision of ground truth observations. These processing technics aim at identifying the associations and integrated indicators that can reveal new information or at least allow its validation. They are currently being used more and more by geologists who consider them a means of improving maps and qualifying certain anomalies, especially those observed on spatial images. These technics have developed in two directions according to the data utilised:

— Superimposition of different remote sensing datasets of multispectral, multiview and multitemporal acquisitions.

— Superimposition of multisource data of geological and geographic types.

7.6.2.1 Superimposition of different datasets of remote sensing

Such superimpositions have been carried out either with the data acquired by a single radiometer at regular intervals over a period of time, or with the data of different radiometers acquired at the same time or at different times.

Data of a single radiometer

The technics of superimposing the datasets of a single radiometer are utilised to detect significant variations in either the distribution of objects or in their responses under the influence of seasonal effects or to determine certain physical parameters.

A. Visible domain

- Example of the Rhône valley (France): MSS images of different dates, whose superimposition indicates variations of certain features (King, 1985). The final image was prepared using multiband Landsat data of the same scene (211.29) acquired at different dates and geometrically corrected by ensuring an accurate location of control points. Band 5 of each season, processed by linear stretching, was used for preparing a colour composite in which the data of 15 May 1976 was coded as blue, that of 18 September 1976 as green and that of 13 January 1977 as red. The output colours on this image depicted such temporal phenomena as forest fires, cloud cover and plumes, which are not of geological importance. Contrarily, terrains with permanent features showed a particular signature and some of the permanent features occurring in the Rhône valley are of geological interest. This is particularly the case of gravelly alluvial terraces which are characterised on the image by a very distinct white colour, which facilitates their mapping.

- In certain contexts the temporal phenomena revealed from superimposition of datasets have a geological significance of high economic value. This refers to desertification or advancement of sands. In this regard Port Sedar in Senegal exemplifies the contribution of this type of processing to the study of port management. Undertaken for the Ministry of Public Works of Senegal, the mission entrusted to the BRGM (Boncompain *et al.*, 1977) comprised several objectives, of which the final goal was the accurate siting of port works and tracing the communication paths. This work was started with a study of the morphological evolution of the site during 1954 and 1974. Two aerial coverages were undertaken for this purpose, one in 1954 in the 1:50,000 scale on a panchromatic emulsion, and the other in 1974 on colour film in the 1:35,000 scale. The interpreted map of aerial photography of 1954 was enlarged to a 1:35,000 scale and compared by superimposition with the interpretation of the 1974 coverage. A number of ground reference points, especially in the vegetation zone, which had not changed between the two dates of data acquisition provided an accurate verification of the two interpretations. This study showed without ambiguity that a large active dune had been displaced locally over 30 to 70 m and required stabilisation to ensure the safety of planned excavations. This result, initially obtained by manual superimposition of interpreted maps was confirmed by a comparison of colour-coded images of the two sets of aerial data. An investigation of this method using the digital data of Spot is presently envisaged for further confirmation.
- More complex in concept, implementation and thematic validity, multidate analysis of the albedo index or the biomass index was undertaken to identify reflectance anomalies (Lefevre, 1983). After applying the technics of vegetative index processing (7+5/7–5) to the MSS data of the Echassières region (France) gathered in March, April and June, the author generated images from a multidate principal component analysis. Component 1 represents the sum of the biomasses of the three dates and enhances the objects of low-vegetative index. Component 2 shows the variations between April and June, emphasising the permanent features and the zones of strong enhancement in the biomass. Component 3 depicts the variations between March and April and in particular the zones in which the biomass has increased significantly.

B. Thermal infra-red domain

- The examples given above represent data of the visible domain of the electromagnetic spectrum. The acquisitions made in the thermal domain during the HCMM mission (Short and Stuart, 1982) provided data for the application of specific superimposition technics to determine certain physical parameters, such as inertia and day/night temperature difference.
- Thermal inertia characterises the response of a material to change in temperature in the diurnal cycle. The calculations for quantifying thermal inertia were carried out on three types of data acquired by the HCMM satellite

either at the same moment-visible and thermal diurnal images-or during the same daily cycle-thermal infra-red images at night.

By limiting one's self to the simple visual approach, it is possible to extract geological information from this type of representation (Scanvic, 1982), as shown in particular by the example of France. The image of thermal inertia resulting from the combination of day and night data of 17 September 1979 depicted several geological formations. Their relative values of thermal inertia on the image (visual impression) were compared with the values obtained from laboratory measurements. It was found that the major granitic rocks had a rather high thermal inertia, conforming to calculations, while Causse limestones had a relatively low thermal inertia, similar to that of the Berriconian Champagne formations. Further, the alluvial deposits of the Seine and Aube rivers were characterised by a higher thermal inertia, undoubtedly because they consist of sands saturated with water. Contrarily, certain terraces of the Rhône River basin showed a low thermal inertia. Lastly, the forests of Orléans, Sologne and Othe showed a high thermal inertia, which implies a response influenced by other features.

- So, in general, a correlation does exist between values estimated from images and values measured in the laboratory, but it would be an exaggeration to say that rocks can be identified on this basis. That is why some authors (Watson and Hummer-Miller, 1981) have developed accurate coding technics for obtaining the numerical value of thermal inertia from the image. This technic was applied to data of the Tongue River area (Wyoming, USA) where the thermal inertia values corrected for topographic variations facilitated differentiation of highly varied geological formations, under the conditions of low occurrence of outcrops. The efficiency of the technic was found to be good because differences in thermal inertia of the order of 5% and definitely of 10–15% were detected. These differences provided criteria for mapping the rocks and soils. In this case the vegetation did not affect the magnitude of thermal inertia but an extension of this result to other areas would need to be verified by other complementary studies.

- The concept of temperature difference results from the daily variation, especially between day and night, as a function of the thermal conductivity of the terrain. The magnitude of this effect can be visually gauged by means of the data of HCMM. Such a study was carried out for France and the Armorican Massif (Scanvic, 1984). Visualisation of the effect was improved by certain additional digital processing technics such as the equidensity and equipopulation methods. The equidensity analysis carried out using the day image of 15 September 1978, with a 1° × 1° interval, apparently yielded no geological information either due to the large interval chosen or because it attempted to identify features which the geologist is accustomed to observing on the ground by conventional means of mapping. The equidensity processing of night-time data, with a class interval of 4°C revealed more interesting information and ideas. The image in this case (Plate 28) was obtained by regrouping grey levels of the pixels by the thresholding technic into 14 classes. A pseudocolour was attributed to each class and very small temperature differences thereby represented. A formula established by NASA was used to study the correlation between temperature and grey-level density of pixels. The resultant image simplified restitution of information and facilitated a quantitative type of interpretation (zones of equal temperature) from the qualitative type data (landscape component). Geological interpretation of these images is not easy due to a number of factors that contribute to the thermal response, such as climate, topography, rock outcrops, vegetation, water content of soils as well as geological nature of the bedrock, and thickness and degree of alteration. Nevertheless, some of the information obtainable from these maps is unambiguous and hence can be explained and used for important applications. Such a situation is observed in the case of a low temperature anomaly extending over a length of 50 km and a width of 2 km (point 1, Plate 28) aligned in 140–150° N, cutting across the Quintin granite. This anomaly is bounded by the isotherm of 2.1°C. The temperature within the anomaly varies from 0.5 to 2.1°C and in the regional surroundings from 2.1 to 4°C. The thermal anomaly, observed only at night and not during the day, indicates a differential cooling after a uniform diurnal situation between this zone and its surroundings. Viewed in a regional geological context, this anomaly can be explained by a hydric phenomenon associated with a lithostructural event. These processing programs facilitated detection of this structure with a higher probability. It might consist of a small Tertiary graben (such structures are common in the region) but this has yet to be confirmed.

The equipopulation (histogram equalisation) image depicts a distribution of the grey levels of the HCMM into 11 classes. Pseudocolour coding provides a better distinction of the classes because a given colour corresponds to a group of pixels whose effective grey level is about 9% of the total grey levels. This type of processing was applied to the HCMM thermal data of the day and night of 15 September 1978 and the difference between the temperatures recorded during the day and at night.

The equipopulation image of the night shows certain zones distributed directionally, indicating some fracture axes, such as the Armorican trough, Montagne Noire, 140–150° N and the Quintin anomaly in particular. The equipopulation image of the day also depicts certain linear belts, representing the directions of fracture zones such as the Armorican trough (less distinct than in the night image) and the Montagne Noire (clearer than in the night image). These belts also represent geological boundaries, such as the western margin of the Rennes basin, south-west margin of Chateaulin basin, southern border of the Quintin granite etc. The equipopulation image corresponding to the day/night temperature differences enhances most of the details observed separately on the original images, with the exception of the southern border of Quintin granite, which is obliterated by the Quintin anomaly of 140–150° N.

By and large, interpretation of this type of image is difficult because a thorough knowledge of the significance of thermal belts is lacking and so the analyst of remote sensing data tends to search for those details of the surface geology that are familiar to him, with an increasing sense of failure. Interpretation of this data should thus be carried out through a fresh approach and by modelling, which *a priori* appears to be fairly difficult to develop.

Data acquired by various radiometers

Superimposition of the data obtained from various radiometers, such as visible, thermal and microwave, acquired simultaneously or otherwise, necessitates a large amount of preparatory work, such as location of characteristic control points common to different types of data, making geometric corrections, resampling for matching the resolutions etc. However, the examples presented here demonstrate the significance of this methodology.

- Example of Vans (Ardeche, France)

This project was co-ordinated by the GDTA and financed by the CEE and its results constituted the subject matter of a memoir (Cluzeau, 1985; Cluzeau *et al.*, 1985). The project involved evaluating airborne data acquired by means of a multispectral radiometer (Daedalus) and a thermal sensor operating in the 10–12 µm spectral band. The thermal infra-red data was resampled so as to obtain a resolution of 20 m, identical to that of the simulated Spot data prepared using the information acquired by the Daedalus radiometer (Plate 29).

The geology of the Vans region is quite simple, consisting of a sequence of limestones and dolomites that outcrop at several places. The results were evaluated using a recent 1:50,000 map (Bessèges geological map) and a ground control of new observations. Analysis of the two types of data led to the following conclusions.

- — Information extracted from the visible band provides new structural data but such is restricted relative to ground observations. This is due in part to the fact that a number of variations observed in this region of particularly wide occurrence of outcrops are of stratigraphic nature and are most often manifested as morphological variations that are difficult to perceive in the absence of stereoscopic views. In this case the use of stereoscopic data furnished by Spot satellite might give better results.
- — Thermal data generally provides more useful and additional information compared to the above.
- — Superimposition of the interpreted maps of the two images modified the recent 1:50,000 geological map.

Images resulting from the digital superimposition of the data of visible and thermal images confirmed these observations, particularly when they were obtained by thematic processing, e.g., principal component analysis or spectral ratioing, which reveal certain details more clearly.

- Example of the Armorican Massif (France)

As part of a research project financed by the CEE, an image of the Armorican Massif was prepared (Scanvic, 1984) using the multispectral data of bands 5 and 7 of Landsat and the day/night temperature difference data of HCMM.

The data was acquired by the respective satellites on different dates but during more or less similar periods to be representative of the same climatic condition. The Landsat data was acquired on 20 August 1976 and the HCMM data on 15 September 1978. Before making the geometric corrections necessary for carrying out superimposition of the data, pixels of different resolutions were resampled at 200 m intervals. Locating support points required for geometric correction of the HCMM data proved difficult partly because of the 'nature of the path' of each satellite and partly due to the low original resolution (600 m) of the HCMM radiometer. Nevertheless, an image was obtained based on the principle of colour composition. Several versions of the images were prepared: MSS4-HCMM day-HCMM night; MSS4, MSS7 day/night temperature difference of HCMM, MSS5, HCMM day and HCMM night. After evaluation it was found that the image obtained from the second combination (Plate30) provided new results which validate the separate interpretations. The result was coded as green for the temperature difference and red and blue for the Landsat bands 7 and 4. The major observations from this image are as follows:

- The leucogranite of West Baud is revealed by a circular structure on the Landsat image in which its radiometric level is identical to that of East Baud. Its low ΔT in HCMM, probably induced by differences in grain size, hydric state and surficial alteration, facilitates discrimination. This is an example of synergistic study.
- The East Baud granite is indicated by a circular structure on the Landsat image and is differentiated by a strong ΔT anomaly on the HCMM data (point 1, Plate 30).
- The Ploumagoar granite (point 2) is not observed on the Landsat reference image and its ΔT anomaly alone facilitates its identification, as its apparent temperature is very low during the night.
- The Quintin granite (point 3) is clearly identified on the Landsat image by its morphology and soil type. Its north-south boundary is indicated on the day image of HCMM-IR by a temperature contrast induced by the existence of a cliff and a forest.
- The boundary of Palaeozoic formations is revealed by a temperature contrast enhanced by the morphological expression (point 4).
- The Armorican Massif, a major regional fault, is marked in the morphology by the Lanvaux Barrens. It is identified on the HCMM-IR day image by a low temperature thermal discontinuity whose response is enhanced by the morphological margins distinctly expressed on the Landsat MSS image (point 5).

This series of observations calls for the following comments.

- Integration of visible and thermal images enhances the response of granite massifs.
- As these granites rarely occur as outcrops, the response observed is due to alterites or other surficial deposits and soils whose response is determined by the hydric state of the surface.

- Example of superimposition of Landsat MSS and microwave images

Images of this type were prepared for Algeria and Utah (Rebillard *et al.*, 1983). As the SIR A and Seasat data used were not in the digital mode, they were subsequently digitised into an intensity scale of 256 levels comparable to that of the Landsat MSS data. The SIR A data was of the L band wavelength (25 cm) acquired in a cross-polarisation mode at an angle of 47° relative to the vertical. In this case the backscatter signal was essentially influenced by the surface roughness in the scale of the wavelength. Contrarily, in the case of Seasat L band and cross-polarisation data acquired at an angle of 20° relative to the vertical, the topography had a preponderant effect on the magnitude of the backscatter signal.

A processing program of linear discriminant analysis was applied to the variables, viz., radar SIR A, radar Seasat and spectral bands of Landsat MSS, to evaluate their efficacy to differentiate well-defined geological features. This procedure helped to quantitatively determine the relative significance of each radiometer. In Algeria, 18 targets classified into 8 groups of geological objects were used in this experiment, each with a size corresponding to 225 pixels. The accuracy obtained for each radiometer, expressed as a cumulative percentage, was found to be excellent.

Other significant results have been described for San Rafael (USA). They show that the integrated analysis of Landsat and Seasat images provides an improved discrimination of different types of rocks (Stewart *et al.*, 1980).

- Example of superimposition of MSS (Landsat, Spot) and microwave data (SIR A)

The Petroleum and Mineral Resources Ministry of Saudi Arabia undertook a project of resource inventory of industrial rocks and minerals. The inventory was supported by 1:10,000 scale mapping of the lithological units around the major urban centres. This survey, conducted in the desert terrain (Chevrel, 1984), formed a large part of the investigation and mapping of surficial deposits which generally constituted a significant part of reserves of construction materials (sands, natural aggregates) that were readily available and exploitable. Multispectral remote sensing is the preferred supporting tool for lithological mapping, which is the basis for prospecting industrial rocks and minerals. This was demonstrated in the Yambu region of Saudi Arabia through the interpretation of images and study and identification of spectral responses on the ground (Chevrel, 1985, 1989).

As earlier experience had shown the limitations of digital approaches such as spectral classification in the field of geology, Chevrel investigated procedures based on other parameters. Taking inspiration from the photointerpretation approach, wherein the image is analysed using three parameters, viz., tonality, morphology and texture, he searched for remote-sensing data that would be amenable to eliminating the inadequacies of spectral classification and his choice fell on radar images. The multivariate approach in this study was based on the following images:

— The Landsat MSS scene was chosen as it covers the whole of the zone of study and provides access to the natural state of regions that have been strongly altered recently by spectacular industrial development, which disturbs the landscape.

— The Spot scene which covers the central and western parts of the area was preferred because of its high spatial resolution (20 m) which makes detailed mapping in the 1:100,000 and 1:50,000 scales possible. Moreover, the date of recent acquisition (1986), subsequent to industrial development, allowed a ready and precise reference to asphalt roads, urbanised zones etc., of a region considerably altered during the preceding years.

— The scenes of radar SIR A (Shuttle Imaging Radar) acquired in November 1981 by the space shuttle 'Columbia'.

The working procedure was as follows:

— Preliminary interpretation of images, consisting of discrimination of photofacies, their lithological interpretation and mapping. This phase preparatory to the ground mission helped in planning the surveys and provided initial data on the regional geology.

— Ground identification and definition of the lithological units to be mapped, tracing their total extent and verifying the validity of preceding interpretations.

— Optimisation of choice of ground control points leading to a significant reduction in surveys.

This method facilitated identification of lithological units defined on the ground and their mapping, tracing of extensions of rock outcrops that present spectral characteristics similar to the known outcrops on the ground and detection of variations within a given lithological group that may have economic potential (Plate 31). This result was achieved due to a permanent interactive approach between the processing of data, interpretation of images and ground studies and a study of mutual correlations. The significant results obtained allow us to envisage that the utilisation of spatial remote sensing and in particular Spot data for the exploration of industrial rocks and minerals could develop in the near future as an essential component of regional or national inventory projects.

- Additional information

Integration of the data obtained from different radiometers facilitates visual discrimination of geological objects and considerably enhances the accuracy of classifications. This improvement results from the original characteristics of the various data utilised, viz., reflectance, temperature, thermal inertia and surface roughness. Numerous authors agree with this statement and in particular the importance of integrated digital correlations of active microwave data. This synergistic effect is emphasised by Daily and Elachi (1978) in the conclusions to a study of Quaternary formations in Death Valley (USA). Such data was gathered by means of airborne radar by Blusson (1985) and Carrere (1987). They found it successful in improving classifications which today are handicapped because based solely on the spectral properties of rocks; image texture and morphology which invigorate photointerpretation are totally ignored. Proceeding with this concept, Carrere prepared new images by digital integration of Landsat MSS and Seasat microwave data, which provided a morphological reference. The classification obtained from this new image, still incomplete from the geological point of view, represents a definite improvement. These results are very promising as they were obtained in adverse conditions, viz., rare occurrence of rock outcrops, vast soil and vegetation cover in the experimental zone, and medium quality of the microwave data due to quasi-verticality of the view. Rabu (1985) also arrived at similar results in other regions.

7.6.2.2 Multicriteria analysis

General

- Integration and simultaneous analysis of the various components of the natural environment are the necessary functions for formulation of the developmental strategy of a country or a region. Some of these components, such as topography, hydrology, pedology, geology and meteorology from the basis of such a strategy, and remote sensing is the sole common factor of these disciplines of earth and environmental sciences. Given its specific characteristics, it contributes quite naturally to the acquisition and evaluation of diverse thematic information, provided it is utilised in conjunction with other conventional means of investigation.
- The procedures for integration and simultaneous analysis were developed individually for some time, according to the needs of each discipline. Today remote sensing is playing the role of an essential driving force in the development of coherent interdisciplinary systems and geographic information systems. In accordance with this objective, specialists of different disciplines are developing specific methodologies.
- As for prospecting subsurface resources, renewable or non-renewable, procedures for integration of remote-sensing data and a number of geological variables have been under development for several years. Methodologies concerning thematic transformation of basic data, the problems of multicriteria analysis and the functioning of geographic information systems have been devised (Bonnefoy and Guillen, 1985).

Thematic transformation of original remote-sensing data is essential. It is necessary to define the optimal form in which such data is to be integrated because this largely determines the result and is the most important problem confronted by geological remote sensing. If one has to confine oneself, as done earlier, to integration into these systems of interpreted data alone, digital or otherwise, such would be highly penalising. This is because the amount of data is continuously increasing unabated and one would be deprived of this vast source of information. That is why it is imperative to develop intelligent digital procedures. Current research on the geological significance of spectral signatures, classifications based on reflectance criteria, analysis of texture and structure and computer-aided photointerpretation is being carried out with this objective in view.

It is also important to investigate the problem of multivariate analysis with regard to the choice of variables and the expected functions of correlations between them. The choice of variables depends on the assigned objective and the examples presented below indicate some of the possibilities. But present developments in earth and environmental sciences are guiding us towards a multi- and interdisciplinary approach. While a good number of natural processes are understood in detail, we have just commenced foreseeing their associations and the deterministic relations they generate. Many questions are now finding answers in the border zone of several disciplines and geographic information systems (GIS) conceptualised to facilitate acquisition of appropriate data. Studies are in progress, for example, on the effects induced by climatic, hydrological and topographic factors, as also on the influence of vegetation and geological substratum on the origin and degradation of soils. They involve a general approach to the problems which can be solved by remote sensing provided it is not used in isolation. This is one of the fundamental principles of the strategy of future earth observation systems (EOS). As the key to understanding various phenomena, it would facilitate obtaining solutions.

The present approach in remote sensing is to correctly gather the data by considering the diversity of each variable used for mapping, with a view to refining the efficacy of geographic information systems and to determining the necessary functions. Some of the essential functions are as follows.

- Development and storage of databases.
- Integration: This is a new function of GIS, which implies a common hierarchy of data and a compromise between complexity and coherence. It also includes means of acquisition (mode of definition of data, systems of co-ordination, accession, reasonable response time), preparation of images and editing.
- Corrections for different effects, the basis of modelling.
- Statistical analysis of various parameters, in the vector and raster forms.
- Synergistic studies, a source of high thematic value.
- Management.
- Forecasting and guiding decision-making, which is the real ultimate objective of GIS.

Examples of Study

The examples presented below still represent a modest contribution of the ongoing research. One group of examples concerns the digital integration of remote-sensing data with non-geological variables, viz., digital terrain models and vegetation maps. The second group deals with digital integration of remote-sensing data with geological variables.

- Example of Reunion Island

This island of volcanic origin is characterised by sharp variations of topography culminating in an altitude of 3100 m at Piton des Neiges. The study consisted of mapping the fracture systems and interpretation in this landscape was not easy due to lack of stereoscopic data. That is why attempts were made to create artificial stereoscopic images by combining altimetric data with MSS Landsat data, and representing the relief from an oblique and anaglyphic view (Malon, 1984). The elevation contours used for creating the digital model were obtained from the 1:50,000 map prepared by the BRGM, NGI and ORSTOM. The contour interval was 200 m. The Landsat MSS data of 20 September 1981 was used because all the images before this date were affected by cloud cover, in spite of several attempts to acquire data through a different procedure. The application of a real-time acquisition mode using the receiving station of Pretoria (South Africa) produced better results. This data, corrected geometrically and radiometrically, was resampled at an interval of 50 m. The digital elevation model was transformed into an image by changing from vector mode to raster mode and by interpolating the intervals between two curves so as to code the altimetric data into grey levels. The images obtained by mixing the two types of data provided the perception of topography either directly on the oblique views or with optical accessories. The advantages of this new mode of observation emerge from the comprehensive results it offers from the interpreted data. Examples of such integrated perception are relations between topography, vegetation and soil type, geomorphological expression of lineaments and faults, lava flows and subsidences in the major caldera of Piton des Neiges. A model (Stieltjes, 1985) that enables forecasting of location of vents of volcanic flows in the eruption base has been proposed using the digital elevation model and the geological map. This forecasting model can be updated in quasi-real time.

- Example of La Paz, Bolivia

The methods for detecting past or present landslides have been based thus far on photointerpretation, used in conjunction with ground surveys and measurements. Remote sensing has been used little because spatial data until the launching of the Spot satellite into orbit lacked the necessary characteristics as its spatial resolution was not compatible with the size of the targets (Landsat MSS) or the systems were not capable of obtaining stereoscopic data (Landsat TM).

At present, detection of these natural hazards is well developed and efforts are under way to prepare maps of the zones that are sensitive to landslides, from the point of view of forecasting and prevention on the basis of information that can be preliminarily extracted from certain remote-sensing data.

The possible use of stereoscopic Spot data for preparation of landslide maps was evaluated in La Paz, Bolivia. This site was selected because the number and variety of known slides make it an exceptional zone for experimentation. The region had already been studied in 1976 by the BRGM, the Central Bureau of Overseas Studies and the Municipality of La Paz, who then prepared a map of engineering construction. This study, largely supported by interpretation of aerial photographs, served as a historic reference (Bles, 1977; Vargas, 1977). Subsequently, this feasibility was demonstrated as various datasets necessary for the preparation of sensitive-zone maps using certain risk factors were acquired. This data included the lithology, tectonics and soil types, which were directly obtained from interpretation and classification of Spot data or derived from digital terrain models computed by an automatic correlation technic (ISTAR). Other parameters studied were drainage pattern, drainage density, slope maps, topography and rock outcrops. All these indicators and maps, prepared with the image processing system 12S, were transferred onto the system 'Synergis'. This program, similar to that of a geographic information system, helped in preparing the map of zones sensitive to landslides, after solving the problems of interfacing and coherence (Scanvic *et al.*, 1989). This map was compatible with the scales of 1:50,000 and 1:25,000, which was the objective envisaged. A similar map was also prepared for Paz del Rio, Colombia (Plate 32).

- Remote Sensing and Depth Geology of France

The program Depth Geology of France enabled development of multivariate methods of analysis for the investigation of the granite domes of Echassières (Rousselin *et al.*, 1984) either by a combination of images or by processing an image by means of a mathematical function.

Several maps were digitised to serve as the base for this development. They included a geological map whose geological units were regrouped to facilitate their restoration, a soil map which showed 30 types of different soils in the Echassières region, a vegetation map and a topographic map. The remote-sensing data used for this analysis was obtained from the Landsat satellites, TIROS N and HCMM. Combinations of different variables were employed to evaluate the effect of certain parameters such as relief, vegetation etc. on the overall response recorded by the spatial sensor. Using the topographic map, pixels whose elevation was higher than 600 m (value 1) and those whose elevation was lower (value 0) were identified. The product of the value of each pixel in the pseudotopographic channel by its temperature was calculated and visualised. It was found from this analysis that elevation generally does not explain the points of low temperature during the day except very locally. Contrarily, the correlation between elevation and the night temperature was much better for all the high temperature points.

The same combination of maps was used to study the correlation between the vegetation cover determined from Landsat data and thermal images. A good correlation was observed between the day-time thermal image and the vegetation, but no correlation could be established between the night-temperature image and vegetation map. The results of correlations between temperature and elevation and between temperature and vegetation cover are significant for geological interpretations because, if these correlations do not explain the linear thermal anomalies, in particular those trending 55° N, they support the assumption of their geological origin.

- Remote sensing, geophysics and geochemistry

A second group of examples concerns the integration of remote sensing with geophysical and geochemical data. The application of these technics necessitates a preliminary definition of the integration strategy and evaluation of their significance. This problem is of primary importance because the interpreter is required to work with at least three types of data: raw data, processed data and interpreted data. He, together with the geologist, has to determine which of the data is correlatable because of homogeneous geological significance. This problem can be solved rapidly, if not already done fully or in part, and the main thing is to understand the significance of integration. In this regard, the examples presented below indicate the progress achieved.

A project of geological, geophysical and geochemical correlations was conducted in Ireland by the Geology Department of Trinity College (Dublin) with the support of the CEE (Critchley and Phillips, 1984). The principal data used in this study corresponded to geology, metallogeny, subsurface geology, geophysics, geochemistry and remote sensing. All the non-digital data was quantified and digitised onto a 1 km^2 grid. Each dataset was subjected to diverse statistical analyses carried out separately. As for the remote-sensing data, preference was given to the structural information interpreted from Landsat images enhanced by filtration. This choice was based on the assumption that the mineralisations were associated with fractures. The image lineament maps were transformed into maps of lineament density, their pattern and types. These values were presented as isodensity contours, which could be interpreted in terms of geology and correlated directly with other sources of data. The integration of different types of data was based on the discriminant analysis technic. This analysis facilitated construction of a classification model, from which the mineralised and non-mineralised zones could be distinguished. The integration of geochemical, geophysical and structural data led to this result, which is expected to reduce exploration costs and to enhance the chances of success.

— Trautwein (1982) developed a model aimed at defining the potential zones of 'porphyry copper' using a database consisting of 16 types or original datasets. Datasets of gravity, aeromagnetics, geochemistry, spectral reflectance (Landsat), topography, geology (petrography and tectonics are known indicators of mineralisation) were used to prepare this database on which different types of processing technics were employed. The model thus determined consisted of 10 parameters. Applied to a test zone, this model revealed three potential zones of mineralisation. Two were already well known but the third was not. Subsequent ground verification showed that it represents a significant copper occurrence.

— Guiness (1983) described a simultaneous analysis of geophysical and geochemical data, the result of which enhanced the value of the components and facilitated interpretation.

— The Remote Sensing Centre of the Paris School of Mines carried out a multivariate analysis as an aid to decision-making in uranium prospecting (Leymarie *et al.*, 1983). Financed by the Atomic Energy Commission (France), the authors developed a geological databank using COGEMA data, which facilitated computerisation of data manipulation, calculations, image processing, classification and identification of landforms. Having defined the characteristics (the variables) combining several initial datasets, the indicators were classified by means of analysis of the characteristics. The data available and utilised in the test zone (La Crouzille, Massif Central, France) was that of remote sensing (Landsat MSS and airborne ARIES-visible and thermal), ground studies (petrography, grain-size analysis, geochemical facies analysis etc.), topography, radiometry and geochemistry.

On the basis of these results, investigations were made to automate the procedure for determining the characteristics that would be favourable or unfavourable to mineralisations, by combining several available datasets. The aforesaid communication does not indicate the state of results, but subsequent experimentation of this system of multivariate analysis as an aid to decision-making associated with an interactive procedure has led to a number of interesting conclusions, particularly regarding granitoids (Durandeau *et al.,* 1984). Thus the existence of a correlation between the linear structures on satellite images and known mineralisation of uranium was demonstrated. This method, based on the calculation of thermal inertia, also helped to differentiate the lithological facies on a regional scale, the mineralised veins on a local scale and the fractures.

— The methodological studies undertaken in the Bresse region (France) by Gess (1983) and Dutartre and Weecksteen (1984) represent a significant contribution to the investigation of a sedimentary basin. A work group consisting of the IFP, BRGM, SNEAP, NGI and CFP proposed to demonstrate the structural importance of remote-sensing data in a region whose depth structure was fairly known but the surface geology masked by recent deposits. The data utilised for this study was Landsat scenes 212.27 and 212.28 and gravity data reproduced in the form of an image index, integrating the whole region of investigation. Several types of output data products were tested, such as the colour composite of Landsat bands 4, 5 and 7, the colour composite of the first three principal components resulting from statistical analysis (PCA) of Landsat data, colour composite of gravity with relief (vertical gradient of gravity on a Landsat image base), colour composites of the first three principal components resulting from statistical analysis of part of the Landsat data (bands 4, 5 and 7) and gravity data (vertical gradient). The digital integration of different variables revealed excellent correlations between remote sensing and gravity data. Some of the most significant features detected from this study may be cited:

— The Saint-Etienne-du-Bois gravity high indicated on the image by discontinuous arcs of circles, some of which represent the eastern margin of the Bourg 'basin' affected by the image lineaments N 20° E and N 140° E, parallel to two axes of gravity 'highs'.

— The perfect directional correlation between the system of N 70° E lineaments and the Bourg gravity ridge, which appears to divide the basin into two subbasins.

— The Cormoz zone of gravity 'highs', perfectly indicated on the image by a series of oval structures whose alignment coincides with the gravity axis of the same name.

— The curvilinear disposition of the structural lines which indicate, in their details, a gravity basin.

— The two gravity axes of Saint-Nizier and Pont-de-Vaux-Lons-le-Saunier on the north-western flank of two structures (3 and 4), which indicate a possible deep structural significance for the system of parallel lineaments between the two.

This study demonstrated the existence of a remarkable correlation between the image lineaments and the deep structures and revealed a new phenomenon at least for a sedimentary basin, viz., the concordance between the curvilinear structures on the image and the structures interpreted from gravity data. This correlation is manifest between the surface effects that integrate varied elements for remote sensing, and the deep structural information of importance in sedimentary terrains (gravity data). The mechanisms of this action are little understood. However, in order to mark the ground surface with the imprints of palaeogeographic and structural phenomena passing through the recent deposits, it is probable that structural reactivations and isostatic readjustments play a significant role. These indicators of reactivation should be searched in the grain-size distribution of alluvium, heavy mineral concentrations and other such details.

- The application of digital integration of different datasets to geological mapping in Tanzania was reported by Bardinet and colleagues (1988). The classification made using space data was validated by an interactive analysis in which it was integrated with other sources of geological information by a special algorithm (Geomulti). The methodology developed in this case helped to determine, in favourable conditions, the different lithologies, structural details and surface boundaries and to detect geobotanical indicators.
- Experiments on simultaneous processing of geological data have been conducted in the equatorial zone, which by definition is difficult to explore.

The problems specific to different geographic zones necessitate an appropriate choice from among the available technics of remote sensing. Such a situation was encountered in Gabon in 1980. A country with high mineral potential, this territory had largely been prospected and its geology was known locally. However, only a general reconnaissance map of 1:500,000 scale existed for a major part of the country. As part of a developmental plan for 1980-1981, the government of Gabon Republic and its geological service decided to undertake an extensive programme of systematic inventory of natural resources based on the acquisition of side-looking airborne radar data over the entire country, followed by geophysical and scintillometric surveys in local areas. The Sodeteg society assisted the geological service of the Ministry of Mines and Hydrocarbons in this major task. It recommended the choice of radar as indispensable because of its all-time acquisition characteristic as the cloud cover is very strong and persistent in this country, and because of the pseudorelief it provides since it was necessary to obtain a 3-D simulated model in the equatorial forest zone in order to interpret the geomorphology (Baelz-Manière and Gilli, 1988).

As part of the mineral inventory programme entrusted to the BRGM, the sections of Mitzic, Boué and Mouila, situated in the control of Trans-Gabon railways, were mapped and prospected in detail. An integrated geological map (Prian *et al.*, 1988) of each region was prepared in the 1:200,000 scale by the work of a French-Gabonese team using different methods. Use of radar data in this context proved very fructuous for the project, particularly in the Mouila section where the Archaean basement constitutes the most extensive zone in a group of very diverse formations characterised by the Pan-African sequence in the south-west and the Lower Proterozoic in the north-east.

A combined study of radar images and aerial photographs was carried out, the former in the 1:200,000 scale and the latter in the 1:50,000 scale, each of which was subsequently reduced.

In this climatic zone the radar mosaic served as a high-quality topographic support (Brosse and Scanvic, 1986). Generally, aerial photography provides powerful discrimination and accurate demarcation of the morphological units. It also facilitates preparation of an exhaustive inventory of image lineaments. Radar images are useful for making correlations of microfeatures that are not entirely clear on aerial photographic mosaics. They also allow a proper representation of fractures according to magnitude and manifestation in morphology. The correlation between the two maps is highly variable depending on the type of morphological model.

— For a high-relief model, the radar image provides a better differentiation of the topography, as in the crystalline terrain situated west of the Ikobé fault.

— For a low-relief model, very fine differentiations are possible with photogeology, as in the sedimentary formations of Nyanga syncline or locally in the Ogooué sequence.

— In the case of an 'average' model, the identifications are very distinct, as in the Chaillu region.

Some lacunae of observation can also be noticed, which are due to the technological characteristics of different sensors. They are:

— Absence of aerial photographic observations in regions where the cloud cover is persistent.

— Lack of observation by radar images when the morphological features are badly oriented relative to the direction of radar beam. The characteristic example is that of the boundary between calcareous schists and siliceous schists represented by a cuesta whose face is oriented in the direction of the radar beam.

In the reconnaissance scale the radar image provides information relatively similar to that furnished by aerial photography but at a considerably lower cost.

A comparison of the schematic geological map (Prian *et al.*, 1988) and interpretation map of radar image clearly highlighted the capacity of the active microwave remote sensing in the conditions prevailing in the region, viz., thick (30 to 40 m height) forest cover, persistent and dense cloud throughout the year, thick zone of alterites (40 to 60 m), as also the quite significant lithological variations that induce characteristic morphologies, perceptible due to the effect of the resultant topography.

A comparison of images of magnetic and scintillometric (Plate 33) data with radar images (Plate 34) enhanced the importance of certain information. Examples of such results with scintillometric data are the Bouvondo and Moukandza gneissic domes indicated by their volcanosedimentary peripheries and greenish rocks, the Bapindji granites and the Waka migmatites, the Diany Myole migmatites, the faults of Ikoy and Ogoulou Offoué and the Chaillu granitoids.

Magnetic data, on the other hand, magnifies the structural characteristics of the Chaillu granite in the region south of Iboundjé and reveals the eastern margin of the Moukandza dome and the Ikoy fault (Fig. 33).

These observations have a dual methodological significance. They indicate the continuity of geological representation of the electromagnetic signal recorded in scintillometry and remote sensing and justify, by their complementary nature, the digital integration of different data.

A comparison of geochemical images obtained from an analysis of stream sediments and alluvial material with radar data also gives an idea of the importance of the integrated approach for geological mapping in regions in which the lithological data inferred from sparse outcrops is too limited. Each map of elemental distribution contributes to geological mapping. Reinterpretation of radar images using all the

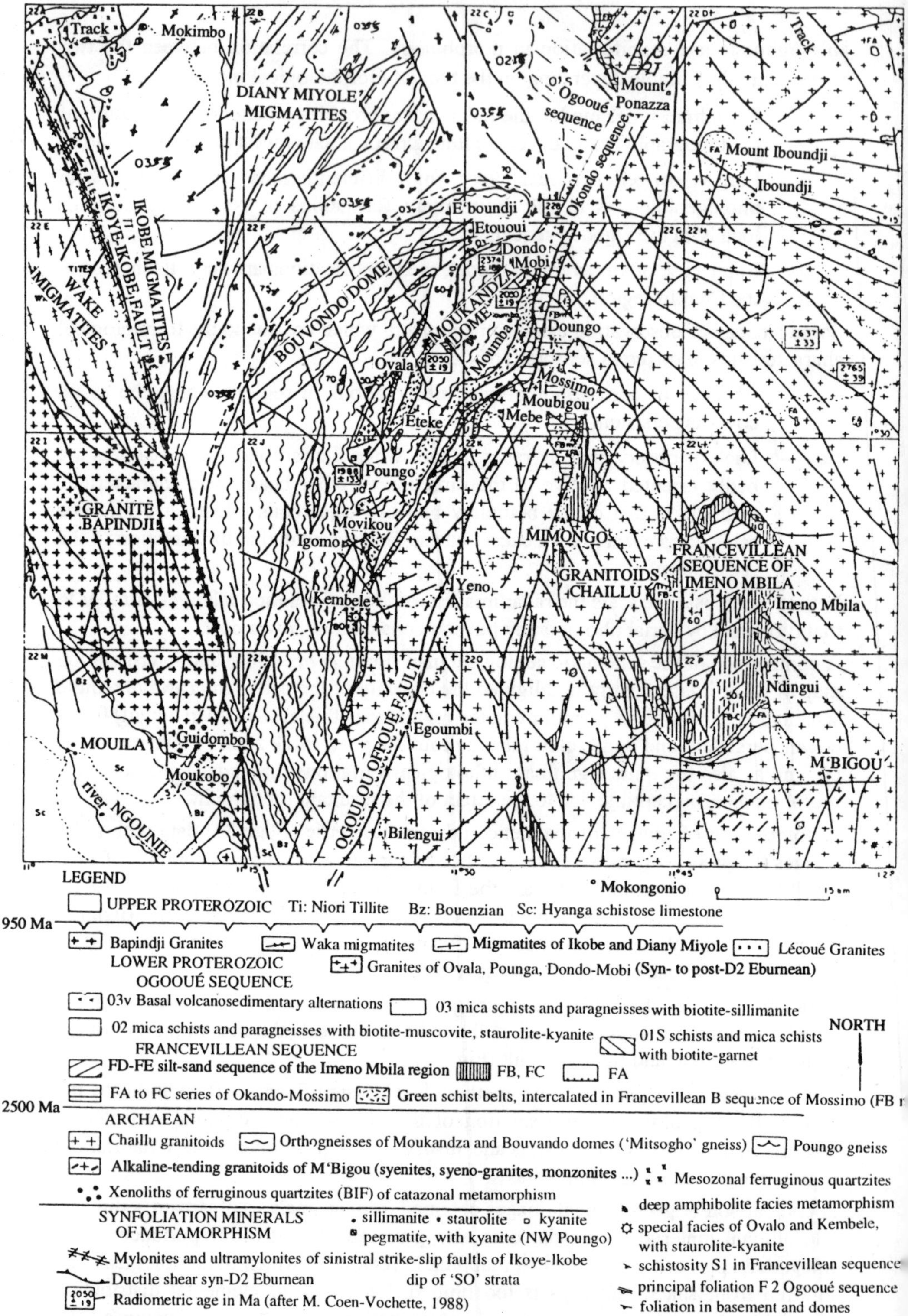

Fig. 33. Schematic geological map of the 1° × 1° region of Mouila (after Prian *et al.*, 1988).

methods employed for optimising geological mapping of these difficult zones proves the importance of this methodology, even beyond what has been presented above, prior to a direct and detailed ground investigation. For example, an integration and simultaneous processing of mutually compatible datasets was carried out in a particular study zone in the Abamie region of Boué using the special algorithm 'Synergis' (Coste, 1985). Interpretations of aerial photos, Goodyear (3 cm) radar images, airborne magnetics and gamma-ray spectrometry and geochemical and stream sediment prospecting data were used for this integration. Going beyond the stage of simple superimposition of the images, which is more of descriptive interest than interpretative, an analysis of correlations was used in this project, preceded by a logical coding of the selected variables. Synthesis of this data enhanced the value of the large amount of information available and usefully complemented ground observations and petrographic studies. Thus it facilitated integrated mapping and demarcation of zones favourable for the occurrence of some mineralisations, even though elaborate processes of integration were not used.

Similar integrated studies using varied datasets were carried out for the Mitzic region (Sebban, 1984).

- A prospecting programme for mercury, lead and zinc around a lease area in Almaden (Spain) was conducted by multicriteria analysis of diverse datasets (Pacheco and Puigarnau, 1986). The various maps were prepared from Landsat TM data acquired on two different dates, processed by IHS. Some zones of abnormal base metal concentrations were traced on the IHS images of the summer, based on the different colours of the soils. IHS processing of the winter images enhanced the textural expressions of some mineralised horizons. The images of the two dates with preferred acquisition of infra-red bands (TM4, TM7) revealed the contact zone of the granitic domes concealed below the surficial deposits.

The integration of different datasets, viz., remote sensing, geophysical and geochemical, has been carried out using different methods. It is clear that the quality of the results depends largely on the quality of processing technics applied in each stage of data manipulation before the process of integration. Good results have been obtained by a method based on IHS and SSP (Synthetic Stereoscopic Processing) technics for integrating the data of gamma intensity, TM4 and aeromagnetics.

- An investigation based on multicriteria analysis was conducted in two experimental zones situated in Portugal and Tunisia (Bonnefoy *et al.*, 1986) by the BRGM, cosponsored by the CNES. Its aim was to study the relation between lithology, geochemistry and spectral signatures to evaluate the importance of synergistic exploration of subsurface resources.

Two methods of correlation were used in Tunisia. The first, essentially a visual approach, consisted of comparing the various maps in pairs. The interpreted maps of images obtained by the application of different processing technics were compared with the lithological map. In spite of thick surficial deposits which strongly influenced mapping by remote sensing, the correlation between the two maps was found to be good. The best correlation was observed between the multielemental geochemical map and the lithological map, because the former is free from the effect of surficial deposits. Lastly, only geochemical data indicated the mineral deposits and occurrences, which are not represented in the spectral or textural variations, at least in the preliminary analysis.

The second method of correlation was based on the digital multicriteria analysis carried out by means of the special algorithm 'Synergis'. The raw remote sensing data, degraded in resolution to ensure compatibility with other datasets, was combined with the multielemental geochemical data according to the principles of factor analysis. As the new image thus obtained was not of much use, the raw remote-sensing image was replaced by a variability index map resulting from the combination of two Landsat images acquired on two different dates, which was partially free from the effects of vegetation cover. Use of this method significantly improved the degree of correlation, even though the results are similar to those obtained from visual interpretation, thus indicating the limitations of multicriteria analysis based only on spectral criteria as far as remote sensing is concerned. However, a visual reinterpretation of the classification considerably improved the metallogenic understanding of the region. This analysis indicated most, if not all of the known mineral deposits with the same spectral response.

Integration and statistical analysis of different datasets along the south Armorican lineament resulted, starting from a model, in the preparation of a map of target zones favourable for shear-zone type gold mineralisation. Each zone is several square kilometres in area, which corresponds to the average size of a lease.

The methodology developed (Braux *et al.*, 1990) is based on the following sequence of operations.

- Interpretation of Landsat TM images generated as colour composites of bands 2 (blue), 5 (green) and 4 (red). In the geomorphological conditions of the Armorican formations, characterised by thick alterites and their strong relationship with the substratum, band 5 provides a signature of the bare soils and their water content, which improves the possibility of lithological identification, especially of granites.
- Development of a geocoded database, integrating the interpreted results of Landsat, gravity, geology, metallogenic and structural data into a single reference base.
- Acquisition of complementary structural, lithological and metallogenic data by ground investigations.
- Establishment of a reference training set using a geometric model of ore control for gold, defined on known deposits (La Chatelet, Villeranges, La Bellière) in the Central and Armorican massifs. The model used is defined as the intersection of two preferential directions of image lineaments (spatial observation), trending 50-70° N and 100-120° N, mainly associated with the Variscan granites which have a distinct gravity expression.
- Investigation of zones exhibiting characteristics of the model by means of the program 'Synergis' (developed by BRGM) in several stages, viz., mapping the intersections of fractures, calculation and identification of their density and selection of promising zones by comparing with the available geophysical, lithological and structural data.

This method resulted in identifying and classifying 11 targets over an area of 20,000 km^2 corresponding to the western part of the south Armorican lineament. These targets were taken up for systematic prospecting on the ground.

7.7 CONCLUSIONS: PRESENT STATUS OF DIGITAL PROCESSING OF REMOTE SENSING DATA IN GEOLOGY

At present, photointerpretation of images optimised by the application of digital processing technics based on their interactive mode of selection, and its digitisation if necessary, are the preferred methods of processing remote sensing data in geology.

These digitised maps are compatible with certain statistical technics, leading to directional rose diagrams or density representations of fractures, and with multicriteria analysis through special algorithms.

Certain programs developed in recent years enable the direct extraction of lineaments of all scales, without the need for visual interpretation. However, they cannot differentiate between natural and anthropogenic lineaments. The intervention of the human eye and brain is still necessary to validate this information in geological terms but these results conform to current research on the digital identification of forms.

Digital classifications based on the spectral criterion alone are not suitable for geological mapping because the surface information generally reflects phenomena that are often masked by vegetation and soils. Investigations are currently in progress for improving the methods of geological classifications by incorporating into them all the criteria used by the photointerpreter, viz., colour or tonality, morphology and texture grouped into a landscape unit, or by making classifications based on the temporal variations of spectral responses.

Another approach currently being explored is that of direct digital photointerpretation performed on a 3-D vision screen or, more ambitiously, based on the identification of shapes in a Digital Elevation Model using expert systems.

The synergistic interpretation by innovative methods based on the digital integration of data obtained from different modes of geological observations has not yet passed the stage of superimposition of datasets. While the capabilities of digital processing technics facilitate demarcation of zones of interest by statistical analysis, we are still not in a position to make use of them to extract a synergistic picture from a group of datasets, as illustrated through a few examples. We had better wait for a concept based on a calculation in which 1 + 1 would make 3.

All this research determines the importance of the role of geological remote sensing in geographic information systems. Already being a source of databases of GIS, with digital terrain models and derived products such as the topographic map, slope map, brightness map, drainage pattern map, river basin map etc., remote sensing, with its geometric conformity, has proved its capability at the level of certain basic technics necessary for the exploitation and management of large-size databases. It is a bank of tools necessary for multicriteria analysis as it provides a synoptic view for all the earth and environmental sciences.

Chapter 8
CONCLUSIONS

The conclusions deal first with mineral exploration through drawing up an appraisal of what can be expected from remote sensing. Emphasis is placed on certain aspects, such as selection of areas that are a priori favourable to a certain type of mineralisation, location of a determined metallotect, understanding lithostructural controls, outlining gossans and perspectives concerning the nature of hydrothermal alteration.

A few proposals for continued research are then formulated concerning improvement in the performance of digital analysis, development of computer-aided photointerpretation, modelling (particularly in the field of structural remote sensing), thoughts concerning the mapping of tomorrow that would be best suited to the needs of environmental management, development of specific instrumentation and, finally, reflections concerning future developments in remote sensing.

Remote sensing technics, having grown through several stages, from panchromatic colour and false-colour aerial photography to highly improved satellite images, viz., coloured, filtered, resampled and classified, have acquired an important place among geological exploration methods. This gradual evolution, rather specific to geological applications, has taken place due to several favourable factors:

- History of a long tradition of photointerpretation both for the interpreter and the user.
- Adaptability of remote-sensing technics to mapping applications.
- Complementary character of the conventional mapping approach, which integrates point observation data into a coherent geological model, and the remote sensing approach, which extracts the geological component of the landscape, for which it gets a synoptic view.
- Rational utilisation of digital processing technics.
- Informative nature of the images (Plate 35).

The extent of utilisation and efficacy of remote sensing in applied geology vary depending on the availability of sensors, climatic conditions, degree of occurrence of rock outcrops and type of application. The data acquired by operational or experimental satellites, accessible from existing databases, has spectral and spatial characteristics that are not always ideally suited for the problems to be studied. Data acquisition is also limited by the number of receiving stations or the recording capabilities of satellites. However, significant improvements are foreseen in these domains. Meteorological conditions are another limiting factor in many regions of the world, since clouds, sand-drifts and atmospheric haze are often impenetrable obstacles. The capability of satellites for repassing over a given point of the globe may partially overcome this handicap but only active microwave systems such as radars can obtain data under all meteorological conditions, and several launchings of such satellites are foreseen.

The efficacy of remote sensing depends on the experience of the interpreter and the occurrence of rock outcrops. Very often, rock formations are masked or distorted by the environment and interpretation in such cases is based on second-order indicators, viz., morphology, vegetation, water content of surficial deposits etc.

Lastly, the application of remote sensing for different geological studies also depends on the specific themes. It varies as a function of the presence of indicators specific to the given problem and their response to remote sensing. Thus, success depends on the compatibility between the spectral, spatial and textural characteristics of the objects investigated and technical specifications of the sensors.

Given these conditions, the practical utility of remote sensing depends on demonstrating its operational capability and on the initial state of knowledge of the region investigated. In general, the results of such studies have been positive in the case of small and medium-scale geological and structural mapping and for different applications.

Remote sensing is presently utilised almost invariably in the preliminary phase of geological surveys. In the phase of decision-making, the data of first-generation satellites is gradually being replaced by those of TM or HRV of Landsat 5 and Spot sensors.

8.1 EXAMPLE OF MINERAL PROSPECTING

In the case of mineral prospecting the results of remote sensing, carried out at different stages of investigation, viz., from formulation of mineral policy and regional exploration up to selection of target points or tracing the extensions of known deposits, offer quite a balanced picture.

Aerospace remote sensing has thus gradually assumed over the past few years a relatively important place in mineral prospecting. Some results are convincing. Still it is difficult to make an objective general statement since prospecting is more often based on the use of several methods. Given this situation, it is not always possible to accurately evaluate the original contribution of remote sensing.

Without going into the metallogenic aspects, spatial observation primarily leads to a simplified classification of mineral deposits into two major groups-those which outcrop at the surface of the earth and are well understood, and those that are concealed at depth, most of which have yet to be discovered. As the electromagnetic energy recorded by the sensors generally comes from the surface or very close to the surface, the first group of deposits can be observed and interpreted from space, provided the environmental conditions, size of the deposit and acquisition technics are favourable. This is not so for concealed deposits. If their prospecting is to be optimised, it is necessary to preliminarily define the metallogenic exploration model for focussing the search on specific occurrences.

An analysis of the literature pertaining to remote sensing appearing in various reviews, the *Canadian Journal of Remote Sensing*, JPL publications or presented in special symposia, revealed that when the principle of this methodology is followed and when the general conditions are not too unfavourable, remote sensing provides new information which is complementary to that obtained on the ground by other methods or technics and which has great importance for prospecting. These publications indicate the efficacy of remote sensing depending on the stage of exploration and the type of deposit investigated.

- In the phase of integrated geological mapping, which forms a preliminary base for mineral planning, aerospace remote-sensing technics have become an indispensable tool. For about a decade now improvements in small-scale mapping have been very significant, as in the case of the 1:3,000,000 geological map of Sudan (3 million km^2) published in 1981.

- In the regional reconnaissance phase remote sensing has importance for improving the geological map when the occurrences of outcrops are favourable, in order to see the exploration region in a 'different' manner. Combining ground surveys and spatial observations, such maps have been prepared in the 1:200,000 scale for the Togo region (the Directorate General of Mines and Geology and the National Bureau of Mineral Exploration, 1986), along the Trans-Gabon rail route (1986-1990, Geological Service of the Ministry of Mines, Gabon) and in Mozambique (the Directorate of Mines and Geology); in the 1:250,000 scale for Sudan and the 1:100,000 scale for Madagascar (OMNIS) for the purpose of mineral and petroleum exploration (1989).

The limitations of visible/near-infra-red spatial remote sensing, the most frequently employed technic, became apparent during these studies and were essentially due to the highly unfavourable meteorological conditions in some regions of the world. These limitations have been overcome by using active microwave remote sensing, that is, radar. Examples of such applications are Brazil (Radam project), Venezuela, Equatorial Guinea and Gabon (Bassot, 1988), where preliminary observations were made with airborne radar in the absence of special satellites, as the first one, ERS-1, was launched only in 1991. Interpretation of this data, highly useful for geological mapping of inaccessible regions, also led to a better understanding of some metallotects of gold (Gabon), iron (Equatorial Guinea) and diamonds (Venezuela).

- In the detailed prospecting phase, remote sensing facilitates the preparation of large-scale (1:50,000) maps suitable to mineral prospecting. It is thus useful for the following studies:

— Selection of favourable mineralised zones using the multivariate analysis technic and geographic information systems. An example is presented by Barr (1990) for the Royaume-Uni zone of the Pennines region, where the method helped in successfully demarcating a lead prospect (Alston block), though the limitations of the method were revealed in other lease areas. Another example has been described in the Armorican Massif (Braux *et al.*, 1991) where the application of remote sensing targeted the search for gold deposits to favourable zones.

— Precise location of metallotects and the demarcation of their geometry. Such studies have led to the discovery of new occurrences (Pb, Zn) in Zimbabwe (Scanvic, 1986; Scanvic and Bache, 1986) and to the mapping of extensions of favourable lithostructures based on a lithological and tectonic model developed over a known Pb-Zn deposit (Christmann, 1989). Digital image processing, supported by reference training sites identified on the ground, proved a decisive factor in these studies.

— Investigation of lithostructural controls of a known deposit by consolidating the inferences obtained from a good knowledge of the terrain. Such was the case at Zeida (Morocco) where the disposition of mineralised lenses of galena and cerusite around a circular structure visible on the Landsat images (Jebrak and Dutartre, 1984) suggested the presence of a high-grade zone in the Triassic formations due to a granite intrusion.

— Spectral detection of gossans, for example, in Zimbabwe (Scanvic *et al.*, 1986) or the oxidised zones associated with uranium mineralisation (Peters, 1983).

— Reinterpretation of regional structure and its relation to known mineral deposits. This approach, based essentially on the mapping of image lineaments, is more often employed for gold exploration (Harris, 1991).

- At the detail stage, satellite remote sensing is handicapped by the limitation of spatial and spectral resolutions of the available acquisition systems. To overcome these limitations, particularly in arid zones that are preliminarily favourable, the mining industry is preferring the airborne systems whose performance has been excellent (spatial resolution better than 5 m, narrow spectral bands). Several airborne studies of mineral districts in the USA are presented in the 'Joint NASA/GEOSAT Test Case Project Report' (Abrams, 1985).

However, satellite remote sensing has also been used at the detail stage for exploration of surface mineral deposits using a local model or for estimation of reserves during exploitation. A typical example of the former is the discovery of a bauxite deposit (reserves of 150 million tons) in Venezuela (Brojanigo and Delpont, 1989; Delpont and Brojanigo, 1991). An investigation in Songolon, New Guinea (Scanvic, 1991) represents an example of the application of remote sensing for estimation of reserves. Here, special digital processing technics applied to Spot satellite data revealed a correlation between the spectral response recorded from space and variations of iron oxide concentrations in the bauxite. Though the degree of correlation is approximate, it is sufficient to determine, in combination with ground analysis, the purity level of bauxite.

A correlation between the reflectance of Spot data and nickel content was observed for deposits in Thio, New Caledonia (Lajoinie and Weecksteen, 1985). These differences, not detectable on the ground by their coloration, are partly related to the specific characteristics of their response in the infra-red band.

Thus the prospecting guides that can be observed and identified from remote-sensing data are morphological, spectral and textural in nature.

Among the morphological features, linear and circular forms, which often reveal tectonic phenomena and intrusions of various nature, serve as exploration indicators. The role of image lineaments in these studies is always significant and the development of fracture models at scales that allow fractal analysis enhances their importance. Certain specific features of the fracture pattern, such as the zones of lineament intersections, are also of interest. All this data can be digitised and statistically analysed. Finally, although visual analysis still remains the general rule, digital extraction algorithms have been developed (Deslandes and Gwyn, 1991) for a more objective mapping of lineaments in regions where the effect of human activity is negligible. These programs include procedures for analysing landforms based on digital elevation models obtained from remote-sensing data.

— Thus the approach of remote sensing is often indirect and its contribution to prospecting depends on the existence of specific indicators of different types of deposits. Past experience has shown that remote sensing can be utilised efficaciously for prospecting under the following conditions.

- Shear zones in perigranitic terrains (gold)
- Mineralisations in volcanosedimentary formations (lead, zinc, silver etc.)
- Mineralisations in alteration products (bauxite, iron, nickel)
- Sulphide deposits (copper, gold)
- Kimberlites (India)
- Uranium, especially in sedimentary formations

8.2 GIS AND MULTIVARIATE ANALYSIS

— The evolution of technologies that enhance the capabilities of remote sensing is accompanied by methodological developments. A major methodological approach in this regard is multivariate analysis which has become operational due to Geographic Information Systems (GIS). However, we are still in the initial stage of conceptualising the strategy (type of processing, scale) and the final trend of correlations between geochemical, geophysical and remote-sensing data, which are the principal imaging methods. Technically, input layering, superimposition and processing of multisource remote-sensing datasets amongst themselves or their integration with other systems, functioning in a raster or vector mode, help in processing the data, mainly to extract synergistic information. Today one can demarcate, based on a metallogenic model, the boundaries within which prospecting efforts should be concentrated. This is definite progress compared to the probabilistic analysis of lineaments. However, development of multivariate analysis technics that reveal the true integration of the various methods is still awaited. Integration does not merely imply determination of the importance of geological phenomena as a function of the frequency of their identification by the various methods employed, but also the extraction of new geological information. The methodology of multivariate analysis using a single technic (such as geochemical, geophysical or remote sensing method) has been perfected but methodologies for intertechnic analysis have yet to be investigated (Rheault *et al.*, 1991).

However, the universality of acquisition of data, its thematic potential and its impact on the development of digital image processing technics have made remote sensing one of the research variables of GIS and an important stimulant for ongoing methodological investigations. Excellent results have been obtained by this methodology in the fields of hydrogeology, natural hazards and mineral prospecting.

8.3 FUTURE EFFORTS FOR INTEGRATING REMOTE SENSING INTO GEOLOGICAL INVESTIGATIONS

Compared to the conventional sources of geological information, remote sensing, which plays a complementary and substitutive role, encounters difficulties in seeking an independent status. Moreover, its utilisation is limited to the initial stages of survey by the user because the data is often difficult to verify and evaluate by conventional means alone. Even if used at various other stages, the qualitative, probabilistic and inductive approach of remote sensing might not always be appreciated in the deterministic approach of the field geologist. This is further complicated by the relative humility imposed on the remote-sensing interpreter by the indirect nature of most of the observations obtained by remote sensing and the need for their correlation with ground truth, the rightful domain of the geologist, and with the subsurface, which is generally explored by the geophysicist or drilling engineer.

Given these conditions, one can anticipate methodological developments directed towards optimal integration of remote sensing with the conventional methods of the geologist. This would prompt an 'expectant' approach to the terrain by the remote-sensing interpreter. It would gradually transit from an integrated interpretation of the photointerpretation technic, which at present is the most operational because it incorporates the efficacy of exploration and identification of the experimental scientist, to comprehension of the landscape aimed at classifying, identifying or quantifying certain descriptive parameters for an objective characterisation.

Concomitantly, the geologist would change his approach to field investigations and be willing to plan them based on the knowledge of the phenomena revealed by remote sensing.

Should this evolution be expedited by the greater availability of sensors and new developments in information technology, it would enable us to utilise remote sensing not merely as a complement to what can be done on the Earth's surface by other proven technics. Structural geology is a good example. The inclusion of remote sensing in a logical sequence of investigations and employment of various methods have proved very successful in this field. In this regard popularisation of spatial geodetic systems such as the GPS, which provide excellent positioning of the image on the ground, would facilitate such an integration, as demonstrated in the investigations in Oman.

However, this still would not suffice. Too often, too much is expected of remote sensing and the anticipated results are estimated on the basis of a specific and independent investigation. It is now recognised that remote sensing can be advantageously utilised in parallel with other technics of geological exploration, each of which has something to draw from the synergy induced by multicriteria analysis. The technological means of exploiting multicriteria analysis are developed on the basis of geographic information systems but as for geological remote sensing, development is hindered by lack of an elaborate and reliable digital mode.

8.4 IMPROVING THE PERFORMANCE OF DIGITAL ANALYSIS

Research on digital analysis must be accelerated and its scope of application widened; up to now it has focussed only on the spectral signature of objects and with varying degrees of success. Technological developments in sensors have evolved high-performance spectroradiometers that enable preparation of maps of relative spectral signatures of objects. Were a reference library of spectral signatures compiled, the identification of some surface objects from satellite remote sensing data would be practicable in favourable cases. Such databases of signatures are being created for laboratory measurements but are not very representative because the favourable cases are limited to regions with good rock outcrops. Even here the ***in-situ*** signature cannot be the same as would be obtained from a sample taken out of its natural environment. In other climatic conditions the situation is much more complex and hence other methods of investigation need to be considered.

One such method belongs directly to the class of spectral analysis. It is based on the analysis of a landscape which is perceived in its totality and whose spectral signature results from the relations between rocks, alterations, surficial deposits, vegetation and soil type. Some of these relations have been studied and sometimes proven but the laws governing them have yet to be established. While each of the elements of the relations is fairly well known, the effects they generate or which they determine are not understood. Hence it is necessary to study the relation between vegetation, soil and geological substratum using the climatic, hydrological and topographic factors and their effect on the formation of landscapes. This general deterministic approach should be multidisciplinary and the interrelations coherent because the problems pertinent to the boundaries of several disciplines are numerous. This is the price for understanding the phenomena observed from space and for the ingenuity of spatial strategy to be evolved by EOS, an international earth observation system, during the 1990s. Structuring such a system would require a concerted effort.

A second method of investigation draws from the photointerpretation technic based on spectral, morphological and textural criteria. Mathematical tools for the analysis of texture exist but they need to be tested on remote-sensing data to evaluate their capability for improving classifications. However, this approach is not realistic even though the thematic totality of landscapes has been demonstrated in the form of a landscape unit and we shall revert to this problem later.

The third approach is based on widening the range of utilisation of the electromagnetic spectrum. This has already become possible in the range between 1.6 and 2.2 mm even though the conditions are not always optimal. This has also become possible in the active microwave range since 1991, when the ERS-1 satellite was launched. However, no development is foreseen in the domain of thermal scanning in spite of the interest sustained by the Thematic Mapper of Landsat V and the exploitation of data of meteorological satellites for geological investigations, whose potential has been indicated by several research studies, viz., Baelz-Manière (1980), Rousselin (1985), Daudon (1985), Bienvenu (1986) and Rouzeau (1989).

Lastly, the spectral resolution, spatial resolution and spectral range utilised, which vary widely for different operational or experimental sensors, provide varied and complementary information that must be managed for optimising it in terms of mapping, resource exploration and investigation of the environment. The tools (GIS) exist and are being adapted, whereas the technological efforts of Japan (JERS-1 project), India (IRS programme), France (Spot-Radar 2000) and the USA (polar observation platform) progressing in this direction, are already promoting thematic applications even though they need to be developed much more.

8.5 DEVELOPMENTS IN UTILISATION OF SPATIAL DIGITAL ELEVATION MODELS

The current potential development of spatial digital elevation models (DEM) is high, since the methods of calculation by correlation are reliable and the acquisition technologies have become efficient. The DEM will form the basis for computer-aided photointerpretation which has developed midway between visual methods and digital classification methods (see Sec. 8.6).

The methodology concerning thematic utilisation of DEM is presently under development and represents the research direction recommended by the exploration conference 'See, Measure and Model'organised by the Ministry of Research in 1990. The products derived from DEM are essential in certain domains, such as the study of erosion and landslides. They will play a very important role in investigating the laws of generalisation of certain models (erosional models, for example) by establishing their interrelations with satellite images, which alone are capable of generating spatial models.

DEM are used to study geochemical dispersion and also for structural interpretation, especially in the case of neotectonics. They enable measurement of certain parameters that are useful and necessary for the geologist, viz., slopes, bed thickness, horizontal or vertical displacement etc. In this domain spatial geodesy has become an essential supplement since it facilitates measurements that are important to plate tectonics, volcanism and ground movements.

Lastly, DEM form the basis for recognition of landforms. Combined with the methods of mathematical morphology and coupled with a classification database, they may define expert systems, at least for some characteristic features such as circular structures, image lineaments and landslides.

Developments in the utilisation of geomorphology add to the applications of DEM. Geomorphology has until now been rather qualitative for want of means. However, the problems posed in the management of our planet necessitate perfecting our knowledge and appreciating the Man-Nature interactions in an environment of fragile equilibrium, wherein tectonics plays a specific role in creating ground instability. These problems have accelerated the evolution of geomorphology, which seeks to develop technics and methods in which remote sensing may and should have an important place.

The perspectives offered by radar interferometry also enhance the significance of this research.

8.6 DEVELOPING COMPUTER-AIDED 3-D PHOTOINTERPRETATIONS

Photointerpretation facilitates the preparation of thematic maps, mostly based on ground identifications. Using aerospace photographic supports, with or without stereoscopic views, thematicians call upon their experience in a given theme, say geology, pedology, to interpret various landscape features. The map thus obtained is used to guide field surveys.

It was once thought that this work would become a perfectly automated task. Nature, however, is a complex system and identification of its different components is not always easy to achieve by means of automatic extraction or digital classification technics.

Rocks and minerals, which constitute the third level of observation from space, are generally difficult to observe directly since the rock outcrops are often masked by soils, alterites or lichens.

The digital approach to utilising remote-sensing data for geological applications has evolved gradually and logically since 1972 because of improvements in the performance of the systems. This evolution can be summarised schematically in four phases:

— Importance of interactive dialogue: The geologist, depending on the theme to be investigated and his knowledge of the terrain, supervises the generation of photographic products using the various processing programs available, which he optimises on an observation screen.

— Classifications on spectral basis: Even with the improved spatial and spectral resolutions, it is difficult to find in the image those pixels representative of rocks and more so of the minerals they contain. Given these conditions, the result, though improved, still remains uncertain.

— Enlarging the number of criteria taken into account for spectral classifications alone: Morphological analysis has presently been developed using mathematical tools. Further improvements may be expected.

— Multivariate analysis and geographic information systems have presently altered the concepts. The dispute between visual interpretation and the pure digital approach has been laid to rest. Each indicates that the product resulting from photointerpretation, while it still remains the optimum method, is handicapped by its analogue form which requires cumbersome and tedious digitisation. Yet, it represents a necessary transition towards multivariate analysis and integration into geographic information systems.

This analysis leads us to hereafter prefer a third approach known as computer-aided 3-D photointerpretation (CAPI 3d). It combines the experience of the photogeologist and his methods of observation with the power of information technology, which together facilitate the preparation of a directly digital product.

CAPI 3d is a program primarily consisting of a stereoscopic pair of remote sensing data, a package of algorithms and an integrated system of interpretation.

Satellite stereoscopic technology is grdually becoming established. Spot satellite has opened the path. The OPS of JERS-1 satellite has facilitated simultaneous stereoscopic acquisition and made it a common product, which was otherwise quite rare.

Specific algorithms are being developed and will facilitate quantitative interpretation. These algorithms are already enabling preparation of digital elevation models which form the basis for producing vertical and oblique orthophotos. The possibility of using these photos for three-dimensional interpretation is being explored. Though limitations exist in the stereoscopic quality of orthophotos, this will be a preferred method in future.

CAPI 3d will develop to the extent that true user-friendly interactive systems permit. At present, the stereoscopic view on the screen is based on the use of anaglyphs that require a special spectacle having a red-tinted glass (left eye) and a green-tinted glass (right eye). Accessible at a lower cost on an image-processing system, the anaglyph has the limitation of black-and-white vision and is not very comfortable.

An alternative approach is based on the liquid crystal seal technic developed by Tektonix. This technic provides a comfortable environment due to vision through polarising glasses.

One can at this stage envisage that the use of annotation algorithms on an image presented on the screen may offer fairly well-developed functions (Leymarie *et al.*,

1991). Such functions include: creation of broken lines, close-spaced or not; modifications of lines already scaled; earmarking the attributes at the time of tracing; selection of optimised local dynamic range in real time and in the parametric neighbourhood; selection of appropriate filters; transformation of annotations into mapping algorithms; carrying out geometric corrections; dual view displays with the facility of superimposing on a reference base such as an existing geological map.

This approach is important as it enables us to quickly take into account the existing databases and to integrate or correct them during the interpretation of colour and stereoscopic data on the screen.

The final product would be a more elaborate map than that resulting from conventional photointerpretation; furthermore, it would be digital and hence easily integrable into the systems of multivariate analysis.

Thus the true efficacy of the CAPI is gradually being established. It is evident that in addition to the preparation of different maps that are fairly informative and the preliminary guidance and optimisation it provides to ground surveys, remote sensing analysis also seeks to demonstrate the originality and specific aspect of the information it reveals. Henceforth easily achievable through CAPI 3d, the integration into geographic information systems would enable the remote-sensing interpreter to exchange different types of data, compare them and optimise them by establishing the synergy between them.

Remote-sensing analysis, which so far has been only a supporting tool for ground surveys, will become a real mapping tool of the future because of the variety of information it generates.

8.7 FORECAST MODELLING

Geological objects are rarely directly visible from space. Objects which cover or mask them almost completely are very frequent and geological remote sensing is generally qualitative. Efforts have been made to model the spectral responses of rocks, minerals and alterations in the visible and near and middle infra-red bands. These studies have contributed to the improvement of the technic since they have facilitated optimisation of sensors and data processing, thus enhancing the discriminatory possibilities. However, the objective of identification of ground objects is still far from realisation.

In the thermal domain a large number of factors that contribute to the response make modelling of geological objects much more difficult and the interpretation of results more complex, though they are often encouraging. In pursuance of this promising area, efforts should be directed simultaneously to technology and basic research.

In the domain of microwave remote sensing, much still remains to be done in the field of geology, for which the radar technic has proved its importance in spite of its empirical utilisation. Basic studies on modelling the signal, backscattered from geological objects, as a function of roughness and electrical permittivity must also take into account the reality of field conditions.

There is, however, a domain, viz., that of fracture studies, which is *a priori* amenable to modelling as structural geologists have developed such possibilities. Is it possible and realistic to transpose them to remote sensing?

The problem of understanding tectonics primarily from remote-sensing data is a least explored direction of research. It emerges from a number of earlier studies (Bouchot, 1989; Caia, 1984; De Fraipont, 1982) which have demonstrated the occurrence of various forms in remote-sensing data of different scales, corresponding to the basement formations in temperate regions with scarce rock outcrops. These forms appear to reflect fracture models, shear models in particular, defined in outcrops. This fractal dimension, suggested or partially demonstrated in some examples, must be generalised. Some research teams are already engaged in such studies.

8.8 DEVELOPMENT OF MAPPING TECHNOLOGY IN FUTURE

This review of present trends should not lead to the thought that the future is in the hands of laboratory and information technologists. On the contrary, major developments are foreseen in the domain of on-the-ground validation of information extracted from present or future satellite data, which is rapidly increasing.

A significant change is currently being observed in the concept of geological mapping and remote sensing will assume an important role in this development.

The observed or indicated changes in geological mapping, which form the subject matter of research, occur at at least three levels.

- Firstly, rapid development has occurred in the last few years in computer-aided mapping but the tools of information technology are essentially used in the stage prior to map production. This model and its usage are closely related to the needs of printing a unique and specified map, such as the 1:50,000 map in France. In this case remote-sensing data is sometimes used when the map is superimposed on a satellite image base. More has still to be done.
- After collecting data by all the available methods of geological mapping, viz., mapping the outcrops, petrographic and geochemical analysis, geophysical surveys and remote sensing, efforts are made to integrate these convergent and complementary datasets. At this stage the methods of multicriteria digital analysis are used to detect the ultimate synergy. Special algorithms are employed for integrating data of any nature, viz., qualitative or quantitative, of different sources such as satellite imagery and graphic databases, and of different formats, viz., vector and raster. Remote sensing has already become an important component of multicriteria analysis but it is generally integrated in the form of digitised graphic maps.
- Finally, research is primarily oriented towards the preparation of maps using regular and highly diversified data with the perspective of generating specific thematic maps. This is the role expected of geological and geographic information systems in which remote sensing may play a significant part. Being a common factor for several disciplines, remote sensing enables us to understand the Earth's surface in its total perspective by integrating vegetation, soil and geology in a complex response which cannot always be readily deciphered. This original perception corresponds to the domain of management and environment. It thus holds great promise provided interdisciplinary specialisations are developed.

The major obstacles to these developments emerge from the multiplicity of participants (the producers and the users) and their objectives, the complexity and diversity of the content of data, the number of materials and programmes utilised. While these problems are basically similar for any information system, in the case of remote sensing they are complicated by the very fact that the data acquired by this technic reflects a large diversity of themes pertaining to earth and environmental sciences. It is therefore imperative to structure this information in a mode suitable to various thematic objectives.

8.9 PORTABLE RADIOMETERS FOR *IN-SITU* MEASUREMENT OF PHYSICAL AND CHEMICAL PROPERTIES OF GEOMATERIALS

Nowadays, *in-situ* measurements are being done to determine the physical and chemical properties of geomaterials in the range of wavelengths used in remote sensing. Data is acquired in the passive mode for the middle infra-red and thermal infra-red bands and in the active mode for the microwave region. Soviet geologists have experimented with these two technics for mineral prospecting (Gorobets *et al.*, 1989). The Research Center of the University of Boulder Colorado, (USA) has developed a light (3 kg) portable radiometer (Puma) connected to a database, which facilitates direct *in-situ* identification of different clay materials.

8.10 CONCEPTS CONCERNING THE EVOLUTION OF REMOTE SENSING INTO A CRAFT

Obviously, remote sensing is not a conventional scientific approach as one cannot dissociate a technic from the human beings who employ it. One should think of its evolution into a craft because remote sensing is concomitantly a largely open field of research and a service tool for rapid developments, especially in geology.

The experience of the last twenty years induces us to reflect on a question: Will remote sensing remain in future an attribute of the specialist or, contrarily, become commonplace?

Each of these two propositions has a realistic aspect when considered from the point of view of a service society. The first-generation remote-sensing geologists, essentially emerging from photointerpretation, mainly concentrated on demonstrating the validity of the phenomena observed over large and more or less known areas, in spite of the limitation of the poor spatial resolution of MSS data. Lacking the ability to walk everywhere by himself, the remote-sensing geologist had earlier to find evidence in what was or had been mapped. He remained a specialist whose positive observations were appreciated, whereas findings without positive response engendered scepticism. Later, under the pressure of a new and vague belief in the total advantage of digital technology, remote sensing became a creed supported and amplified by the myth of a certain informatic vision that would replace all other forms of investigation of the Earth.

The above notwithstanding, an equilibrium has been established between visual and digital methods, each striving to progress towards the other, as also between geologists, irrespective of whether they specialise in remote sensing or not. The mapper or structural geologist provides access to the field and opens the clues to interpretation of satellite observations, which he takes into account for what they are and where they were found. This progress has become possible through the availability of second-generation satellite data, which facilitates the development of applications and more users. Unlike the various specialists for whom it is intended, the remote-sensing scientist, primarily a generalist, has experienced a loss of self-esteem albeit the former have begun to appreciate the importance of satellite information in their respective domains. To rectify this situation, it is preferable to develop new profiles characterised by a dual competence combining remote sensing with a specific field of geology or with the interface between different disciplines. These new specialists might develop either from remote sensing or from the various branches of applied geology. Since this formula is already bearing fruits in mapping, hydrogeology, mineral exploration or in the problems of natural hazards, one may expect more universality, promoted by the independence offered by microinformation technology. In due course it may be imagined that every geologist, entrusted in this organisation with a work, would take total charge of it. But, then, what would be the consequences for remote sensing should it become a commonplace craft?

Even though the outcome would not be the same, one may refer to photogeology, which witnessed rapid development until the 1960s, then subsequently became too common at the cost of efficacy. Aerial photography is often considered a support for field observations and photointerpretation is no longer as systematic and hence is less efficient. This is a regression which needs to be examined so as to avoid the possibility of remote sensing reaching the same state, should it become too commonplace.

Modern technological developments are outstriding developments in thematic applications, which are hindered by the difficulties of field investigations. At present, due to the effects of dual expertise, the handicap is persisting or even slightly increasing, but the changes in the technological front are too intense to be totally reduced. Moreover, inherent to its two facets, there exists in dual expertise an excess at one pole which one fears may blunt the other.

Still, at the end of this discussion, one's response is to undoubtedly affirm the importance of dual expertise because it enables us to efficiently integrate remote sensing with other applied studies, to manage a pool of specialists engaged in leading or following the technological evolution developed by space agencies and industries, and to ensure its adaptation to the needs of the market. However, it is the duty of professional societies to consolidate this desirable evolution through investments in materials because the competence of man alone cannot offset the absence of efficient means.

LITERATURE CITED*

Abrams, M.J. (1983). Remote sensing for porphyry copper deposits in Southern Arizona. *Econ. Geol.*, vol. 78, no. 4.

Abrams, M.J. (1985). *JOINT NASA / GEOSAT Test Case Project. Final Report.* American Association of Petroleum Geologists, Tulsa, Oklahoma.

Abrams, M.J. and Brown, D. (1985). Silver Bell Arizona Porphyry Test Site Report. *The Joint NASA / GEOSAT Test Case Project. Final Report* (H.N. Paley, ed.). AAPG, Tulsa, Oklahoma, pt. 2 **(1)**: 4.1-4.70.

Abrams, M.J. and Goetz, A.F.H. (1985). Imaging spectrometry: past, present, future. Spectral signatures of remote sensing objects. 3rd Colloquium, Les Arcs, France.

Abrams, J.M., Conel, J. and Lang, H. (1985). *The Joint NASA / GEOSAT Test Case Project. Final Report* (H.N. Paley, ed.). AAPG, Tulsa, Oklahoma, pt. 2 ( ): 1.1-9.110.

Ahmadzadeh, M. (1984). Le district aurifère de Saint-Yrieix (Haute-Vienne) [The gold-bearing district of Saint-Yrieix (upper Vienna)]. 3rd cycle thesis, Clermont Univ. II.

Alsac, C., Arene, J., Breton, J.P., Scanvic, J.Y., Ferrandes, R. and Mennechet, C. (1980). Méthodes d'investigation d'un massif granitique [Methods of investigation of a granite massif]. *J. CEA sur le Stockage des Déchets* (unpubl.).

Aye, F. (1982). Contróles géologiques des gîtes volcanosédimentaires de Pb, Cu and Zn de la bordure du bassin de Chateaulin (France) [Geological controls of volcanosedimentary deposits of Pb, Cu and Zn on the perimeter of Chateaulin basin (France)]. Doct. thesis, Univ Paris VIII.

Baelz-Manière, S. (1980). Utilisation de thermographies infrarouge recueillies par satellite pour le calcul de l'évapotranspiration a l'échelle régionals [Utilisation of satellite thermal infra-red images for calculating evapotranspiration on a regional scale]. 3rd cycle thesis, Univ. Paris VI.

Baelz-Manière, S. and Gilli, J. (1988). Programme d'acquisition des données aéroportées et leur interprétation [Programme of acquisition of airborne data and its interpretation]. *Chron. Rech. Min.*, no. 491, pp. 3-12, June 1988.

Bailey, G. and Dwyer, J.L. (1985). Evaluation of Landsat thematic mapper data for geologic mapping in semiarid area. ERIM symposium, April 1985.

Barbier, J. (1984). Télédétection des gisements par mise en évidence de l'empoisonnement des sols par les métaux. La France vue à travers les images Landsat 1 and 3 [Remote sensing of mineral deposits by detecting enrichment of soils by metal. France seen through Landsat 1 and 3 images]. *Chron. Rech. Min.*, no. 477, pp. 15-22.

Bardet, M.G. (1963). Géologie du diamant [Geology of diamonds]. *Mém. BRGM*, no. 83, vol. 1-2.

Bardinet, Cl., Gabert, G. and Monget, J.M. (1988). Application of multisatellite data to thematic mapping. *Geol. Jahrb.*, vol. 67, pp. 3-74, Hannover.

Barr, M.W. (1990). Pennine exploration dataset re-examined. "Remote sensing: an operational technology for the mining and petroleum industries". Conf., London 29-31 Oct. 1990, Institution of Mining and Metallurgy, pp. 3-14.

Bassot, J.P. (1988). Apport de la télédétection à la comprehension de la géologie du Gabon [Role of remote sensing in understanding the geology of Gabon]. *Chron. Rech. Min.*, no. 491.

Baudoin, A. (1988). Rapport du groupe de travail IV.3 de la commission IV de la SIPT [Report

*Some entries were incomplete in the French original—General Editor.

of the working group IV.3 of the Commission IV of ISPR]. *Bull. SFPT*, no. 111, 1988-3, Paris.

Becker, F. and Raffy, M. (1987). Problems related to determination of land surface parameters and fluxes over heterogeneous media from satellite data. *Adv. Space Res. Rev.*, 4(11): 45-57.

Becker, F., Nerry, F., Ramantsizehena, P. and Stoll, M.P. (1985). Emissivité dans l'infrarouge thermique: condition de mesure par rétrodiffusion et comparaison avec une méthode de mesure passive [Emissivity in the thermal infra-red band: condition of backscatter measurement and its comparison with a passive method]. Proc. 3rd Int. Coll. on Spectral Signatures of Objects in Remote Sensing. Les Arcs, France, LSA 247, pp. 241-244.

Bérard, P. (1982). Interprétation des images satellites en complement à la photo-interprétation traditionnelle pour la définition des structures hydrogéologiques au Niger et en Haute-Volta [Satellite image interpretation as a complement to conventional photointerpretation for tracing hydrogeological structures in Niger and the Upper Volta]. Jubilé Castany Orléans, Nov. 1982. BRGM Publ.

Bérard, P., Castaing, C., Gombert, P. and Scanvic, J.Y. (1990). Alimentation en eau de la mine d'or d'Hassai (Soudan); impact de l'étude télédétection [Water supply to the Hassai gold mine (Sudan): impact of remote sensing study]. BRGM Publ., Orléans, France.

Berthiaux, A. and Marot, A. (1989). Etude structurale de l'unité de Saint-Georges-sur-Loire par télédétection [Structural study of the Saint-Georges-sur-Loire region by remote sensing]. BRMG Report 89 DT 006 TED.

Berthiaux, A., Girault, F. and Scanvic, J.Y. (1988). Cartographie géologique a l/100 000 du Karoo du bassin de Morondava (Madagascar) par télédétection [The 1:100,000 geological mapping of Karroo region of the Morondava basin (Madagascar) by remote sensing]. BRGM Report 88 MGD 151 TED.

Berthiaux, A., Clement, J.P., Minoux, L. and Wynes, R. (1987). Programme d'évaluation Spot. la region des Mauges. Analyse et comparaison des données spatiales Spot et TM-1987. [Program of Spot evaluation. The Mauges region. Analysis and comparison of space data of Spot and TM-1987]. BRGM Report DT 043 TED.

Bertraneu, J. (1973). Stratégie de la recherche minière [Mineral exploration strategy]. In: *Geology* 2, *Encyclopédie de la Pléiade*. Gallimard ed., Paris, pp. 323-328.

Bienvenu, J. (1986). Detection et études de phénomènes géologiques en Aquitaine. Analyse et synthèse de données satellitaires thermiques. Expérimentation radiométrique *in situ* [Detection and study of geological phenomena in Aquitaine. Analysis and synthesis of satellite thermal data. *In-situ* radiometric experiment]. 3rd cycle thesis, Univ. Orléans, 207 pp.

Biju-Duval, B., Devaux, E., Gonnard, R., Latarche, C. and Rivereau, J.C. (1975). Apport de l'étude des images du satellite Landsat 1 à la connaissance du la structure du domaine méditerraneen [Contribution of the study of Landsat 1 satellite images to understanding the structure of the Mediterranean region]. *Rev. Inst. Fr. Petrole*, 30 (5): 841-853.

Birnie, R.W. and Francica, J.R. (1979). Remote detection of geobotanical anomalies: a method for mineral exploration. Int. Symp. Remote Sensing Enviornment 12 h Manila, 1978, pp. 795-804.

Blanchet, P. (1957). Development of fractures analysis as an exploration method. *Bull. Amer. Assoc. Petrol. Geol.*, 42 (8): 1748-1759.

Bles, J.L. (1977). Cuadro geologico-Plano de desarollo urbano de la Ciudad de la Paz. Rapport Honorable Alcadia Municipal de La Paz.

Blom, R.G., Crippen, R.E. and Elach, C.E. (1984). Detection of subsurface features in SEASAT radar images of Means Valley, Mojave-Desert, California. *Geology*, no. 12, pp. 346-349.

Blusson, A. (1985). Etude structurale du faisceau de failles des Cévennes (France) à l'aide de données multiples: télédétection, terrain géophysique [Structural study of the Cévennes fault system (France) using multiple data: remote sensing and ground geophysics]. Thesis, Univ. South Paris.

Boirat, J.M. (1979). Synthèse sur la répartition des gîtes de fluorine et barytine dans le Massif central (France) a l'Est du Sillon houiller [Integrated study on the distribution of fluorite and baryte deposits in the Massif Central (France) to east of the Coal Belt]. *Mém. Univ. Orléans*, Diplôme Ingéniorat, Ecole Supérieure des Transports, de l'Énergie et des Ressources Minérales.

Bolviken, B. and Lag, J. (1978). Natural heavy-metal poisoning of soils and vegetation: an exploration tool in glaciated terrain. *Inst. Mining and Metall.*, vol. 86, pp. 8173-8180.

Bolviken, B., Honey, F., Levine, S. R., Lyon, R.J.P. and Prelat, A. (1977). Detection of naturally heavy-metal poisoned areas by Landsat 1 digital. *Data. J. Geochem. Exploration*, vol. 8, pp. 457-471.

Boncompain, R., Driesen, D. and Pasquet, R. (1977). Port Sedar. Etude geologique, géophysique et géotechnique du site portuaire de Kheur-Farah Pahlavi. République du Senegal [Port Sedar. Geological, geophysical and geotechnical study of Kheur-Farah Pahlavi port site, Republic of Senegal]. BRGM Report 88 AGE 029.

Bonham-Carter, G.F. (1984). Statistical association of gold occurrences with Landsat derived lineaments Timmons, Kirkland Lake area, Ontario, *Can. J. Remote Sensing*.

Bonnefoy, D. and Guillen, A. (1985). Le traitement des données multisource [Processing multisource data]. BRGM Report 85 DAM 040 GMX.

Bonnefoy, D., Chevrel, S., Dutartre, Ph., Minoux, L. and Orgeval, J.J. (1986). Télédétection: discrimination d'objets géologiques et miniers. Analyses muyltitemporelles et multispectrales de données spatiales et de leur relation avec la géochimie [Remote sensing: discrimination of geological and mineral objects. Multitemporal and multispectral analyses of space data and its relation with geochemistry]. Part 1: Portugal, BRGM Report 86, PRT 162 GEO, 68 pp., Part 2: Tunisia, 86 SGN 163 GEO, 73 pp.

Bothorel, A. (1983). Télédétection des ressources terrestress: signature spectrale des roches et de minerais en zone aride [Remote sensing of earth resources: spectral signatures of rocks and minerals in arid zones]. *Mém. Sciences de la Terre, Univ. P.M. Curie, Paris*.

Bouchet, Ph. (1985). Classements et contrôles sur le terrain des données Landsat sur la cartographie automatique et la recherche de signatures spectrales des gîsements minéraux au Sud du Maroc [Classification and ground verification of Landsat data using automatic mapping and investigation of spectral signatures of mineral deposits in southern Morocco]. Thesis. *Mém. Sciences de la Terre, Univ. P.M. Curie*. 116 pp., no. 8513.

Bouchot, V. (1989). Contexte géologique et structural des minéralisations aurifères du district de Saint-Yrieix (Limousin, Massif central français) [Geological and structural context of gold mineralisations of Saint-Yrieix (Limousin, French Massif Central)]. Orléans BRGM, 280 pp. (Thesis: Sciences de la terre: spécialité: géologie structurale et métallogénie, Orléans, 1989).

Bouchot, V., Dutartre, Ph. and Gros, Y. (1988). Contribution de l'imagerie satellite Landsat Thematic Mapper à la caracterisation et à la cartographie de zones et cisaillement aurifères du district de Saint-Yrieix (France) [Contribution of Landsat Thematic Mapper imagery to the characterisation and mapping of gold-bearing shear zones of the Saint-Yrieix district (France)]. BRGM Report 88 DAN 008 DEX/TED.

Bouchot, V., Dutartre, Ph. and Gros, Y. (1991). Analyse multicritere des données satellitaires pour la cartographie prévisionnelle des minéralisations aurifères de type discordant. Exemple du district de Saint-Yrieix (Massif central français) [Multicriteria analysis of satellite data for preliminary mapping of discordant type of gold mineralisations. Example of the Saint-Yrieix district (Massif Central, France)]. *Chron. Rech. Minière*, Dec. 91, no. 505. BRGM eds., Orléans.

Boyer F., Routhier, P. and Bouvier, H. (1975). Mise en évidence de quelques ceintures métallogéniques d'Europe; conséquences théoriques et pratiques [Detection of some metallogenic centres in Europe: Theoretical and practical consequences]. *C.R. Acad. Sci. Paris*, vol. 280, 17 March 1975, series D 1133-1136.

Braux, C., Delpont, G., Bonnefoy, D., Cassard, D. and Bonnemaison, N. (1990). Interpretation of remote sensing data over the south Armorican shear zone (Brittany, France): A contribution to exploration of gold minerlization. Remote Sensing Internat. Conf. and Exhibition, London.

Braux, C., Delpont, G., Bonnefoy, D., Cassard, D., Corpel, J. and Lagny, P. (1991). Identification de secteurs a vocation aurifère dans un couloir tectonique règional par interprétation de données satellitaires, géologiques et géophyiques intégrées dans un système d'informations géographiques. L'exemple du cisaillement sud-armoricain (France) [Identification of probable gold-bearing sectors in a regional tectonic zone by interpretation of satellite, geological and geophysical data integrated into a geographic information system. Example of the south Armorican shear zone (France)]. *Chron. Rech. Minière*. Dec. 1991, no. 505. BRGM eds., Orléans.

BRGM (1978-1982). Geological Survey of Sudan. Co-operation Programme of Mineral Research in Sudan. Red Sea Hills Mission. Internal Reports 1978-1982.

Brohan, M. (1981). Prospection systématique en Guinée équatoriale. 1re et 2e phases [Systematic prospecting in Equatorial Guinea. 1st and 2nd phases]. BRGM Report 81 RDM 016 AF and 81 ROM 061 AF.

Brojanigo, A. and Delpont, B. (1989). Uso de una imagen Spot para la prospeccion de bauxita-region de Los Pijiguaos-Guyana Venezolana. IV Simposio Latino Americano de Percepcion Remota IX Selper-Baricloche Argentina, Nov. 1989.

Brosse, J.M. (1975). La télédétection en géologie structurale [Remote sensing in structural geology]. Thesis, Doct. Appl. Geol., Univ., Orléans, 13 pp.

Brosse, J.M. and Scanvic, J.Y. (1976). Rôle de l'échelle dan l'apport de la télédétection en géologie structurale [The importance of scale in the contribution of remote sensing to structural geology]. *Journ. de Télédétection du GDTA, Toulouse,* pp. 99-117.

Brosse, J.M. and Scanvic, J.Y. (1985). Etude morphogéologique des radargraphies et des photographies aériennes de la feuille de Boué (Gabon) [Geomorphological study of radar images and aerial photographs of the Boué region (Gabon)]. BRGM Report 85 SGN 032 GEO.

Brosse, J.M. and Scanvic, J.Y. (1986). Etude morpholithologique et structurale des radargraphies et des photographies aériennes de la feuille Mouila (Gabon) [Morpholithological and structural study of radar images and aerial photos of the Mouila region (Gabon)]. BRGM Report 86 GAB 003 GEO.

Brosse, J.M. and Scanvic, J.Y. (1989). Etude morpholithologique et structurale des photographies aériennes et des radargraphies de la feuille Franceville (Gabon) [Morpholithological and structural study of radar images and aerial photographs of the Franceville region (Gabon)]. BRGM Report 89 GAB 137 TED.

Brosse, J.M., Chevrel, S., Dutartre, P., Girault, F., Malon, J.F. and Scanvic, J.Y. (1987). Rôle de Spot dans le développement des applications géologiques de la télédétection [Role of Spot in the development of geological applications of remote sensing]. Coll. Spot-Images, Paris Nov. 1987, pp. 773-785.

Buckingham, W.F. and Sheldon, E. (1983). Mineralogical characterization of rock surfaces formed by hydrothermal alteration and weathering-application to remote sensing. *Econ. Geol.*, vol. 78, pp. 664-674.

Burnol, L. (1974). Geochimie du béryllium et types de concentration dans les leucogranites du Massif central français [Beryllium geochemistry and types of concentrations in leucocratic granites of the French Massif Central]. *Mém. BRGM,* no. 85.

Caia, G. (1984). Etude des minéralisations à sidérite et sulfures du secteur de Gausson et de leur environment géologique. Permis d'Uzel, Cotes-du-Nord (France). DEA soutenu en septembre 1984 à l'Université d'Orléans. UER des sciences fondamentales et appliquées. Département des sciences de la Terre [Study of siderite and sulphide mineralisations of the Gausson region and their geological environment. Uzel permit, Cotes-du-Nord (France). DEA undertaken in September 1984 at Orléans University. UER of basic and applied sciences. Department of Earth Sciences]. BRGM: Département des Gîtes Minéraux, 115 pp.

Caire, A. (1975). Les règles de la fracturation continentale et le rôle des géofractures dans l'évolution de l'écorce terrestre [Laws of continental fracturing and the role of geofractures in the evolution of the Earth's crust]. *Rev. Geogr. Phys. Geol. Dynamique, Paris,* 17 (4): 319-354.

Campy, M. and Macaire, J.J. (1989). Géologie des formations superficielles [Geology of Surficial Deposits]. Masson Publ., Paris, 433 pp.

Cannon, H.L. (1955). Geochemical relations of zinc-bearing peat to the Lockport Dolomite, Orleans Country, New York. *U.S. Geol. Surv. Bull.*, no. 1000d-D, pp. 119-185.

Cannon, H.L. (1960). Botanical prospecting for ore deposits. *Science,* 132 (3427): 591-598.

Carrere, V. (1987). Exploitation des données multisources et traitement d'images en cartographie géologique [Application of multisource data and image processing in geological mapping]. Thesis, Paris VI.

Carrere, V. (1990). Mineral identification and hydrothermal alteration mapping using airborne visible/infrared imaging spectrometer (Aviris) data. Examples from the goldfield mining district, Nevada. In: JPL Openfield Report, Pasadena, California.

Carrere, V. and Abrams, M.J. (1988). An assessment of Aviris data for hydrothermal alteration mapping in the goldfield mining district, Nevada. *Proc. Airborne-visible-infrared Imaging Spectrometer (Aviris) Performance Evaluation Workshop.* Jet Propulsion Laboratory, Pasadena, California, publ. 88-38 (G. Yane, ed.), pp. 134-154.

Carte Géologique de la France et de la marge continentale à 1/1 500 000 (1978-1979) [Geological Map of France and the Continental Margin in the 1:1,500,000 Scale (1978-1979)]. BRGM publ.

Castaing, C., Dutartre, P., Gouyet, J.F., Loiseau, P., Martin, P. and Pointet, T. (1989). Etude pluridisciplinaire d'un réseau de discontinuités images Spot en milieu granitique couvert [Multidisciplinary study of discontinuities on Spot images over a covered granitic terrain]. *Bull. SFPT,* no. 113 (1989-1).

Cazabat, C. (1975). Topologie ertzienne de la France. Premiers résultats [Ertzian topology of France. Preliminary results]. *Soc. Fr. Photogramm. Bull.*, no. 60.

Cervelle, B. (1989). Spot des yeux braques sur la Terre [Spot with Its Eyes Focussed on the Earth]. CNRS, Paris, 214 pp.

Cervelle, B. and Bahu, C. (1974). Mesure au microscope de la reflectance d'une poudre monominerale, en fonction de la taille des grains et de leur absorption [Microscopic measurement of the reflectance of a monomineral powder as a function of the grain size and absorption]. *C.R. Acad. Sci., Paris*, vol. 278, series D.

Challans, Ters, M. and Viaud, J.M. (1983). Notice de la carte géologique a 1/50 000 [Notes for the 1:50,000 geological map]. BRGM, Orléans.

Champetier de Ribes, G. and Scanvic, J.Y. (1983). Résultats préliminaires à l'étude des images SIR A de la region de Yaounde (Cameroun) [Preliminary results of the study of SIR A images of the Yaounde region (Cameroon)]. *Bull. SFPT,* no. 92, pp. 33-42.

Chang, S.H. and Collins, W. (1983). Confirmation of the airborne biogeophysical mineral exploration technique using laboratory deposits. *Econ. Geol.*, 78 (4): 33-42.

Chen, K.S. and Fung, A.K. (1992). A comparison between backscattering models for rough surfaces. *Proc. IGARSS* 92, vol. **1**, pp. 907-909.

Chen Song Lin (1987). Télédétection-traitement d'images, confrontation des données: contribution à la compréhension du contexte métallogenique de la Montagne noire et de l'Albigeois (France) [Remote sensing-image processing and data interpretation: contribution to the understanding of metallogenic context of the Montagne Noire and Albigeois regions (France)]. Thesis, P. Sabatier Univ. Toulouse.

Chevrel, S. (1984). Cartographie lithologique et prospection des minéraux industriels en milieu désertique. Apport de la télédétection et evaluation de quelques traitements numériqués de données Landsat, étude methodologique appliquée à la région du Qasim (Arabie) [Lithological mapping and prospecting of industrial minerals in desert terrains. Role of remote sensing and evaluation of some digital data-processing technics. A methodological study applied to the Qasim region (Arabia)]. BRGM Report 84 SGN 249 GEO.

Chevrel, S. (1985). Industrial mineral resoures map of Buraydah, Kingdom of Saudi Arabia, scale 1:1,000,000. Kingdom of Saudi Arabia, Ministry of Petroleum and Mineral Resources, Deputy Ministry for Mineral Resources DMMR. Geoscience map GM 110, 1985, explanatory notes, 14 pp.

Chevrel, S. (1986). Etude par télédétection d'objets géologiques d'après leur signature spectrale [Remote-sensing study of geological objects using their spectral signatures]. BRGM Report 86 SAU 069 GEO, 42 pp.

Chevrel, S. (1989). Remote sensing, a tool for lithological mapping and industrial minerals and rocks prospecting. Yanbu region, Kingdom of Saudi Arabia. Proc. 1st Symp. on Remote Sensing in the UAE, Al Ayn, V.A.E., February 1989.

Chevrel, S., Chevremont, P., Wyns, R., Lemetour, J., Altoba, A. and Beurrier, M. (1989). The use of digitally processed Spot data in the geological mapping of the ophiolite of Northern Oman. Mascat Symposium 1989 and BRGM Report 530412 TED SGN 20.

Chiron, J.C. (1980). Introduction à la carte tectonique de la France [Introduction to the tectonic map of France]. *Mém. BRGM,* no. 10.

Chorowicz, J. (1989). Reconnaissance automatique de formes géologiques et géomorphologiques [Computerised identification of geological and geomorphological patterns]. *Bull. SFPT,* no. 114-1989-3.

Chorowicz, J. and Mukonki, M. (1979). Linéaments anciens. Zones transformantes recéntes et géotectoniques des fossés de l'Est africain d'après la télédétection et la microtectonique [Ancient lineaments. Recent transform faults and geotectonics of the East African trench from remote sensing and microtectonic study]. *Mus. Roy. Afr. Centr. Tervuren Belg. Rapp. Annuel* 1979, pp. 143-167.

Chorowicz, J., Kim, J., Manoussis, S., Rudant, J., Foin, P, and Veillet, I. (1989). A new technique of recognition of geological and geomorphological pattern in digital terrain models. In: *Remote Sensing Environment,* pp. 229-239. Elsevier, New York.

Christmann, P. and Scanvic, J.Y. (1987). Apport de l'imagerie Spot à la cartographie géologique en pays aride. Exemple de la Republique Arabe du Yémen. [Role of Spot imagery in the geological mapping of arid regions. Example of the Arab Republic of Yemen]. *Bull. SFPT*, no. 106.

Cluzeau, B. (1985). Utilisation de données de télédétection dans le domaine des longueurs d'ondes visibles, thermiques et hyperfréquence sur un site test, les Vans: évaluation, comparaison et complémentarité de ces données pour des applications géologiques [Use of remote-sensing data of the visible, thermal and hyperfrequency wavelength bands in the

Vans test site: evaluation, comparison and complementary nature of this data for geological applications]. Thesis, P. Sabatier Univ., Toulouse.

Cluzeau, B., Kerr, Y., Gess, G. and Scanvic, J.Y. (1985). Les Vans (Ardèche, France) [The Vans (Ardèche, France)]. SAR 580. GDTA Report, Toulouse, Contrat CEE no. 152 MPP, pp. 1-107.

Cole, M. (1982). Integrated use of remote sensing imagery and geobotany in mineral exploration. *Transgeol. S. Afr.* 85, pp. 13-28.

Cole, M. (1984). Geobotany in geological mapping and mineral exploration. In: *Remote Sensing for Geological Mapping.* BRGM Doc. no. 82. IUGS no. 18, pp. 267-283.

Cole, M. (1987). Evaluation of Spot imagery for discriminating vegetation communities associated with surficial and bedrock geology and identifying anomalous and stressed vegetation associated with numeralization. Spot 1 Utilisation des Images-Bilan- Résultats Paris 1987-CNES-Cepadues Ed. Toulouse, pp. 755-772.

Cole, M. and Brown, R.C. (1976). The vegetation of the Ghanzi area Botswana. *J. Biogeo.*, no. 3, pp. 181-185.

Colleau, A. and Lenotre, N. (1990). Une méthode d'analyse numérique combinatoire en néotectonique [A combination method of numerical analysis in neotectonics]. SGN GEO Report R30578.

Collins, W., Chang, S.H., Raines, G., Canney, F. and Ashley, R. (1983). Airborne biogeophysical mapping of hidden mineral deposits. *Econ. Geol.*, 78 (4): 723-782.

Corpel, J. (1990). Traitement et réinterprétation des données gravimetriques disponibles sur la zone de synthèse Haut Bocage vendéen [Processing and reinterpretation of gravity data over the synthesis zone of Vandian Haut Bocage]. BRGM Report R 31765 GPH SGN 90.

Corpel, J. and Debeglia, N. (1982). Valorisation de l'aéromagnetisme en Bretagne. Action concertée CEE MPP 137 F [Evaluation of aeromagnetic data of Brittany. Concerted action of CEE MPP 137 F]. BRGM Report.

Coste, B. (1985). Essai de traitement simultané de l'ensemble des données de l'inventaire transgabonais: le secteur d'Abamie [Experiment in simultaneous processing of different datasets of Trans-Gabon region: the Abamie sector]. BRGM 85 GAB 20.

Cottard, F. (1982). Mise en évidence par l'autocorrélation optique et l'interpretation sur images spatiales de guides structuraux ayant présidé à la répartition des gîtes d'étain et de tungstène dans le Nord-Ouest de la péninsule ibérique [Detection of structural guides controlling the distribution of tin and tungsten deposits in the north-western part of the Iberian peninsula using optical autocorrelation and interpretation of space images]. *Bull. Soc. Géol. Fr.*, XXIV (1): 117-126.

Coulon, M. and Frizon de Lamotte, D. (1988). Les craies éclatées du secteur d'Omey (Marne, France): le résultat d'une bréchification par fracture hydraulique en contexte extensif [The split chalks of the Omey sector (Marne, France): the result of brecciation by hydraulic fracturing in a tensile context]. *Bull. Soc. Géol.* Fr., IV (1): 177-185.

Critchley, M.F. and Phillips, W.E.A. (1984). The correlation of geological, geochemical and geophysical data with satellite imagery in North Central Ireland. Final Report EEC contract MSM 111 EIR (H), Bruxelles.

Crowley, J.K. (1986). Visible and near infrared spectra of carbonate rocks: reflectance variations related to petrographic textures and impurities. *JGR*, vol. 31, no. 85.

Crowley, J.K. and Vergo, N. (1987). Visible and near infrared (0.4 to 2.5 microns) reflectance spectra of selected mixed layer clays and related minerals. *Photogrammetric Engineering Remote Sensing*, vol. LIII, no. 6.

Daily, M.L. and Elachi, C. (1978). Application of multispectral radar and Landsat imagery to geologic mapping in Death Valley, JPL publ. 7819.

Darch, J.P. and Barber, J. (1983). Multitemporal remote sensing of a geobotanical anomaly. *Econ Geol.*, 78 (4): 770-782.

Daubree, A. (1979). Etudes synthétiques de géologie expérimentale [Integrated Studies of Experimental Geology]. Dunod Publishers, Paris. Librairie des Corps des Ponts et Chaussées, des Mines et des Télégraphes.

Daudon, Ph. (1985). La fracturation du socle en bas Limousin et Quercy, ses répercussions sur la couverture sédimentaire. Données de télédétection, en particulier infrarouge thermique [Basement fractures in lower limousin and Quercy and their repercussions on the sedimentary cover. Remote sensing data with special reference to thermal infra-red bands]. 3rd cycle thesis, Univ. Orléans.

Dawson, M.S., Olenera, J., Fung, A.K, and Manry, M.T. (1992). Inversion of surface parameters using fast learning neural network. *Proc. IGARSS* 92, vol. **1**, pp. 910-912.

Debeglia, N. and Gerard, A. (1973). Eléments interprétatifs de base obtenus a l'issue de lever aéromagnetique (Massif central 1972) [Basic interpretative elements obtained from aeromagnetic survey (Massif Central, 1972)]. BRGM Report, 21 pp.

Debeglia, N. and Gerard, A. (1975). Apport de l'aéromagnetisme a l'étude du Massif central [Role of aeromagnetic survey in the study of the Massif Central]. *Chron. des Mines*, sect. III, no. 3.

Debrand-Passard, S., Ogier, M. and Medioni, R. (1977). Un exemple de corrélation entre le données obtenues à partir des images ERTS, des études géophysiques et de la cartographie géologique [Example of correlation between data obtained from ERTS images, geophysical studies and geological mapping]. *Journ. de Télédétection,* GDTA, Sept. 77.

De Fraipont, P. (1982). Approche multiscalaire de la fracturation du Massif central. [Multiscalar approach to fractures in the Massif Central]. *Chorn. des Mines,*, sec. III, no. 3.

Delpont, G. and Dutartre, Ph. (1983). Etude du granite de Chemillé par télédétection [Study of Chemille granite by remote sensing]. BRGM Report 83 SGN 406 GEO.

Delpont, G. and Brojanigo, A. (1991). Spot, outil de prospection minière régionale appliqué à la recherche de bauxite en forêt tropicale dense: example de la région de Los Pijiguaos (Venezuela) [Spot-a regional mineral prospecting tool applied to the exploration of bauxite in dense tropical forests: example of the Los Pijiguaos region (Venezuela)]. *Chron. Rech. Minière,* Dec. 1991. BRGM publ., Orléans.

Demange, J., Haubensack, C. and Rancon, J.P. (1983). Reconnaissance géothermique de la République du Rwanda. [Geothermal exploration of the Republic of Rwanda]. BRGM Report 83 SGN 192 GTH.

Deroin, J.P. (1990). Imagerie satellitaire à haute résolution spatiale progrès méthodologiques et résultats géologiques: applications aux Cévennes et au Bas-Languedoc (France) [High spatial resolution satellite imagery methodological developments and geological results: applications in the Cévennes and Bas-Languedoc regions (France)]. Doct. thesis, Paris VI.

Deroin, J.P. and Deroin, Th. (1990). Effet d'une déficience en manganèse sur les feuilles de hêtre: conséquence sur la réponse spectrale [Effect of manganese deficiency on beech leaves: its influence on spectral response]. *C.R. Acad. Sci., Paris,* no. 311, pp. 863-871.

Deroin, J.P. and Rudant, J.P. (1991). Radar Gars 2000: sous-groupe géologie (site du Larzac). Application pour la cartographie des états de surface en relation avec la géologie [Radar Gars 2000: Geology subgroup (the Larzac site). Application for mapping the surface conditions associated with the geology]. BRGM Publ. Report 32504, 35 pp.

Deroin, J.P. and Rogov, A.N. (1994). Comparison of ALMAZ and ERS data. Geological and environmental results in the mountainous area of the Cévennes (France). *Int. Conf. Remote Sensing Methods in Geol. and Ecol. Res.* St. Petersburg, pp. 51-52.

Deroin, J.P. and Rouzeau, O. (1994, in press). A method for interpreting stereo-simulated ERS and Landsat-TM data. Geological application to the Velay area (France). *European Symp. Satellite Remote Sensing*. 26–30 Sept. 1994, EUROPTO, Rome.

Deroin J.P., Rudant J.P. and Wadsworth, A. (1991). Apport du radar imageur SAR au suivi de structures géologiques a terre et an mer. Example de l'Ile de Wight (Manche centrale) [Role of SAR imaging radar in tracing geological structures on land and in sea. Example of wight Island (Central Manche)]. *C. R. Acad. Sci., Paris,* ser. II, vol. 312, pp. 1099-1106.

Deroin, J.P., Cautru, J.P. and Rudant, J.P. (1994). The SAREX-92 experiment in French Guyana: lithological and geomorphological aspects, comparison with ERS-1 satellite data. European Space Agency, Spec. Paper 76, pp. 205-214.

Deroin, J.P., Braux, C., Cautru, J.P., Rudant, J.P. and Toux, L. (1993). The ERSs SAR: an aid for geological mapping and mineral exploration in rain-forest areas. ESA PFF 11 Pilot Project French Guyana. Second ERS-1 Symp. Hambourg, 11-14 October, 1993. European Space Agency, Spec. Paper 361, pp. 897-902.

Deroin, J.P., Chorowicz, J., Deroin, T., Dutartre, Ph., Scanvic, J.Y. and Simonin, A. (1989). The Landscape unit approach applied to geological mapping, on Spot image, in temperate climatic zone. EUG, Strasbourg. *Photogrammetria,* no. 45, pp. 382-403.

Deroin, J.P. and Scanvic, J.Y. (1995) Apport de l'imagerie radar á Pe cartographie geologique: exemples et reflèxions [Role of radar imagery in geological mapping: examples and reflections]. Bul SFPT No. 138 1995/2 Saint Mandé France.

Derre, C. and Roger, G. (1984). Filons à étain-tungstène dans la moitie nord du Portugal. Guides de récherche [Tin and tungsten veins in the northern half of Portugal. Exploration guides]. *Chron. Rech. Minière,* no. 475, June.

Deslandes, S. and Gwyn, Q.H.J. (1991). Evaluation de Spot et Seasat pour la cartographie des linéaments: comparaison basée sur l'analyse de spectres de Fourier [Evaluation of Spot and Seasat for lineament mapping: comparison based on Fourier spectral analysis]. *Canadian Remote Sensing Journal,* April 1991, Ottawa, Canada.

Dewolf, Y. (1987). Interprétation et perception de la morphologie du Bassin de Paris [Interpretation and perception of the morphology of the Paris basin]. *Bull. Inf. Géol. Bass. Paris Mém.* hors ser. Fr. A 1987, pp. 19-38.

Disnar, J.A., Gauthier, B. and Carter, J. (1986). Utilisation des hydrocarbures gazeux en prospection régionale des minéralisations plombozincifères de couverture: application au gîte de Treves (France) [Use of gaseous hydrocarbons in regional prospecting for concealed lead-zinc deposits: application to the Treves deposit (France)]. *Chron. Rech. Minière,* no. 482, pp. 67-76, March.

Dolivo-Dobrovolsky, A.V. and Strelnikov, S.I. (1978). Principal types of the largest linear and ring structures of the Earth's continental crust. *Proc. Int. Conf. Basement Tectonics, USA.*

Dottin, O. (1980). Notice de la carte géologique de la France et de sa marge continentale à l'échelle du 1/1 500 000 [Notes to the geological map of France and its continental margin in the scale of 1:1,500,000]. BRGM publ., Orléans, France.

Duplan, L. (1973). Tectonique - les grands linéaments de la France [Tectonics-the major lineaments of France]. *C.R. Acad. Sci., Paris,* 276 (D): 2947-2949.

Durand, M.A. (1986). Caractérisation morphologique du contexte dans les images de télédétection-utilisation des connaissances externes pour la reconstruction de structures [Morphological characterisation of features in remote-sensing images: use of external sources for reconstructing the structures]. Thesis, P. Sabatier Univ., Toulouse, 189 pp.

Durandeau, A., Monget, J.M. and Rousselin, T. (1984). Computer processing of satellite and airborne thermal data for mapping Hercynian granites in relation with uranium deposits. 27th Int. Geol. Cong., Moscow, 1984 and *Geochr.,* suppl. no. 10.

Dutartre, Ph. (1984). Analyse numérique de carte de fracturation-programme LINEAME

[Numerical analysis of the fracture map the programme LINEAME]. BRGM Report 84 SGN 3921 GEO, 31 pp.

Dutartre, Ph. (1985). Potentiel d'applications des données aérospatiales en relation avec l'acquis méthodologique et schématique de l'équipe télédétection [Scope of application of aerospace data in relation to the methodological and schematic acquisition of remote-sensing equipment]. BRGM Report 85 SGN 553 GEO.

Dutartre, Ph. and Weecksteen, G. (1984). Recherche méthodologique sur la contribution de la télédétection à la connaissance structurale d'un fossé sédimentaire. Exemple de la Bresse (France) [Methodological investigation of the contribution of remote sensing to the structural information of a sedimentary basin. Example of Bresse (France)]. *Chron. Rech. Min.*, no. 477, pp. 29-34.

Dutartre, Ph. and King, C. (1986). Structural and geobotanical contribution of remote sensing to exploration of ore deposits in Rhodope (Greece). CEE MSM Report 037F-BRGM 86 GRC 099 GEO, 85 pp.

Dutartre, Ph., Castaing, G., Chevremont, Ph., Martin, P. and Scanvic, J.Y. (1983). Méthodologie de la recherche par télédétection des coupoles non affleurantes [Methodology of exploration of concealed domes by remote sensing]. BRGM Report 83 SNG GEO 132.

Duvigneaud, P. (1958). La végétation du Katanga et des sols métalliques [Vegetation of Katanga and metal-bearing soils]. *Soc. Roy. Belg. Bull.*, no. 90, pp. 127-286. *Economic Geology* (June-July 1983), 78 (4): 573-796.

Elachi, C., Roth, L.E. and Schaber, G.G. (1984). Spaceborne radar subsurface imaging in hyperarid regions. *IEEE Trans. Geoscience and Remote Sensing,* GE-22, no. 4, pp. 383-388.

Elachi, C., Blom, R., Daily, M., Farr, T. and Saunders, R.S. (1980). Radar imaging of volcanic field and sand dune fields. Publ. 80 61 in report of radar geology workshop, Snowmass, Colorado, 1979.

ERIM Symposium (1989). Michigan University (USA) publ.

Fauconnier, D. (1988). Télédétection et forêts. Cartographie du couvert végétal dans les forêts domaniales de Vouzeron Allogny et Saint-Palais sur la feuille Saint-Martin d'Auxigny a 1/50 000 [Remote sensing and forests: Mapping the vegetation cover in the provincial forests of Vouzeron Allogny and Saint-Martin d'Auxigny]. BRGM Report 88 SGN GEO 161.

Favard, J.C., Friedman, A.L., Thyrested, T. and Scanvic, J.Y. (1981). Airborne remote sensing in East Greenland. ERIM Symposium 1981.

Feybesse, J.L. (1981). Tectonique et microtectonique de la région de Laroquebrou (Massif central français). Rôle de la déformation ductile et évolution du Sillon houiller [Tectonics and microtectonics of the Laroquebrou (Massif Central, France). Role of ductile deformation and evolution of the Coal belt]. Thesis, Clermont-Ferrand.

Flouzat, G. (1988). Morphologie mathématique et analyse des images de télédétection [Mathematical morphology and analysis of remote-sensing images]. *Rev. Photointerprétation,* no. 1988-5, fasc. 1. Technip Publ., Paris.

Flouzat, G. and Merchoub, Y. (1983). Modelisation de l'extraction manuelle d'éléments texturaux par la morphologie mathématique [Modelling the manual extraction of textural elements by mathematical morphology]. *Rev. Photointerprétation,* 6 (6): 59-64.

Fourniguet, J. and Motti, E. (1983). Morpholithologic survey applied to strategic geochemical prospecting in Saudi Arabia through remote sensing method. BRGM Report 84 SGN 134 GEO.

Fourniguet, J., Motti, E. and Pointet, H. (1984). Application de la télédétection à la prospection hydrogéologique en zones arides [Application of remote sensing for hydrological prospecting in arid zones]. Challenges in African Hydrology and Water Resources. Harare Symposium.

Francis, P. (1983). Les calderas volcaniques géantes [Large volcanic calderas]. *Rev. Science,* no. 70, August 1983 (French edition of *Scientific American*).

Froidevaux, C.M. (1980). Radar, an optimum remote sensing tool for detailed plate tectonics, analysis and its application to hydrocarbon exploration. JPL publ. 80 61, Radar Geology Workshop, Snowmass, Colorado, July 1979.

Gaddis, L., Mouginis-Mark, P., Singer, R. and Kaupp, V. (1989). Geologic analyses of Shuttle imaging radar (SIR-B) data of Kilauea Volcano, Hawaii. *Geol. Soc. Amer. Bull.,* no. 101, pp. 317-332.

Galley, J. (1986) Spectral reflectance of carbonate minerals in the visible and near infrared. Calcite aragonite and dolomite. *Amer. Mineralogist,* no. 71.

Gess, G (1983). La télédétection et l'exploration pétrolière [Remote sensing and petroleum exploration]. *Total Information,* no. 96.

Girard, M.C. and Girard, C.N. (1977). Télédétection à la surface des sols [Remote sensing at the surface of soils]. *2nd Coll. Red Sol et Télédétection,* Rome, pp. 55-64.

Gladwell, D.R. (1984). Application of reflectance spectrometry to mineral exploration using portable radiometers. *Econ. Geol.,* vol. 78, no. 4.

Gorobets, B. S., Gaft, M.L. and Podolski, A.M. (1989). *Luminescence of Minerals and Ores Handbook.* Inst. Advanced Studies, Ministry Geol., USSR, Moscow (in Russian).

Greffard, J., Sarcia, C. and Bourg, A. (1985). Etude de la contamination des sols sous les champs d'épandage d'Achères (France)]. [Study of soil contamination under the debris fields of Achères (France)]. *Hydrogeologie,* no. 1, pp. 55-84.

Grolier, J. (1981). Contribution de l'analyse structurale à l'interprétation génétique des bassins houillers: quelques exemples du Massif central, France [Contribution of structural analysis to the genetic interpretation of coal basins: some examples of Massif Central, France]. *Soc. Nat. Elf Aquitaine,* pp. 601-620.

Grubbs, D., Henderson, F. and Penfield, G.T. (1984). Bauxite exploration by satellite. *Proc. 1984 bauxite Symp.,* Los Angeles, pp. 200-242.

Guigues, J. (1984) Prospection stratégique et métallógenique: caractérisation d'auréoles autour des granites hercyniens dans l'Est du Massif armoricain (France) [Strategic and metallogenic prospecting: Characterisation of haloes around the Hercynian granites east of the Armorican Massif (France)]. *Chron. Rech. Min.,* no. 475, June.

Guillemot, J. (1962). Structures circulaires en milieu désertique [Circular structures in desert terrains]. *Rev. Photointerprétation,* no. 4, fasc. 1, 2, 3. Technip Publ., Paris.

Guiness, E.A. (1983). Digital image processing applied to analysis of geophysical and geochemical data for southern Missouri. *Econ. Géol.,* vol. 78, no. 4.

Harris, J.R. (1991). Mapping of regional structure of eastern Nova Scotia (Canada) using remotely sensed imagery: implications for regional tectonics and gold exploration. *Canad. J. Remote Sensing,* 17 (2): 122-135.

Hauff, P.L., Kruse, F.A. and Thiry, M. (1990). Characterization of interstratified daolinite/smectite clays using infrared reflectance spectroscopy (1.2-2.5 μm): Extended Abstract, 2nd Int. Symp. Geochemistry of the Earth's Surface and Mineral Formation, July 2-8, Aix-en-Provence, France. *Chem. Geol.,* vol. 84, pp. 267-270.

Hawkes, H.E. (1957). Principles of geochemical prospecting. *U.S. Geol. Surv. Bull.,* no. 1000-F, pp. 225-252.

Hobbs, W.H. (1904). Lineaments of the Atlantic border region. *Bull. Soc. Geol. Amer.,* vol. 15, pp. 483-500.

Hopkins, W. (1841). On the geological structure of the Weaden district and the bas Boullonais. *Trans. Geol. Soc. London* (2nd ser.), vol. VII, pp. 1-51.

Hornler, D.N.H., Barber, J. and Barringer, A.R. (1980). Effects of heavy-metal on the absorbence and reflectance spectra of plants. *Int. J. Remote Sensing,* 1 (2): 121-136.

Howard, J.A., Watson, R.D. and Hessin, T.D. (1971). Spectral reflectance properties of *Pinus ponderosa* in relation to copper content of the soil-malachite mine, Jefferson County, Colorado. *Proc. 7th Int. Symp. Remote Sensing of Environment,* Ann Arbor, Michigan, pp. 285-296.

Humbert, M., Masure, Ph. and Scanvic, J.Y. (1989). Mouvements de terrain. Aménagement et imagerie Spot [Ground movements: Management and Spot imagery]. *Bull. Soc. Fr. Photogrammètrie et Télédétection,* no. 114.

Hundington, J.F., Gabell, R. and Green, A.A. (1983). The role of Landsat imagery in exploration for some large area mineral deposits. CSRIO, North Fydi, Australia (unpubl.).

Hunt, E.B. and Salisbury, J.W. (1971). Visible and near infrared spectra of minerals and rocks. *Modern Geol.,* vol. 1-2.

Hunt, G. and Ashley, P. (1979). Spectra of altered rocks in the visible and near infrared. *Econ. Geol.,* vol. 74, pp. 1613-1629.

Huppé, P. (1958). Aperçu sur le reseau rhegmatique de la croûte terrestre [Overview of the rhegmatic network of the Earth's crust]. *Bull. Trim. du Serv. d'Inform. Géol.,* no. 40, pp. 2-15.

Jebrak, M. and Dutartre, Ph. (1984). Apport de l'imagerie spatiale à la synthèse gitologique du district plombifère de la haute Moulouya (Maroc) [Contribution of space imagery to metallogenic synthesis of the lead district of upper Moulouya (Morocco)]. *Chron. Rech. Min.,* no. 477, pp. 29-34.

Jourde, G. (1978). Etude et prospection minière du sud-ouest Cameroun, phase préliminaire [Study and mineral prospecting of South-west Cameroon, preliminary phase]. BRGM Report 78 RDM 059 AF.

Kayle, A. (1987). Surface emittance temperature and thermal inertia derived from thermal infrared. Multispecial scanner data from Death Valley (USA). *Geophysics,* pp. 52-88.

King, C. (1985). Contribution de la télédétection à la connaissance des sols et des formations superficielles; l'évaluation de l'apport des données de simulation Spot [Contribution of remote sensing to the investigation of soils and surficial deposits: evaluation of the role of simulated data of Spot]. Doct. thesis in engineering, INA Mém.

King, C. (1985). Etude de sols et des formations superficielles par télédétection. Approche de leurs caractéristiques spectrales, spatiales et temporelles dans le visible et le proche infrarouge [Study of soils and surficial deposits by remote sensing: Approach of their spectral, spatial and temporal characteristics in the visible and near infra-red bands]. BRGM doc. no. 96, 200 pp.

Kosakevitch, A. (1979). Chapeau de fer: problèmes de définition et de nomenclature pratique [Iron hats: problems of definition and practical nomenclature]. *BRGM Bull.,* sect. II, no. 213, pp. 141-149.

Kravtzov, A.J. and Fridman, A.J. (1965). Les gaz naturels des gisements metalliferes [Natural gas of metallic deposits]. *Dokl. Akad. Nauk USSR,* 165 (5): 1168-1170.

Kutina, J., Pokorny, J. and Vesela, M. (1967). Empirical prospecting net based on the regularity distribution of ore veins with application of the Jihlava mining district, Czechoslovakia. *Econ. Geol.,* vol. 62, pp. 390-405.

Labourguigne, J., Megnien, C., Scanvic, J.Y. and Weecksteen, G. (1978). A propos de quelques accidents tectoniques de la craie en Champagne (France) [Remarks on some tectonic faults in the chalks of Champagne region (France)]. *BRGM Bull.,* sect. 1, no. 1, pp. 57-64.

Labovitz, M.L., Masuoka, E.J., Bell, R., Siegrist, A.W. and Nelson, R.F. (1983). The application of remote sensing to geobotanical exploration for metal sulphides. Results from the 1980 field season at Mineral, Virginia. *Econ. Geol.*, 78 (4): 750-760.

Lajoinie, P. and Weecksteen, G. (1985). Télédétection et recherche minière-Simulation Spot en Nouvelle Calédonie-application au nickel oxydé [Remote sensing and mineral exploration: Spot simulation in New Caledonia-application to nickel oxide]. BRGM Report 85 NCL 049 GEO.

Lallier, S. (1977). Etude géologique at géophysique au Sud-Est du massif granitique d'Echassières (Allier), France [Geological and geophysical study south-east of the Echassiéres Massif (Allier), France]. Thesis, Univ. Orléans.

Lambert, P. (1974). La structure impactitique de Rochechouart (Limousin, France) et son contexte structural régional par interprétation des photos-satellites ERTS [The impact structure of Rochechouart (Limousin, France) and its regional geological set-up by the interpretation of ERTS satellite photos]. *BRGM Bull.*, Fr., sect. 1, no. 3, pp. 177-188.

Lavreau, J. (1989) Exploration de l'espace spectral d'une scène Thematic Mapper au Burundi (région Musongati) [Exploration of the spectral space of a Thematic Mapper scene in Burundi (Musongati region)]. *Journ. Télédétection Mus. Roy. Afrique Cent., Belg. Annales*, ser. no. 8, *Sci. Géol.*

Lavreau, J., Theumissen, K. and Trefois, Ph. (1991). Identification per télédétection de métallotectes lithologiques et structuraux dans le sud de Burundi [Remote sensing identification of lithological and structural metallotects in southern Burundi]. *Chron. Rech. Min.*, Orléans, France.

Lefevre, M.J. (1979). Détection aéroportée de phénomènes géologiques de sub-surface révélés par une anomalie de température de brillance (Coat en Noz, Côtes-du-Nord) France [Aerial detection of subsurface geological phenomena revealed by temperature and brightness anomalies (Coat en Noz, Côtes-du-Nord) France]. *Rev. Photointer.*, no. 6, fasc. 3a.

Lefevre, M.J. (1980). Télédétection d'anomalies géochimiques a Echassiéres (Allier, France). [Remote sensing of geochemical anomalies in Echassiéres (Allier, France)]. *Chron. Rech. Min.*, no. 453, pp. 41-64.

Lefevre, M.J. (1982). Tél'edétection d'anomalies géobtaniques appliquée à la recherche miniére par la méthode de traitement d'images à Echassiéres (Allier, France) [Remote sensing of geobotanical anomalies applied to mineral exploration by the image-processing method in Echassiéres (Allier, France)]. *Bull. Soc. Geol.*, Fr., XXIV (1): 127-130.

Lefevre, M.J. (1983). Tél'edétection d'anomalies géobtaniques appliquées à la recherche miniére [Remote sensing of geobotanical anomalies applied to mineral exploration]. BRGM, GTDA, state thesis, Univ. P. Sabatier, Toulouse.

Lehmann, J.P. and Scanvic, J.Y. (1969). Etude géologique du permis Zemoul (Algérie) [Geological study of Zemoul permit area (Algeria)]. Rept. réf. 30 052, Bureau Etudes Ind. et de Coop. de l'Inst. Franç. du Pétrole. Paris-Rueil.

Letouzey, P. (1985). Interprétation géologique du bassin de la basse Magdalena Sierra de San Jacinto (Colombie) [Geological interpretation of the lower Magdalena Sierra de San Jacinto basin (Colombia)]. *Symb. sur la Gĕodynamique des Caraibes*. Technip Publ., Paris.

Leymarie, P., Baelz-Maniére, S., Durandeau, A. and Monget, J.M. (1983). Un systeme multivarié d'aide à la decision en prospection de l'uranium [A multivariate system as an aid to decision-making in uranium prospecting]. *Symp. AEN/AIEA. Méthodes de Prospection de l'Uranium.* OCDE Publ., Paris, pp. 889-901.

Leymarie, P., Pison, N. Renouard, L., Clément, E., Dardel, J. and Ballery, J.-L. (1991). Nouvelles techniques d'interprétation des images Spot. Application à la prospection de l'uranium: exemple du district de Grants (Nouveau-Mexique, USA) [New technics of Spot image interpretation. Application to uranium prospecting: example of Grants district (New Mexico, USA)]. *Chron. Rech. Min.*, no. 505. BRGM Publ., Orléans.

L'Homer, A. (1986). Poster baie du Mont-Saint-Michel [Bay poster of Mont-Saint-Michel]. (for sale by BRGM Publ.).

Lointier, M., Rudant, J.P., Deroin, J.P., Prost, M.T., Charron, Ch., Deffontaines, B., Raymond, D. and Cautru, J.P. (1993). Apport des données SAR-ERS-1 a l'étude de l'environment côtier tropical guyenais [Role of SAR-ERS-1 data in the study of coastal environment of tropical Guyana]. *De l'optique au radar, Paris. Coll. CNES-ESA.* Cépaduhès Publ., pp. 59-69.

Longshaw, T.G. (1976). Application of an analytical approach to field spectroscopy in geological remote sensing. *Modern Geology,* vol. 5.

Lopes, A., Touzi, R. and Nezry, E. (1990). Adaptative speckle filters and scene heterogeneity. *IEEE Trans. Geoscience and Remote Sensing,* 28 (6): 992-1000.

Louhala, S. (1988). Signatures spectrales de roches en milieu tempéré: valeurs réelles et valeurs perçues par le satellite [Spectral signatures of rocks in temperate regions: true values and values obtained from satellite data]. Thesis, Univ. Paris VII. BRGM Doc. no. 168, 205 pp.

Lowell, J.D. and Guilbert, J.M. (1970). Lateral and vertical alteration mineralization zoning in porphyry ore deposits. *Econ. Geol.*, no. 65, pp. 373-408.

Lowman, P.O. (1991). Original shape of the Sudhury structure, Canada: a study with airborne imaging radar. *Can. J. Remote Sensing,* 17 (2): 152-184.

Lyon, R.J.P. (1975). Mineral exploration applications of digitally processed Landsat imagery. U.S. Geol. Surv. Prof. paper no. 1015, pp. 271-292.

Malon, J.F. (1984). Remote sensing processing application: Reunion Island. 18th symp. ERIM, Oct. 1984, Paris. BRGM Report 84 SGN 396 GEO.

Malon, J.F. (1984). Principaux traitements des images satellitaires et associés [Principal and associated technics of satellite image processing]. BRGM Report 84 SGN 396 GEO.

Marconnet, B. (1985). La télédétection spatiale, une méthode pour la prospection minière stratégique et tactique [Space remote sensing, a method for strategic and tactical mineral exploration]. Thesis, Univ. Nancy 1, 244 pp.

Marconnet, B., Gagny, C., Leymarie, P. and Monget, J.M. (1982). Phénomène de transparence d'un leucogranite à étain-tungstène sour couverture métamorphique (Province de Zamora, Espagne). Confrontation entre traitement des signatures spectrales et réalité terrain [The phenomenon of transparency of a leucogranite with tin-tungsten mineralisation below a metamorphic cover (Zamora province, Spain). Comparison between spectral signatures and ground truth]. Soc. Int. de Photogrammétrie et Télédétection. Symp. Int., Toulouse, Sept. 1982, pp. 2-9.

Marcoux, E. (1980). Le district de Pontivy: sa place dans la métallogénie plombo-zincifère du Massif armoricain (France) [Pontivy district: its place in the lead-zinc metallogeny of the Armorican Massif (France)]. Doct. thesis, 3rd cycle, Univ. Clermont-Ferrand, 252 pp.

Marek, K.H. and Schmidt, K. (1994). Preliminary results of the comparative analysis of ERS-1 and ALMAZ-1 SAR data. *ISPRS J. Photogrammetry and Remote Sensing,* 49 (3): 12-18.

Maroncle, J.L. (1985). Synthèse géologique et métallogénique de la région du faisceau de Pontgibaud et de ses marges (Massif central français) [Geological and metallogenic synthesis of Pontgibaud swarm region and its surroundings (Massif Central, France)]. BRGM Report DL 349, Clermont.

Maroncle, J.L., Le Chapelain, J.R. and Perichaud, J.J. (1989). Geological, geochemical and metallogenetic evaluation of the Pontgibaud swarm and its surrounding area. Massif

Central (France). *J. Geochem. Explor.*, 32, 1989, pp. 411-412. Observer Sciences Publ., Amsterdam.

Martin-Kaye, P.H.A. (1984). The U.K. Pennine project. The computer correlation of exploration data including remotely sensed data. Sem. Remote Sensing Geol. Mapping, Feb. 1984 and Action concertée CEE.

Masson, Th. (1986). Le volcanisme sur Mars [Volcanism on Mars]. *Rev. Photointer.*, 86/4. Technip Publ., Paris, pp. 7-17.

Massonnet, D. (1992). Perspective de l'interferometrie par Radar [Perspectives of interferometry by Radar]. SFPT Meeting, Grenoble, December 1992, Télédétection en Montagno. Bull. SFPT, no. 127, 1992-93.

Massonnet, D. and Raboute, T. (1993). Radar interferometry: limits and potential. *IEEE Trans. Geoscience and Remote Sensing,* vol. **31**, no. 2, March 1993.

Massonnet, D., Rossi, M., Carmona, C., Adragna, F., Peltzer, G., Feigl, K. and Raboute, T. (1993). The displacement field of the Lander earthquake mapped by rader interferometry. *Nature,* vol. **364**, pp. 138-142.

Mayo, E.B. (1958). Lineament tectonics and some ore distinct in the Southwest Mining. *Engineering*, Nov., pp. 1169-1175.

Mayo, E.B. and Milton, N.M. (1983). Use of reflectance spectra of native plant species for interpreting airborne multispectral scanner data in the east Tintic Mountains, Utah. *Econ. Geol.*, **78** (4).

Meunier, A., Parneix, J.C. and Beaufort, D. (1984). Les altérations hydrothermales des massifs de granite [Hydrothermal alterations of granite massifs]. *Journ. Granite, Orléans.* BRGM doc. no. 64, pp. 133-141.

Milton, N.M. (1984). Geobotanical techniques for lithologic discrimination and mineral exploration. Sem. Remote Sensing for Geol. Mapping, Feb. 1984.

Milton, N.M., Collins, W., Chang, S.H. and Schmidt, R.G. (1983). Remote detection of metal anomalies on Pilot Mountain, Randolph County, North Carolina. *Econ. Geol.*, 78 (4): 605-617.

Minatome, S.A. (1982). Prospection for uranium mineralization in favourable areas of the Kingdom. Remote sensing studies report. Report MSA CR 03, pp. 1-110.

Minoux, L. (1986). Marais breton et télédétection [Breton swamps and remote sensing]. BRGM Report 86 SGN 723 GEO, 32 pp.

Moon, W.M. (1991). Born inversion of surface scattered SAR wave field. *Can. J. Physics,* vol. **69**, pp. 1256-1260.

Moore, D.B. (1971). Subsurface geology of Southwest Alabama. *Alabama Geol. Survey Bull.*, vol. **99**, 79 pp.

Morris, R.V., Neely, S.C. and Mendell, W.W. (1982). Application of Kubelka-Munk theory of diffuse reflectance to geologic problems: the role of scattering. *Geol. Res. Letters,* vol. **9**, no. 2.

Motti, E., Vasquez Lopez, R. and Bigot, M. (1980). Les minéralisations Zn-Pb-Ba et Cu de la marge du rift de la mer entre Yambu Al Bahr et le golfe d'Aquaba (Arabie Saoudite) [Zn-Pb-Ba and Cu mineralisations of the rift margin of the sea between Yambu Al Bahr and the Gulf of Aqaba (Saudi Arabia)]. *BRGM Bull.* (2) II, 1-2, 1980-1981, pp. 113-134.

NASA (1989). SAR Synthetic Aperture Radar, Earth Observing System, Instrument Panel. NASA Report, vol. IIf, Wash.

NASA J.P.L. (1933). Imaging spectrometry: the next step in remote sensing.

Nezry, E., Beaudoin, A., Lopes, A., Rudant, J.P. and Deroin, J.P. (1993). Preprocessing of multifrequency SAR images: application to geological study of karstic formations (Le Larzac, France). *Proc. IGARSS* 93, *Tokyo,* vol. 4, pp. 2135- 2137.

Nguyen, P.T., Simon, S. and Asfar, L. (1984). An application of pattern recognition for structural analysis in geology. Int. Conf. on Pattern Recognition, Montreal, July 1984.

Nicolini, P. (1980). Méthodes de la Gitologie [Methods of Metallogeny]. M. Deschamps Publ., Haiti.

Norman, J.W. (1977). The World is a bit cracked. *New Scientist*, Feb., pp. 320-322.

Olhoeff, G.R. (1986). Electrical conductivity. In: Proc. 8th Workshop on Electromagnetic Induction in the Earth and Moon. Rev. paper Intern. Assoc. Geomag. and Aeronomy, Neuchatel.

Pacheco, A. and Puigarnau, P. (1986). Mapping contact metamorphic rocks in Extramadura Spain using TM data. 2nd European Workshop on Remote Sensing in Mineral Exploration. EEC Brussels. Eosat *Landsat Application Notes*, vol. 3, May 1988.

Paul, S. (1988). Glossaire des termes officiels de la télédétection aérospatiale [Glossary of official terms of aerospace remote sensing]. *SFPT Bull.*, no. 112, p. 4.

Pavillon, M.J. and Recoing, M. (1977). Notice explicative de la carte geologique a 1/50 000 de Bugeat (France) [Explanatory note to the 1:50 000 geological map of Bugeat (France)]. BRGM Publ., Orléans, France.

Pélissonnier, H. (1974). Vers une métallogenie quantitative [Towards quantitative metallogeny]. *Annales des Mines*, March, pp. 47-58.

Peltzer, G., Armijo, R. and Tapponnier, P. (1987). Estimation des vitesses de déplacement sur la faille de l'Altin Dagh (Nord Tibet Chine). Spot 1: utilisation des images, bilan resultats [Estimation of displacement velocity on the Altin Dagh fault (northern Tibet, China). Spot 1: Utilisation of images and net results]. *CNES*, Cepadues Publ., Toulouse.

Peragallo, J. (1978). L'emploi du radar en geophysique [Application of radar in geophysics]. BRGM Report 78 GPN 030.

Perrodon, A. (1985). Histoire des grandes découvertes Pétroliéres [History of Major Petroleum Discoveries]. Copublication Elf-Aquitaine and Masson, 224 pp.

Peters, C.C. (1983). Use of airborne multispectral scanner data to map alteration related to roll-front uranium migration. *Econ. Geol.*, vol. 78, no. 4.

Pinna, M. (1983). Inventaire de l'extrême Nord de la République Populaire du Mozambique. Reconnaissance géologique. Rapport final [Inventory of the northernmost part of the People's Republic of Mozambique. Geological explorations. Final report]. BRGM Report 83 MOZ 002.

Pinna, P., Motti, E., Dutartre, Ph., Lafage, M., Scanvic, J.Y. and Jourde, G. (1988). New geological data from Mozambique: the role of remote sensing in large-scale mapping, structural interpretation and prospective metallogeny over 120,000 km^2 of polyphase basement. In: 2nd European Workshop on Remote Sensing in Mineral Exploration. EUR 11317 report, pp. 58-69.

Podwysocki, M.H. (1984). Use of multispectral scanner images for assessment of hydrothermal alteration in the Marysvale, Utah, mining area. *Econ. Geol.*, vol. 78, no. 4.

Press, N.P. and Norman, J.W. (1972). Detection of ore bodies by remote sensing: the effects of metals on vegetation. *Inst. Mining and Metal., Trans.*, B, vol. 78, pp. B166- B168.

Pretorius, J.P.G. and Portridge, T.C. (1974). The analysis of angular atypicality of lineaments as an aid to mineral exploration. *Journ. South African Inst. Mining and Metallurgy*, May, pp. 367-369.

Prian, J.P. (1991). Synthese géologique et géochimique, potentialités minières, de l'Archéen et du Protérozoique du degré carré Mouila (Gabon), avec carte géologique au 1/200 000. Synthèse du district aurifere d'Etéké [Geological and geochemical synthesis, and mineral potential of the Archaean and Proterozoic in the 1° × 1° map of Mouila (Gabon) with the

1:200,000 geological map. Integrated map of the Eteke gold district. *Direction des Mines du Gabon et BRGM.*

Prian, J.P., Simeon G., Johan, V., Ledru, P., Piantone, P., Coste, B. and Eko N'Dong, J. (1988). Valorisation géologique de l'inventaire minier de l'Archéen et du Protérozoique inferieur des feuilles Mitzic Boué et Mouila à 1/200 000 (Gabon central) [Geological evaluation of the mineral inventory of Archaean and Lower Proterozoic deposits in the 1:200,000 map of Mitzic Boué and Mouila (central Gabon)]. *Chron. Rech. Min.*, vol. 491, pp. 67-104.

Prost, Gary. (1994). *Remote Sensing for Geologists, a Guide for Image Interpretation.* Gordon and Breach Science Publ., Lausanne, Switzerland, 325 pp.

Rabu, Y. (1985). Traitement d'images multisources. Application à la géologie provençale [Multisource image processing. Application to regional geology]. Thesis, Paris Orsay, 142 pp.

Rebillard, Ph. (1980). Méthodologie du radar latéral aéroporté (bande L) sur les Alpes françaises méridionales [Methodology of airborne lateral radar (band L) over the southern Alps of France]. Thèse Doct. Spec., Univ. Grenoble.

Rebillard, P. and Evens, D. (1983). *Geophysical Research Letters,* vol. 10, no. 4.

Rheault, Michel et al. (1991). *Can. J. Remote Sensing,* 17 (2): 185-190.

Rickard, T.A. (1926). *Eng. Mining J. Press,* 78 (4): 605-617.

Robert, D. (1968). Geologie linéamentaire et richesses minérales [Lineament geology and mineral wealth]. *Rev. Ind. Minérale,* 50 (10): 743-757.

Rock, B. (1984). Remote detection of geobotanical anomalies associated with hydrocarbon microseepage, using TM simulator and imaging spectrometer data. Seminar on Remote Sensing for Geological Mapping, Orléans, France, Feb. 1984.

Rondot, J. (1983). L'astrobleme de Charlevoix-cicatrisation apres l'impact d'une grosse meteorite [Charlevoix astroblem-cicatrisation after the impact of a large meteorite]. *Ann. Sci. Univ. Clermont-Ferrand II* (France), 75 (34): 169.

Roquin, C. (1979). Contribution à l'étude de la structure des données en prospection géochimique [Contribution to the study of data structure in geochemical prospecting]. 3rd cycle thesis, Orléans.

Roquin, C. (1984). Inventaire géochimique de la feuille Nort-sur-Erdre (France) [Geochemical inventory of the Nort-sur-Erdre region (France)]. BRGM Report, Orléans.

Roquin, C., Pion, J.C. and Dandjihou, T. (1987). Spot image and geochemical mapping of lateritic cover in southern Mali. Spot 1: utilisation of images, final results. CNES. Cepadues Publ., Toulouse, pp. 871-879.

Rousselin, T. (1985). Télédétection thermique des granites uranifères du Nord Limousin [Remote sensing of uraniferous granites of northern Limousin]. 3rd cycle thesis, Ecole Nationale Supérieure des Mines de Paris, 81 pp.

Rousselin, T., Bourbon, M. and Monget, J.M. (1984). Thermographie et méthodes multivariées de cartographie de l'apex d'Echassières: aide à la localisation du sondage profond [Thermography and multivariate methods for mapping the Echassières dome: aid for siting a deep borehole]. Ecole Nationale Supérieure des Mines, BRGM convention, Armine, Feb. 1984.

Routhier, P. (1980). Où sont les métaux de l'avenir? [Where are the metals of the future?]. *Mém. BRGM,* no. 105.

Rouzeau, O. (1985). Apport des données satellitaires thermique. Bassin de Paris [Role of satellite thermal data. Paris basin]. DEA Univ. d'Orléans.

Rouzeau, O. (1988). Contribution de la télédétection spatiale à la connaissance de l'anomalie thermique du forage de Condé-sur-Escaut, France [Contribution of spatial remote sensing to investigation of the thermal anomaly of Condé-sur-Escaut, France]. BRGM Report 88 DT 037 ED.

Rouzeau, O. (1989). Problématique de transposition des modèles de distribution de température dans le sol aux interprétations issues de la télédétection thermique spatiale. Application au Bassin de Paris [Problem of transposition of distribution models of ground temperatures to the interpretation of thermal remote-sensing data. Application to the Paris basin]. Thesis memoir, Univ. d'Orléans, 2 vols, pp. 1-162.

Rouzeau, O. and Scanvic, J.Y. (1989). Interprétations structurales de linéaments observés sur des données spatiales thermiques multidates et multicapteurs dans le Bassin de Paris [Structural interpretation of lineaments observed in multidate and multisensor thermal remote-sensing data of the Paris basin]. *SFPT Bull.*, no. 113, pp. 8-12.

Rowan, L., Goetz, A. and Ashley, R.P. (1977). Discrimination of hydrothermally altered and unaltered rocks in visible and near infrared multispectral images. *Geophysics,* vol. 42, pp. 522-535.

Roy, D.W. and Boily, J. (1993). Délimitation de domaines structuraux par classification non-supervisée de cartes de linéaments [Demarcation of structural domains by unsupervised classification of lineament maps]. *16th Symp. Can. on Remote Sensing, 8th Cong. Quebec Assoc. Remote Sensing, AQT/CRSS.*

Royer, J.J. and Laurent, J.L. (1982). Imagerie spatiale et cartographie géochimique. Etude des corrélations entre géochimie en roches et télédétection. Application au massif de la Marche orientale [Space imagery and geochemical mapping: Study of the correlations between rock geochemistry and remote sensing. Application to the East Marche massif]. ATP CNRS Report 80/CNES/279.

Rudant, J.P. and Cervelle, B. (1988). Evaluation des données Varan en géologie et en géomorphologie sur le Sud-Est de la France [Evaluation of Varan data in geology and geomorphology in south-eastern France]. Proc. 4th Int. Coll. on Spectral Signatures, Aussois (France), Jan. 1988. ESA SP 287.

Rudant, J.P., Deroin, J.P. and Polidori, L. Multiresolution analysis of radar images and its application to lithological and structural mapping: Larzac (Southern France) test site. *Int. J. Remote Sensing* (accepted).

Rudant, J.P., Baltzer, F., Deroin, J.P. Lointier, M., Maitre, H., Penicand, C., Prost, T. and Sabatier, D. (1994). Possibilité cartographique offerte par ERS-1 en contexte tropical. Exemple en Guyane française: bathymétrie, formations de la plaine côtière, géologie, hydrologie, urbaniame, iconocarte; apport spécifique des séquences multitemporelles [Mapping possibility offered by ERS-1 in tropical regions. Example in French Guyana: bathymetry, coastal-plain formations, geology, hydrology, urban development, image map; specific role of multidate data]. ERS-1 Pilot Project Workshop, Tolede, June 1994, p. 108.

Rudant, J.P., Cautru, J.P., Lointier, M., Charron, C., Dechambre, M., Deffontaines, B., Deroin, J.P., Prost, M.T., Raymond, D. and Vanderhaeghe, D. (1993). Pilot project PPF12. Coastal and fluvial environment in French Guiana: first results of ERS-1 SAR data analysis. Cannes, November 1992. *ESA Special Papers*, 359, pp. 835-845.

Rudant, J.P., Lointier, M., Prost, M.T., Cautru, J.P., Deroin, J.P., Sabatier, D., Girault, D., Tonon, M., Nezry, E., De Grandi, F., Conway, J. and Sieber, A. (1993). ERS-1 SAR data analysis of the French Guiana coastal plain: some thematic aspects. 2nd ERS-1 Symp., Hamburg, 11-14 October 1993. *European Space Agency, Spec. Paper* 361, pp. 953-960.

Sabins, F.F. (1983). Geologic interpretation of space Shuttle radar images of Indonesia. *Amer. Assoc. Petrol. Geol. Bull.,* 67 (11): 2076-2099.

Sabins, F.F., Blom, A. and Elachi, C. (1980). Geologic use of linear fractures maps derived from small-scale images. Proc. 3rd Basement Tectonics Symp.

Salisbury, J.W., Walter, L.S.and Daria, D. (1988). Midinfrared (2.6-13.5 μm) spectra of igneous rocks. U.S. Geol. Surv. Open File Report 88-686, pp. 1-108.

Salman, A. and Tronin, A. (1990). Satellite thermal survey. A new tool for the studies of seismoactive regions. Res. Inst. Remote Sensing Methods in Geol. USSR, Leningrad Berzevoy.

Saunders, D.F., Thomas, G.E., Kinsman, F.E. and Beatly, D.F. (1973). ERTS 1 imagery use in reconnaissance prospecting. Type 3 Final Report to NASA US department Com. Natl. Tech. Inf. Service E 74-10345.

Sawatzky, D.L. and Rames, G.L. (1980). Geologic use of linear fractures maps derived from small-scale images. Proc. 3rd Basement Tectonics Symp.

Scanvic, J.Y. (1975). Apport de l'imagerie spatiale multispectrale à la comprehension tectonique du Massif central français et de son environment sedimentaire [Role of multispectral spatial imagery in understanding the tectonics of the French Massif Central and its sedimentary environment]. *Chron. des Mines,* sect. no. 4.

Scanvic, J.Y. (1977). The French Atlantic littoral and the Armorican Massif. NASA-ERTS 2. Fralit project. BRGM Report 77 SGN 052 GEO, 13 pp.

Scanvic, J.Y. (1980). Notice explicative à l'interprétation photogéologique du sud-ouest Cameroun [Explanatory note to the photogeologic interpretation in south-west Cameroon]. Publ. Direc. Mines et Geol. Repl. Cameroun Yaoundé, pp. 1-21.

Scanvic, J.Y. (1981). Télédétection spatiale et recherche minière [Space remote sensing and mineral exploration]. *SFPT Bull.,* no. 83.

Scanvic, J.Y. (1982). Perspectives d'application de la télédétection thermique spatiale. Les renseignements de l'expérimentation HCMM sur la France [Perspectives of application of spatial thermal remote sensing. Results of the HCMM experiment over France]. *Géochronique* no.2 pp. 7-12.

Scanvic, J.Y. (1983). Utilisation de la télédétection dans les sciences de la Terre [Utilisation of remote sensing in earth sciences]. *Manuels et Méthodes,* no. 7. BRGM publ., Orléans.

Scanvic, J.Y. (1984). Etude par la télédétection du district de Pontgibaud, Massif central (France) [Remote sensing study of Pontgibaud district, Massif Central (France)]. BRGM Report 84 SGN 330 GEO, 16 pp.

Scanvic, J.Y. (1984). A strategy for implementing remote sensing in exploration geology. Application to the Armorican Massif (France). Symp. Int. ERIM, Paris.

Scanvic, J.Y. (1984). Correlations between spatial remote sensing technique, geochemical and geophysical data in western France: an integrative and orientation technique for geological mapping and ore exploration. In: *Remote Sensing for Geological Mapping.* BRGM doc. no. 82, IUGS publ. no. 18.

Scanvic, J.Y. (1984). Télédétection et prospection minière. Panorama 1984 [Remote sensing and mineral exploration. 1984 panorama]. *Chron. Rech. Min.,* no. 477, pp. 5-14.

Scanvic, J.Y. (1984). Etude par télédétection (radar et photographies aériennes) de la region de Mitzic (Gabon) [Remote sensing (radar and aerial photographs) study of the Mitzic region (Gabon)]. BRGM Report 84 SGN 016 TED.

Scanvic, J.Y. (1986). Apport de la téléanalyse à la synthèse géologique et métallogénique du faisceau de Pontgibaud et de ses marges (Massif central français) [Contribution of remote sensing to the geological and metallogenic synthesis of the Pontgibaud swarm and its surroundings (French Massif Central)]. II Cong. des Sciences de la Terre. Soc. Geol. France, Clermont-Ferrand, p. 162.

Scanvic, J.Y. (1986). Evolution du potentiel d'utilisation de la télédétection en cartographie géologique et minière: exemple du Zimbabwe. Spot 1 Prémiers résultats [Developments in the application of remote sensing to geological and mineral mapping. Example of Zimbabwe. Spot 1 Preliminary results]. CNES. Cepadues Publ., Toulouse, pp. 181-185.

Scanvic, J.Y. (1988). Contribution de la télédétection à la connaissance structurale et à la prospection minière de la Nouvelle Calédonie orientale [Contribution of remote sensing to structural investigation and mineral prospecting in eastern New Caledonia]. BRGM Report 88 NCL 098 TED.

Scanvic, J.Y. (1989). Etude par télédétection de la Montagne noire et de l'Albigeois, synthèse des travaux de Chan Song Ling [Remote sensing study of Montagne Noire and Albigeois. Synthesis of investigations by Chen Song Ling]. BRGM Report R30094 TED DT 89.

Scanvic, J.Y. (1991). Télédétection en recherche minière: champs d'application et perspectives d'avenir [Remote sensing in mineral exploration: fields of application and future perspectives]. *Chron. Rech. Min.*, no. 505, pp. 3-5.

Scanvic, J.Y. and Weecksteen, G. (1980). Notice explicative de la carte des linéaments de la France, échelle 1/1 000 000 [Explanatory note of the lineament map of France, 1:1,000,000 scale]. BRGM doc. no. 14, 30 pp.

Scanvic, J.Y. and Bache, J.C. (1986). Recherche minière au Zimbabwe. Apport de la télédétection [Mineral exploration in Zimbabwe. Role of remote sensing]. BRGM Report 86 RHO 138 SGN, 44 pp.

Scanvic, J.Y. and Girault, F. (1989). Imagerie Spot et inventaire des mouvements de terrain. L'exemple de La Paz-Bolivie [Spot imagery and ground movement inventory. Example of La Paz, Bolivia]. *Revue Photointerprétation,* 11 pp.

Scanvic, J.Y., Delpont, G. and King, Cl. (1985). Structural and geobotanical contribution of remote sensing to exploration of hidden deposits in Brittany (France). Final report CEE MSM 037F-BRGM 85 SGN 583 GEO, 118 pp.

Scanvic, J.Y., Brosse, J.M. and Prian, J.P. (1991). The use of side looking radar for geological mapping of equatorial forest: example of Mouila, Gabon. Earsel workshop 'Radar Geology', Graz, Austria, July 1991.

Scanvic, J.Y., Cluzeau, B., Kerr, Y. and Gess, G. (1985). SAR 580-'Les Vans' site test F5. Contrat CEE-GDTA N 152 MPP-F(G).

Scanvic, J.Y., Castaing, G., Chevremont, Ph., Dutartre, Ph. and Martin, P. (1986). Methodology for locating subsurface domes by remote sensing. Remote sensing in mineral exploration. CEE Report. EUR 10 334, pp. 169-186.

Scanvic, J.Y., Brosse, J.M., Desprats, J.F., Girault, F. and Imbault, M. (1993). Etude structurale en milieu altéritique: apport des données SAR-ERS-1 du plateau des Mauges (Loire Atlantique, France) [Structural study in an alteritic terrain: role of SAR-ERS-1 data in the Mauges plateau (Atlantic Loire, France)]. Coll. Int. CNES 'de l'optique au radar' Paris, 10-13 May 1993.

Sebban, C. (1984). Essai de synthèse de données multivariées dans le cadre d'une banque de données géologiques du Gabon [Experiment in integrating multivariate data into the geological databank of Gabon]. *Mém. Cetel.* GDTA, Toulouse.

Seyler, F. (1986). Recherche méthodologique sur le traitement d'images de télédétection aérienne basée sur l'analyse préalable des formations superficielles avec leur couverture végétale [Methodological investigation on image processing of aerial remote-sensing data based on preliminary analysis of surficial deposits with their vegetation cover]. Thesis, Paris VII, 262 pp.

Short, M.N. and Stuart, M.L. (1982). *The Heat Capacity Mapping Mission (HCMM) Anthology.* NASA, US Govt. Printing Office, Washington, D.C., 267 pp.

Siegal, D.B. (1983). Use of Landsat multispectral scanner data for the definition of limonitic exposures in heavily vegetated areas. *Econ. Geol.*, vol. 78, no. 4

Singhroy, V.H. (1987). An overview of geobotanic remote sensing in Canada. *Proc. 11th Canadian Symp. on Remote Sensing,* Waterloo, Canada, Ontario Centre for Remote Sensing, pp. 259-276.

Singhroy, V.H. and Kruse, A.F. (1991). Detection of metal stress in boreal forest species using the 0.67 μm chlorophyll absorption band. *Proc. 8th Thematic Conf. Geol. Remote Sensing Exploration. Engnrg. and Environ.*, April 1991, Denver, Colorado. Publ. Envir. Res. Centre Inst. Mich., Ann Arbor, pp. 361-373.

Singhroy, V.H., Slaney, R. Lowman, P., Harris, J. and Moon, W. (1993). Radarsat and radar geology in Canada. *Can. J. Remote Sensing*, 19 (4): December 1993.

Sonder, R.A. (1938). Die Lineamenttektonik und ihre Probleme. *Ecol. Geol. Helv.*, vol. 31, pp. 199-238.

Sossa, Simawengo (1980). Contribution à la pétrologie, géochimie, géologie structurale du massif de Montmarault (France) [Contribution to the petrology, geochemistry and structural geology of Montmarault Massif (France)]. Doct. thesis, Univ. Orléans.

Steward, H.W., Blom, R., Abrams, M. and Daily, M. (1980). Rock type discrimination and structural analysis with Landsat and Seasat data. San Rafael, Utah. PJL ubl. 80.61. In: Report Radar Geol. Workshop, Snowmass, Colorado, 1979.

Stieltjes, L. (1985). Carte des coulées historiques de la Fournaise et notice [Map of Historic Flows of Fournaise and a Note]. BRGM Publ., Paris.

Tamain, G. and Deroin, J.P. (1989). Zonalité intra et periplutonique des minéralisations tardi-varisques des Cévennes, France [Intra- and periplutonic zoning of Late Variscan mineralisations of Cévennes, France]. Guide Télé-mégatectonique.

Tapponnier, P. and Molnar, P. (1977). Active faulting and tectonics in China. *Journ. Geophys. Res. USA*, 82 (20): 2905-2930.

Taranik, J.V., Reynolds, C.C. *et al.* (1978). Targeting exploration for nickel laterites in Indonesia with Landsat data. *Int. Symp. Remote Sensing Envir. Proc.*, vol. 11, pp. 1037-1051.

Taud, H. and Parrot, J.F. (1990). Analyse numérique des formes sub-circulaires sur les images satellitaires [Numerical analysis of subcircular forms in satellite images]. *Deux. Journ. de Télédétection de l'Orstom*, Bondy.

Teleki, P. and Weber, C. (1984). Remote sensing for geological mapping. Proc. Sem. IUGS. BRGM doc. no. 81, IUGS publ. no. 18.

Trautwein, C.M. (1982). Digital database application to porphyry copper mineralization in Alaska. US Dept. Interior, Open File Report 82801.

Tricart, J.L.F. (1972). Méthodes de travail: les enregistrements d télédétection source d'information pour l' étude de l'environment éecologique [Methods of investigation: Remote-sensing data, a source of information for studying ecological environment]. *Rev. Gĕomorph. Dyn.*, XXVII (1): 30-41.

Ulaby, F.T., Moore, R.K. and Fung, A.K. (1982). *Microwave Remote Sensing. Active and Passive.* Addison-Wesley Publ. Co., Norwood, USA.

Ulaby, F.T., Moore, R.K. and Fung, A.K. (1986). *Microwave Remote Sensing*. Artech Publ. House, Inc. 2162 pp. vol. 3: *From Theory to Applications.*

Ulaby, F.T., Bengal, T.H., Dobson, M.C., East, J.R., Garvin, J.B. and Evans, D.L. (1990). Microwave dielectric properties of dry rocks. *IEEE Trans. Geoscience and Remote Sensing*, 28(3): 325-336.

Vairon, J. (1977). Etude méthodologique de la distribution des gîtes de flourine-barytine du Massif Central à l'Ouest du Sillon houiller (France) [Methodological study of the distribution of fluorite-baryte deposits of Massif Central in the western part of the Coal Belt (France)]. *Mém, Univ. Orléans.* Diplôme Ingéniorat 'Ressource et Matériaux Minéraux'.

Valencon, F. and Berger, L. (1980). Détermination des zones de circulation des fluides par géochimie et télédétection dans un massif granitique [Determination of fluid circulation zones by geochemistry and remote sensing in a granite massif]. Centre de Télédétection et d'Analyse des Milieux Naturels. Ecole Nat. Sup. des Mines.

Varet, J., Stieltjes, L., Gerard, A. and Fouillac, C. (1980). Prospection géothermique du Mont-Dore [Geothermal prospecting in Mont-Dore region]. BRGM Report 80 SGN 150 GTH.

Vargas, E. (1977). Plano de desarolo urbano de la cuidad de La Paz. Rapport Honorable Alcadia Municipal de La Paz, 1977. Caracteriscas geomorphologicas de la Cuenca de La Paz y Alrededores.

Vennat, G. (1982). Un exemple de relation volcanisme-pluto-nisme dans la region de Gannat-Les-Ancizes, Massif central (France) [Example of relation between volcanism and plutonism in the Gannat-Les-Ancizes region, Massif Central (France)]. Thesis, UER sciences, Clermont-Ferrand.

Vincent, R.K. (1977). Uranium exploration with computer processed Landsat data. *Geophysics,* 42 (3): 536-541.

Vincent, R.K. (1977). *Geochemical Mapping by Spectral Rationing Methods in Remote Sensing and Mineral Exploration* (W.L. Smith, ed.). Dowden, Hutchinson and Ross, Inc., Strasbourg, pp. 251-278.

Vogt, J. (1981). Carte seismotectonique de la France [Seismotectonic map of France]. *Mém, BRGM,* no. 111.

Walter, L.S. and Salisbury, J.W. (1988). The basis for the spectral behaviour of silicates in the thermal infra-red and applications for remote sensing. Signature Spectrale d'Objet en Télédétection. 4th Colloque Aussois, France ESA.

Watson, K. (1982). Regional thermal inertia mapping from an experimental satellite. *Geophysics,* 47 (12): 1681-1687.

Watson, K. and Hummer-Miller, S. (1981). A simple algorithm to estimate the effective regional atmospheric parameter for thermal inertia mapping. *Rev. Remote Sensing of Environment,* vol. 11, pp. 455-482.

Weecksteen, G. (1983). Evaluation de l'apport des radar-graphies SIR A en géologie: comparaison avec les images Landsat et les photographies aériennes. Cas de la République du Mali: exemples en Asie et en Amérique [Evaluation of the role of SIR A radar images in geology: comparison with Landsat images and aerial photos. Case of Mali Republic: examples in Asia and America]. *SFPT Bull.,* no. 92-4.

Weecksteen, G. (1984). Premiers résultats de l'évaluation de l'apport des simulations Spot en recherche minière [Preliminary results of evaluation of the role of Spot simulations in mineral exploration]. GDTA Publ., Toulouse. *Expériences de Simulation du Satellite Spot en Afrique de l'Ouest.*

Weecksteen, G. (1985). Télédétection et recherche minière. Simulation Spot en Nouvelle Calédonie. Application au nickel lateritique [Remote sensing and mineral exploration. Spot simulation in New Caledonia. Application to lateritic nickel]. BRGM Report, 85 NCL 049 GEO.

Weecksteen, G. (1985). Evaluation géologique des clichés de la chambre photogrammétrique des Spacelab I dans le Nord de la France [Geological evaluation of the snaps of the photogrammetric camera of Spacelab 1 in northern France]. *SFPT Bull.,* no. 99, pp. 16-26.

Wertz, J.B. (1963). Le concept linéamentaire aux Etats- Unis [The concept of linements in the United States]. *Rev. de l'Industrie Minérale,* Nov., pp. 708-764.

Wertz, J.B. (1971). Les linéaments et failles majeures de l'Ouest des Etats-Unis. Leur importance au point de vue minier [Lineaments and major faults of the Western United States-their importance in mineral exploration]. *Rev. Géogr. Phys. Géol. Dyn.,* 13 (4): 315-326.

Whitney, G., Abrams, M.J. and Goetz, A.F.H. (1983). Mineral discrimination using a portable ratio-determining radiometer. *Econ. Geol.,* vol. 78, no. 4.

Won, J.S. and Moon, W.M. (1992). Inversion of SAR data for surface scattering. *Geophys. J. Int.*, vol. 108, pp. 423-432.

Wood, J.A., Lasserre, M. and Fedosejeus, G. (1990). Analysis of mid-infrared spectral characteristics of rock outcrops and evaluation of the Kahle model in predicting outcrop thermal inertia. Remote Sensing Environ., 30.169.195/1990.

Wyns, R. (1984). Etude géologique du champ filonien de la Bellière dans le périmètre du P.E.R. de Gesté (Maine-et-Loire) [Geological study of the lode clusters of Belliers in the perimeter of P.E.R. of Geste (Maine-et-Loire)]. BRGM Report 84 SGN 251 GEO.

Wyns, R. and Bechennec, F. (1992). Synthèse régionale Vendeé-Mauges. Etude géologique a 1/250 000 [Regional synthesis of the Vendeé and Mauges areas. A geological study at 1/250,000]. BRGM Report R 36216.

Yost, E. (1971). The reflectance spectra of mineralized trees. *Proc. 7th Int. Symp. on Remote Sensing of Environment,* Ann Arbor, Michigan, pp. 269-297.

Ziserman, A., Boirat, J.M. Lecuyer, P., Vairon, J. and Scanvic, J.Y. (1982). Les gîtes fluro-barytiques du Massif Central. Etude statistique et informatique-mise au point provisoire [The fluoro-baryte deposits of the Massif Central. Statistical and computer study-provisional application]. *Bull. BRGM* 2 (14): 451-462.

Landsat Images

Original data © *Earthnet. Distributed by GDTA. BRGM Processing. Fig. 3 (1973), 4 (1976), 10 (1976), 12 (1976), 13 (1977), 16 (1974), 27 (1984), 28 (1984), 35 (1978), 57 (1985), 58 (1974).*

Original data © *ESA. Distributed by Eurimage/GDTA. BRGM Processing. Fig. 10 (1987), 11 (1988), 17 (1988).*

Original data © *EOSAT. Distributed by GDTA. BRGM Processing. Fig. 24 (1988).*

Original data © *ESA/ESRIN. BRGM Processing. Fig. 21.*

Printed in India

An image of France seen from the NOAA satelite

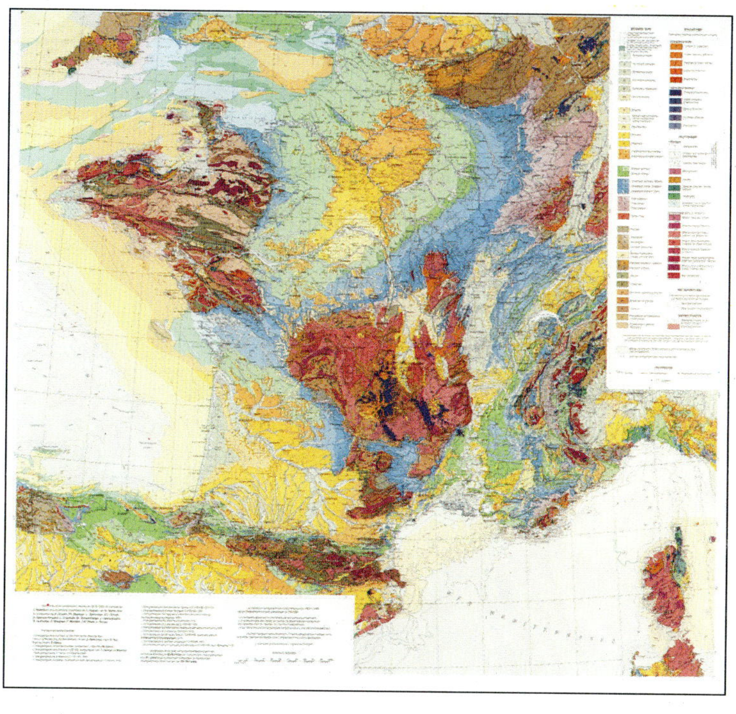

Plate 1. Geological map of France (Original scale: 1/1 500 000) published by BRGM

Plate 2. The Massif Central between the Sillon Houiller formation and the Limagne of Allier

1—igneous intrusions; 2—Aigueperse fault; 3—Limagne formation; 4—Riom trench (major subsidence of Limagne); 5—Puys Range

2
2
1
3
4
5

Plate 3. Spot image of Kanye region. Botswana and its interpretation
1—dolomite: 2—syenite: 3—dolerite
(CNES, 1986. Spot Image Distribution. BRGM Processing)

2
1
2
3
3
2

Plate 4A. Spot image of Yemen
(CNES, 1986. Spot Image Distribution. BRGM Processing)

1

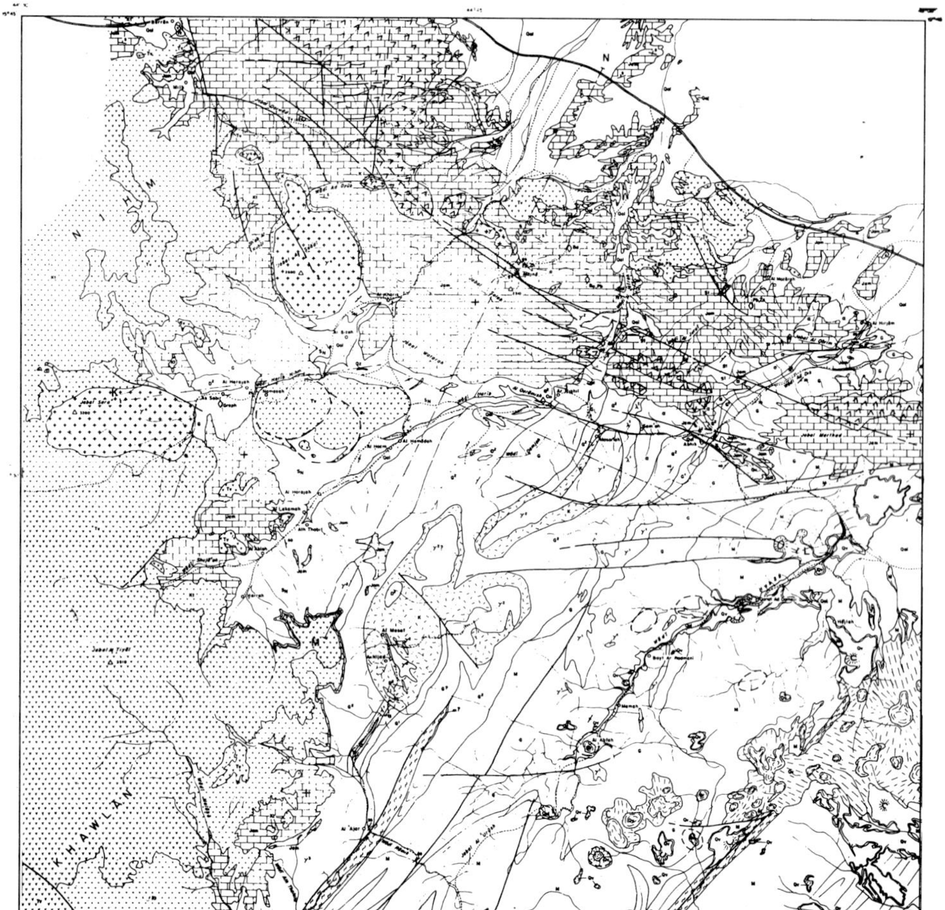

Plate 4B. Earlier geological map (1987 publication)

Legend (from the Nihm geological map)

	Known ground faults	Eb, qa	Quaternary formations
	Other inferred faults	Jam	Sedimentary formations
	Lithological boundaries	Tv	Volcanic intrusions
	Water courses	M, Y2, G1, G2, K, Sn	Basement formations
③, ④...	Particular points (see text)		

0 1 2 3 4 5 Km

Plate 4C. Interpretation of Spot imagery (1987)

1—Ussel-Meymac granite; 2—Sillon Houiller;
3—Commentry Bezenet structure;
4—Roanne structure

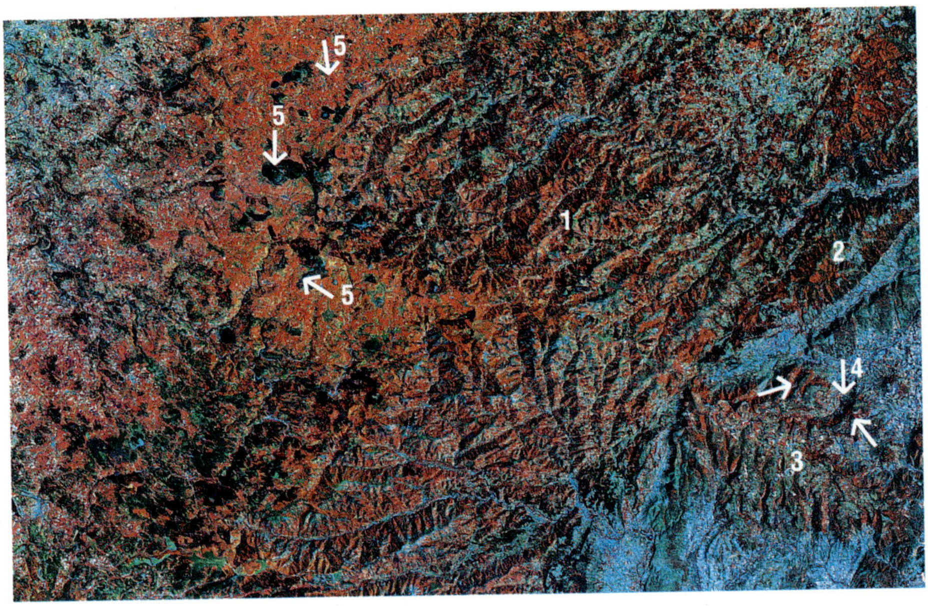

1—Vellay granite; 2—Dumieres granite; 3—Coirons Volcanic Massif; 4—Circular structure;
5—Phonolite occurrences

Plate 5. Granites of Ussel-Meymac (top) and Velay (bottom) as restored Landsat MSS imagery.

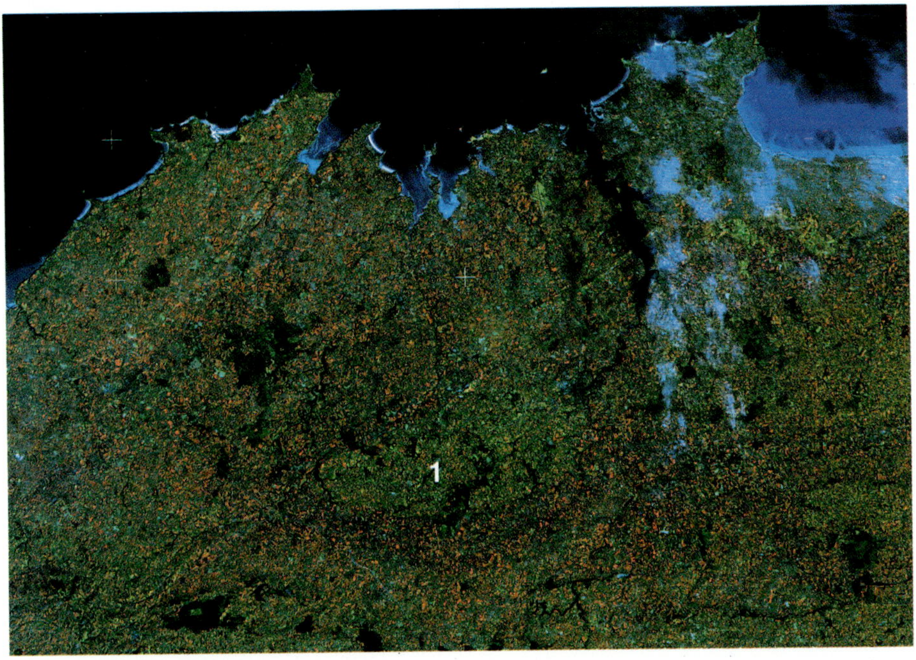

Plate 6. Granite massifs of Dinan (1. top—Landsat TM imagery) and Huelgoat (2. bottom—spot)
(CNES, 1989. Spot Image Distribution. BRGM Processing)

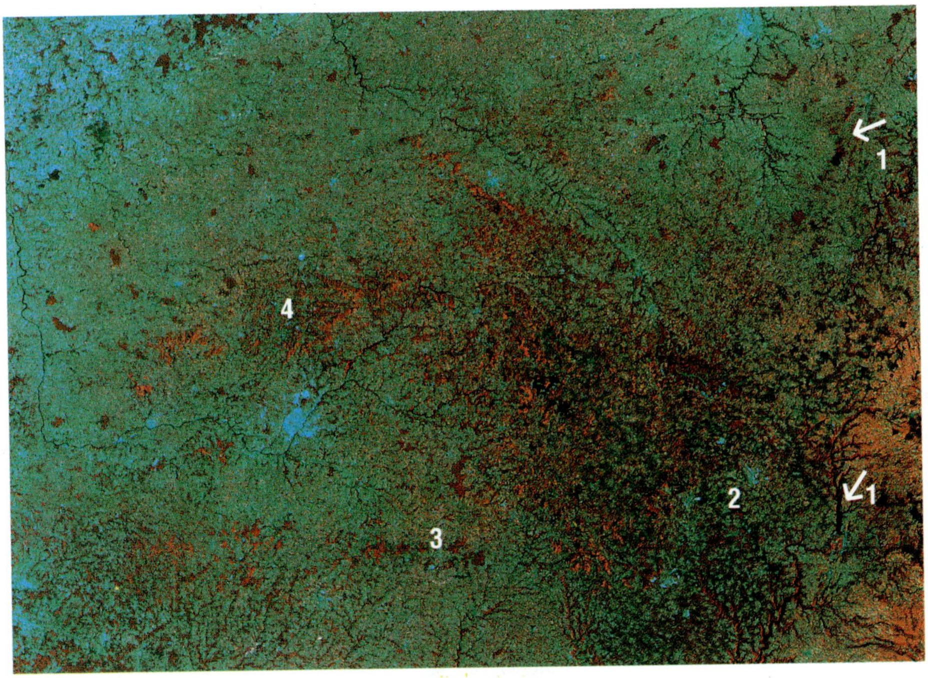

Plate 7. Western part of the Massif Central
1—Sillon houiller clearly visible up to Borte-les-Orgues dam; 2—Millevaches granite; 3—Limousin schists; 4—Ambazac Massif

Plate 8. Remote-sensing map of the hydrogeology of Smara region (Morocco)

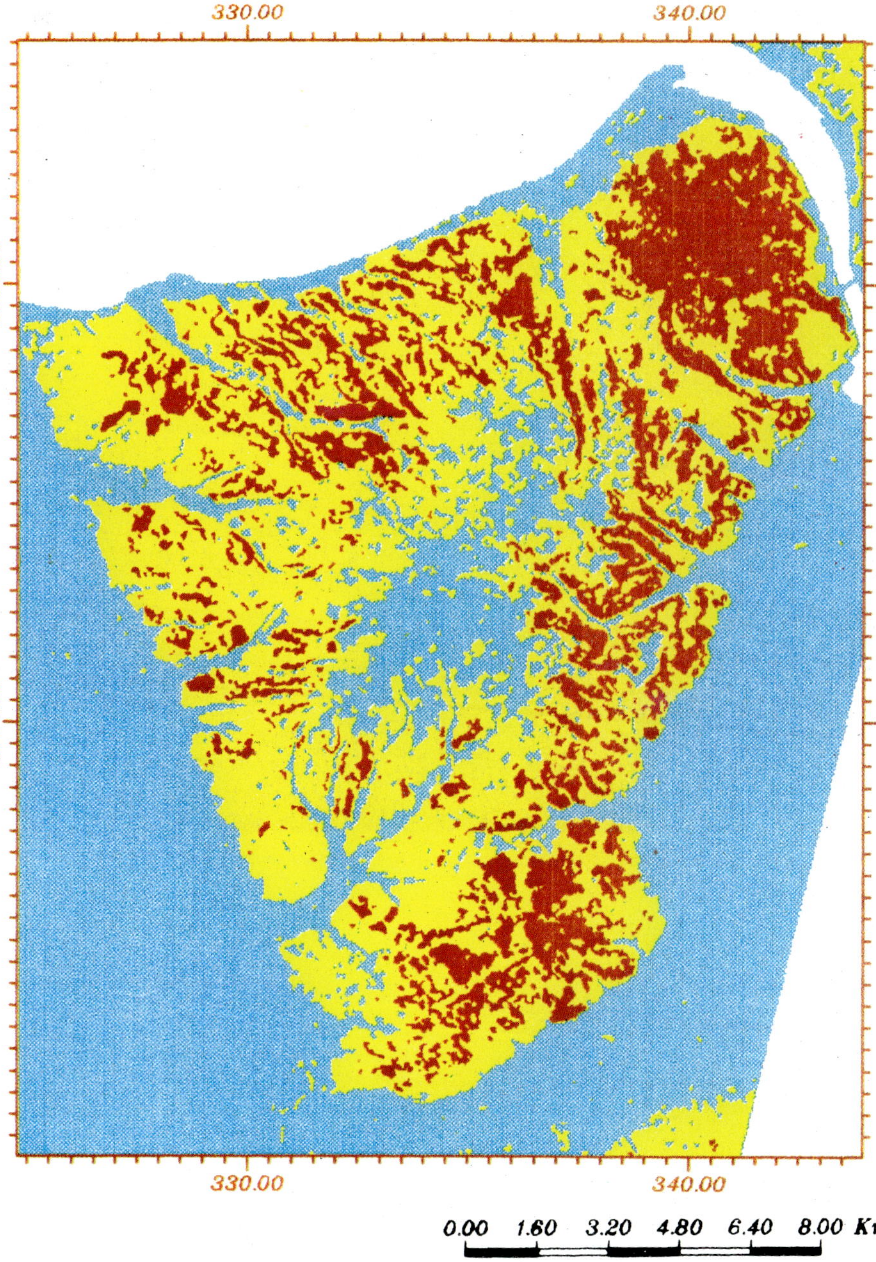

Plate 9. Map showing zones susceptible to landslides (Linkou-Taiwan). prepared from the processing and interpretation of Spot images.

highly susceptible

moderately susceptible

less susceptible

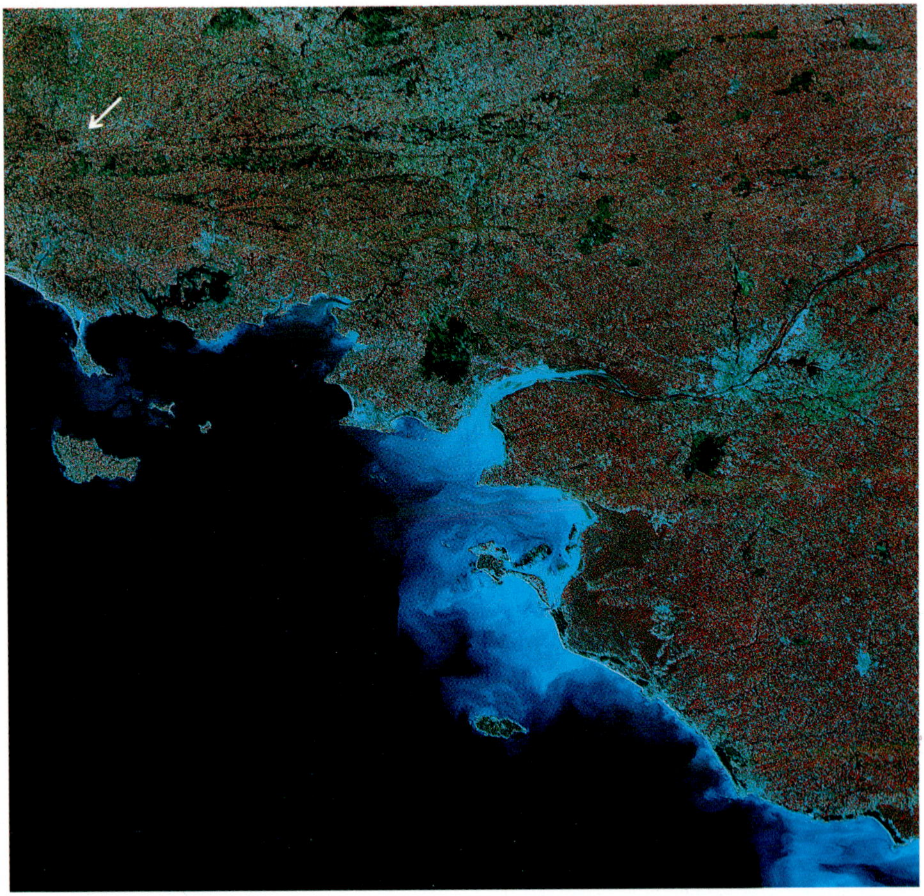

Plate 10. Morphological lineaments. Example of sillon Armorican (Landsat MSS image).

Plate 11. Landsat TM image and its interpretation: example of Chateaulin basin. Armorican Massif (France).

Mappable fault (small-scale geological map)

Lithological boundaries

Lineaments controlling the morphological or drainage elements. Radiometric lineaments.

(0) Circular structures

(1) à (23) Specific points

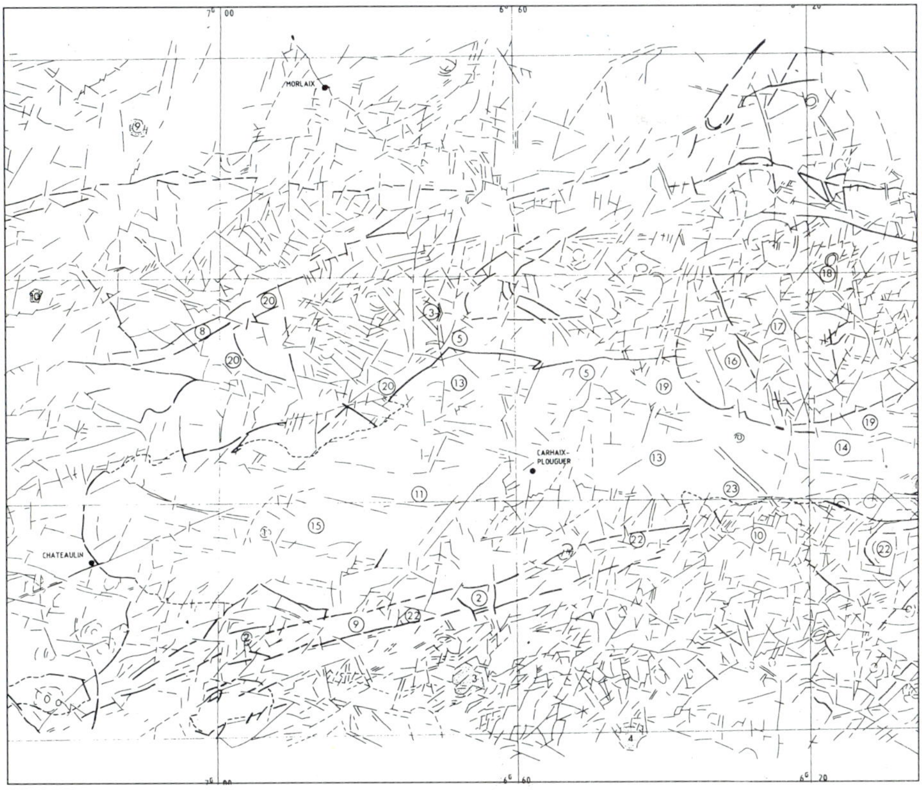

MORLAIX
CARHAIX-
PLOUGUER
CHATEAULIN

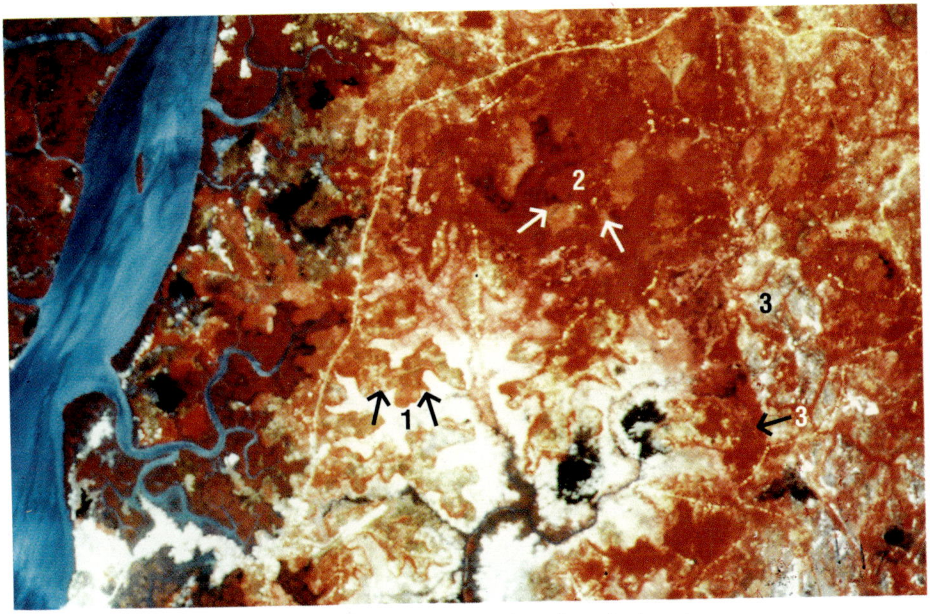

Colour composite: (1) emphasises the boundary of salt-water zones (2) the top points of the bedrock below the fluviodeltaic basin and (3) fractures in the sandstones and quartzites

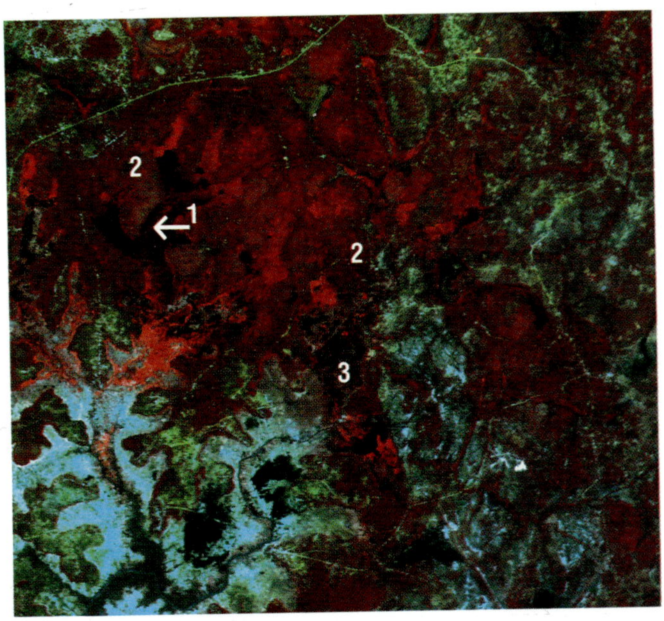

Sinusoidal stretching: enhances the appearance of low ground (1), the boundary of the fluviodeltaic basin and the contrast between the schists (2) and the fluviodeltaic sediment (3), which are not visible on the colour composite.

Plate 12. Spot image of Songlon basin Guinea
(CNES, 1986. Spot Image Distribution. BRGM Processing)

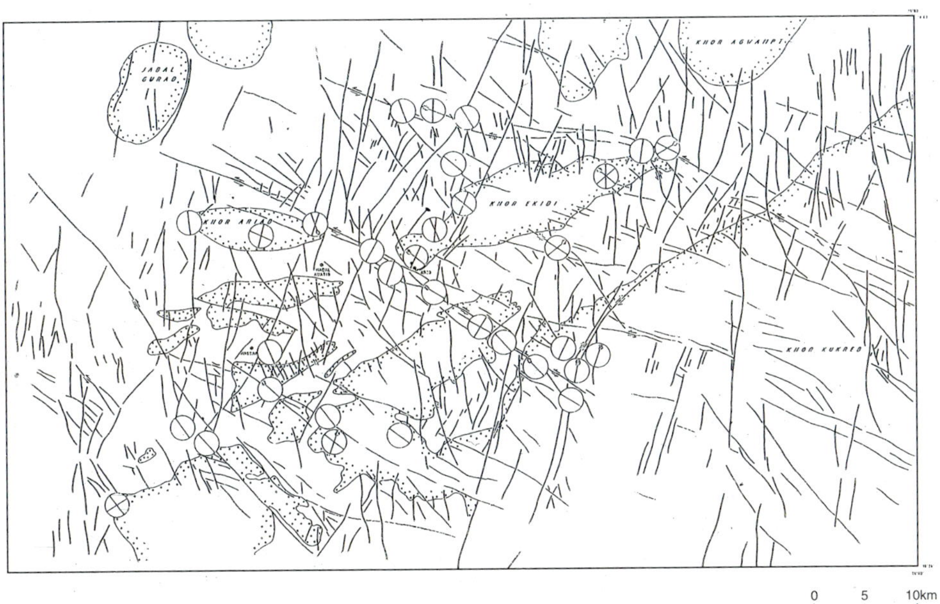

Plate 13. Hassai: colour composite of Landsat TM data and its interpretation

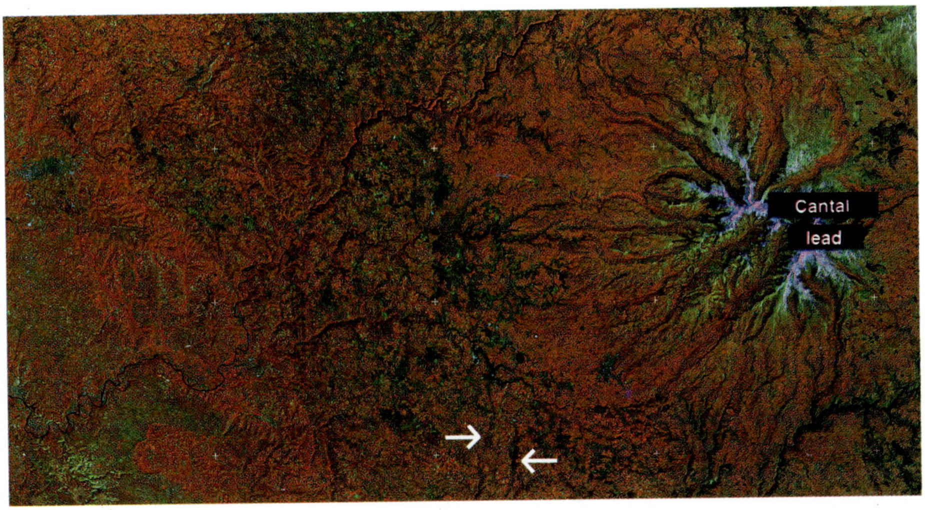

Plate 14. Argentat-Châtaigneraie region. Massif Central (France). Landsat TM

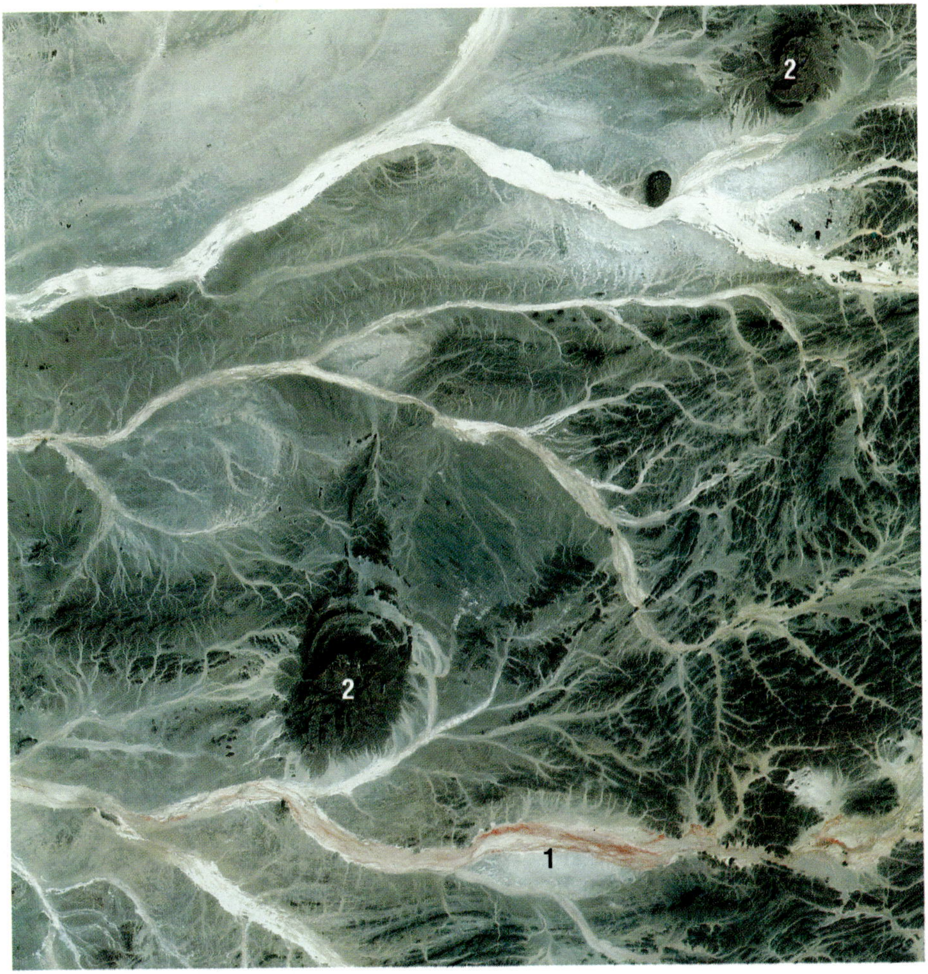

Plate 15. Spot image of circular structures in Sudan, depressed (1) and elevated (2)
(CNES, 1987. Spot Image Distribution. BRGM Processing)

(a)

(b)

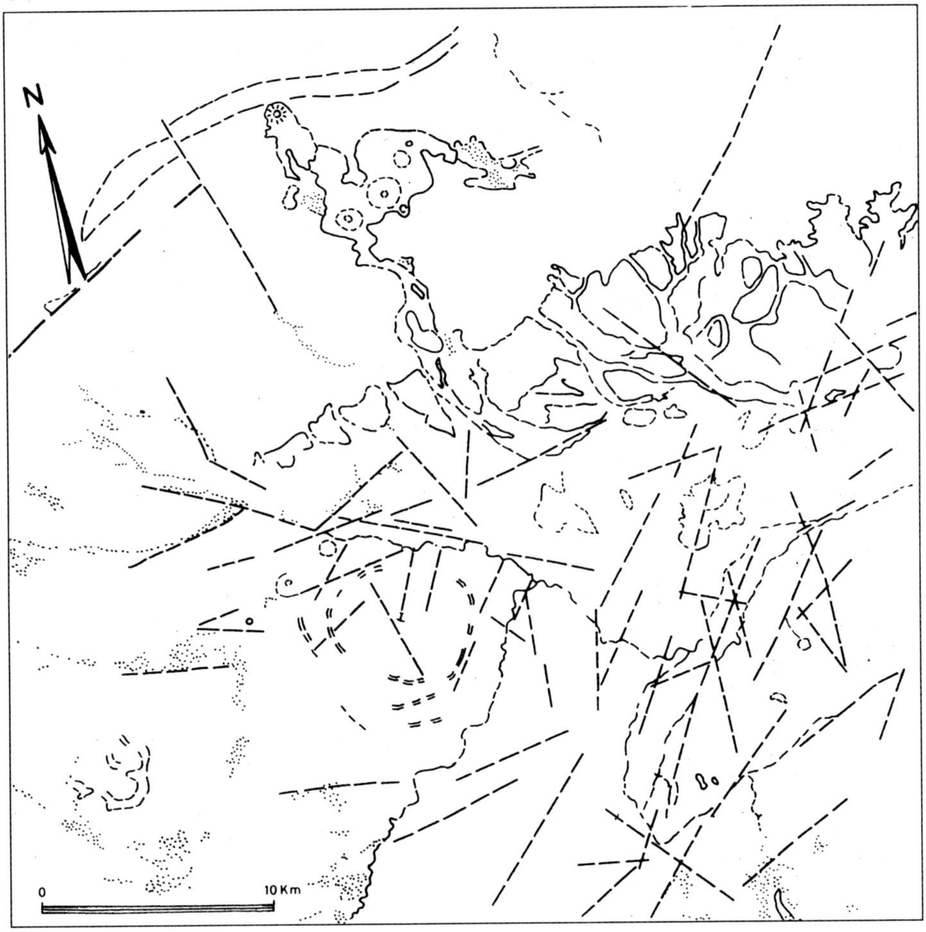

Plate 16. Example of a circular structure visible on Landsat MSS imagery and its interpretation. associated with mineralisation of lead, silver and baryte (after Jebrak and Dutartre, 1984).

Plate 17. Iron cappings in Zimbabwe; Combining spectral response with shape for their identification on a Spot image (© CNES, 1986. Spot Image Distribution. BRGM Processing)

Plate 18. Spot image of Boké bauxite (Guinea).
1—Laterites; 2—Lateritic bauxites
(CNES, 1986. Spot Image Distribution. BRGM Processing)

2

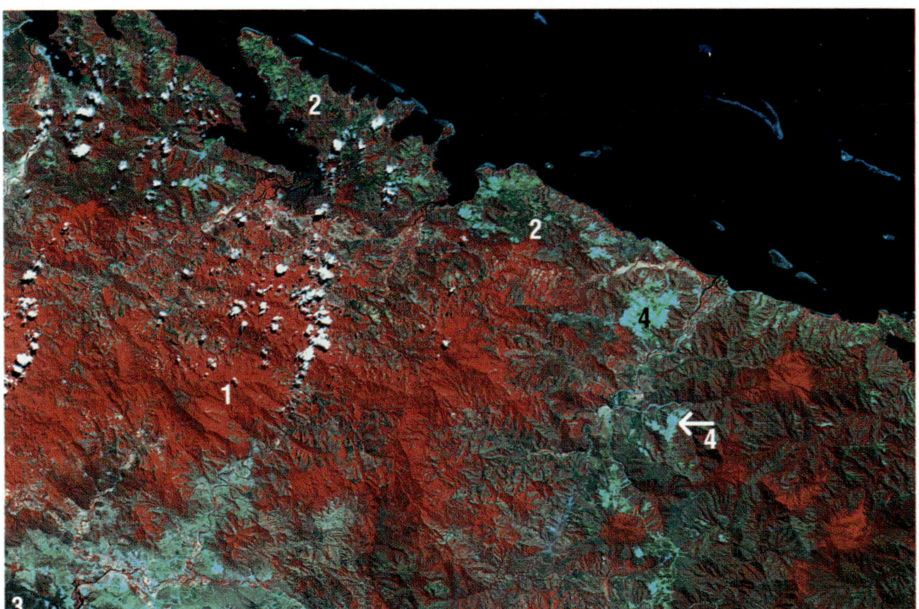

Plate 19. Nickeliferous alterites in New Caledonia; Spot images
(CNES, 1986, 1988. Spot Image Distribution. BRGM Processing).
1—unweathered peridotite; 2—weathered peridotite; 3—basalt; 4—'flows' with nickel mineralisation. Top—southern zone; Bottom—Thio zone

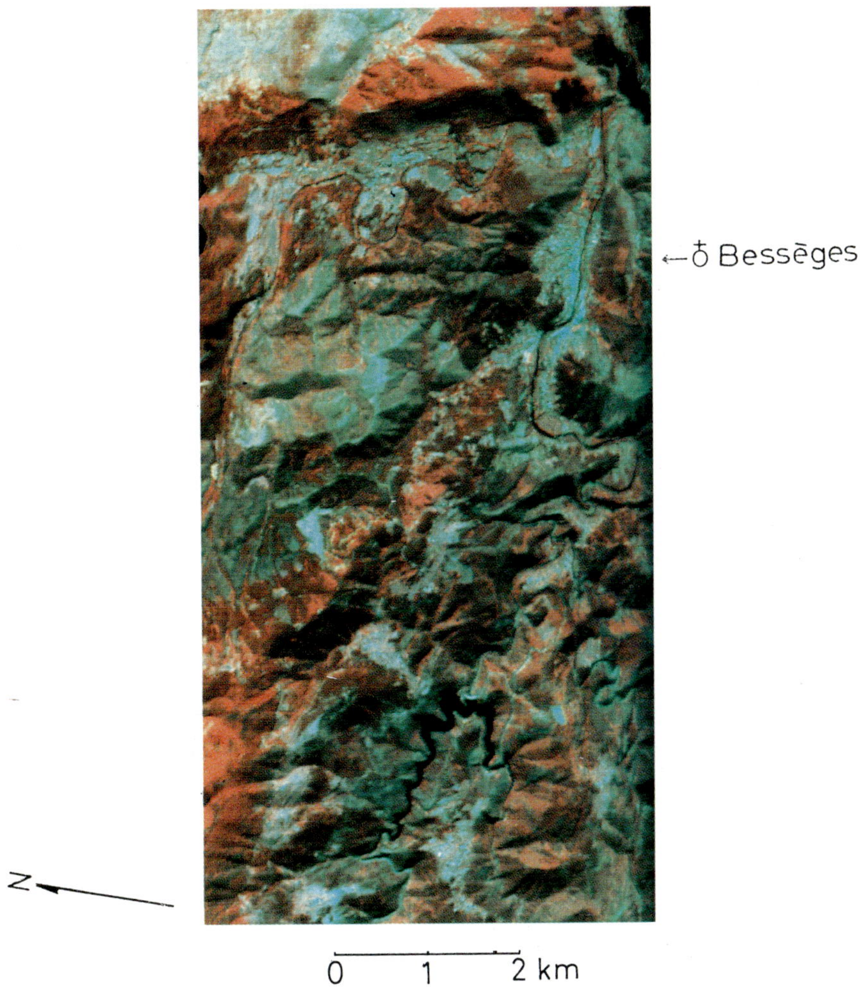

Plate 20. Example of photographic restitution of certain bands of GER (31-42-48) from the Cévennes region (France) (after Deroin, 1990).

(a)

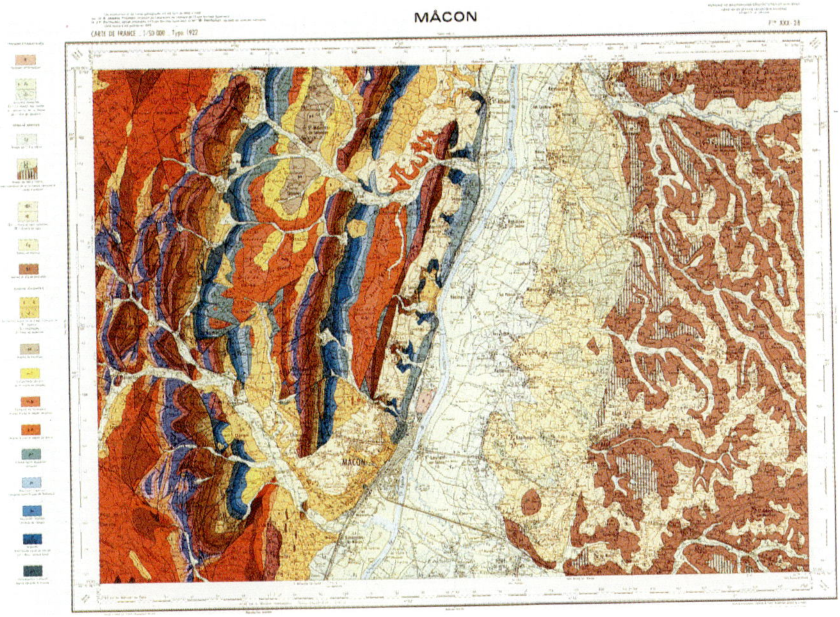

(b)

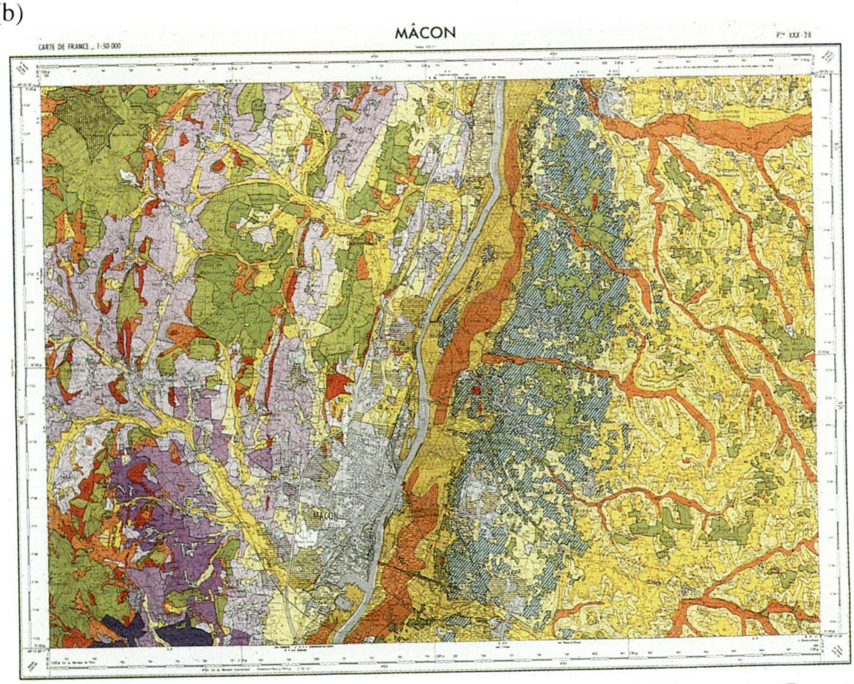

Plate 21. Comparison of geological map and soil-type map. Example of the Mâcon region (France).
a—Geological map published by BRGM; 6—Soil map (SCET AGRI).

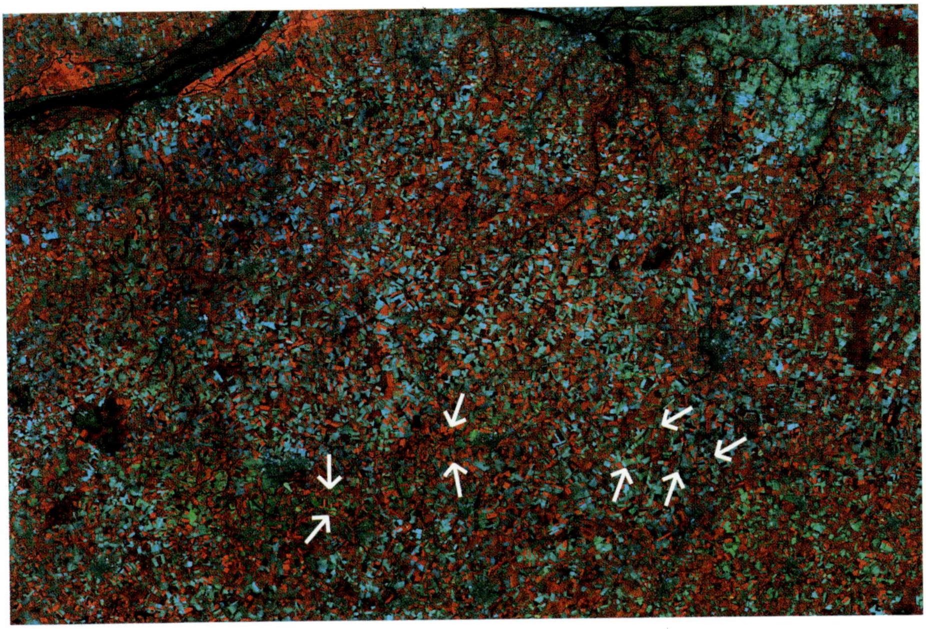

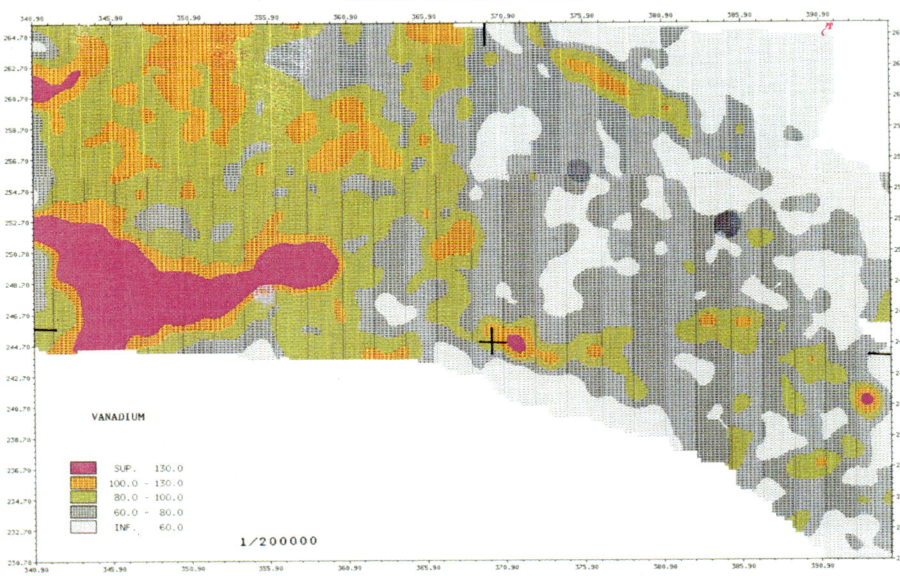

Plate 22. Spectral response of the Beaupréau sequence. Landsat TM image 6-3-1.

A)

Plate 23. Examples of geological landscapes on Spot images. Top: Parthenay-Mirebeau region; bottom: Mont-Saint-Michel bay (CNES, 1986. Spot Image Distribution. BRGM Processing).

B)

(a)

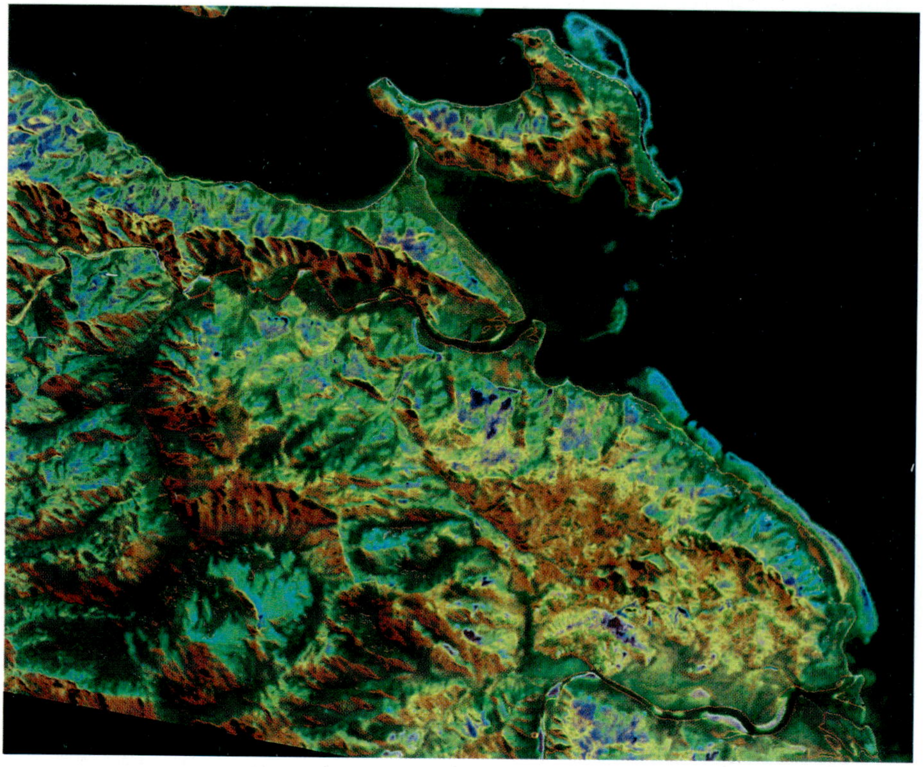

Plate 24. Example of sinusoidal stretching of a Spot image: Poume granite (New Caledonia) (a) and its interpretation (b) (CNES, 1987. Spot Image Distribution. BRGM Processing)

(b)

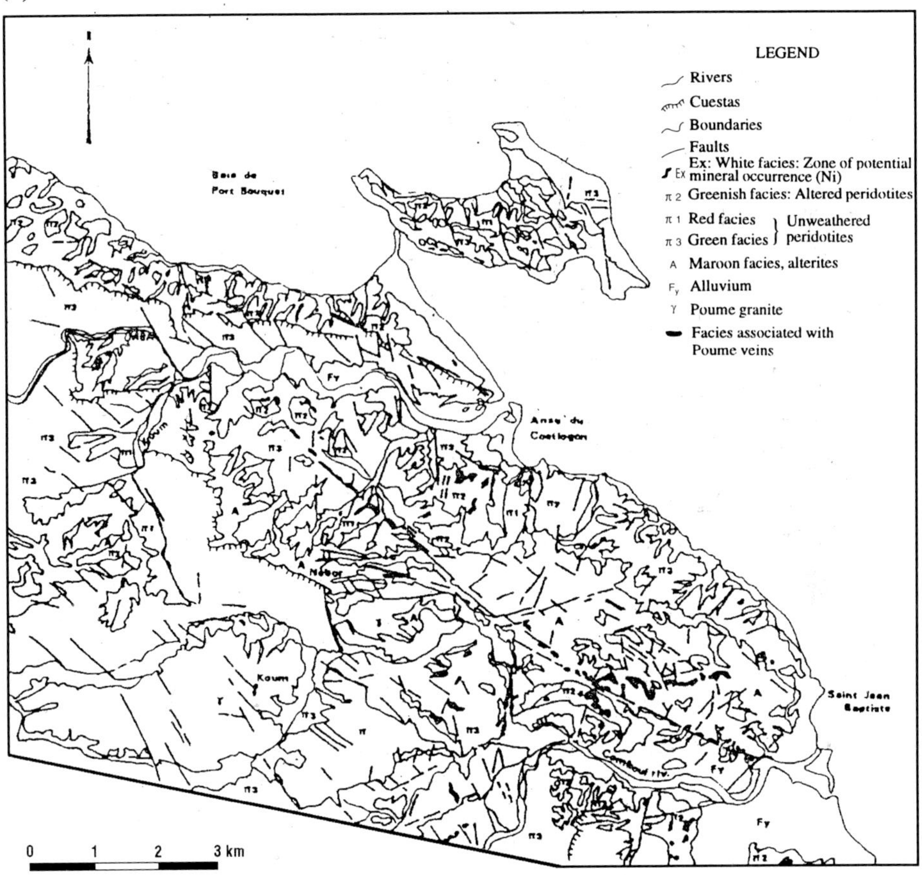
LEGEND
Rivers
Cuestas
Boundaries
Faults
Ex Ex: White facies: Zone of potential mineral occurrence (Ni)
π 2 Greenish facies: Altered peridotites
π 1 Red facies
π 3 Green facies
Unweathered peridotites
A Maroon facies, alterites
F_y Alluvium
γ Poume granite
Facies associated with Poume veins
Baie de Port Bouquet
Anse du Coetlogon
Saint Jean Baptiste
Koum
0 1 2 3 km

Plate 25. Perspective view over La Paz city, integrating the Spot image, its DEM and the plan of the site (after Germain) (CNES, 1988. Spot Image Distribution. BRGM Processing).

A—Altiplano; B—Royal Cordillera; C—Pampajasi fold; D—Rio Irpavi River; E—Llojeta flow; F—La Serana d'Aranguez; G—La Paz city; H—Rio La Paz

A
B
C
G
D
H
F
E

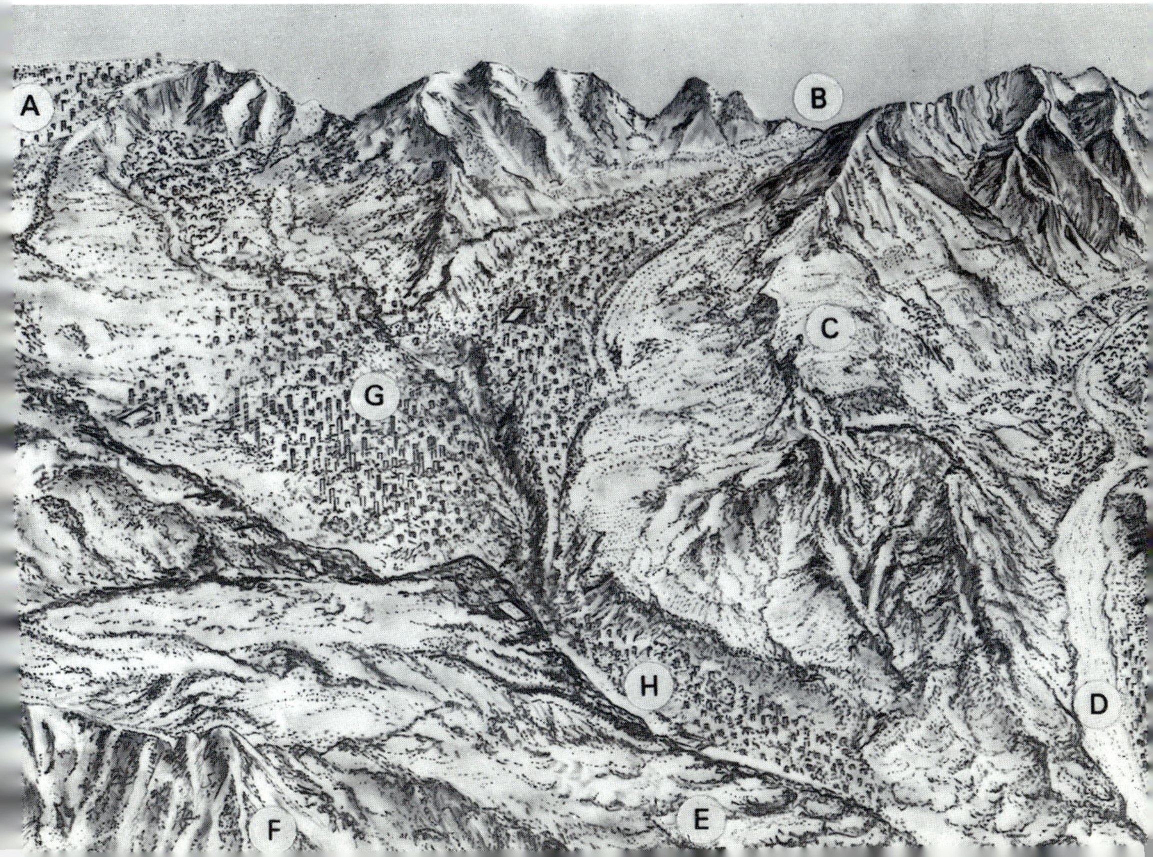
A
B
C
G
H
D
F
E

(a)

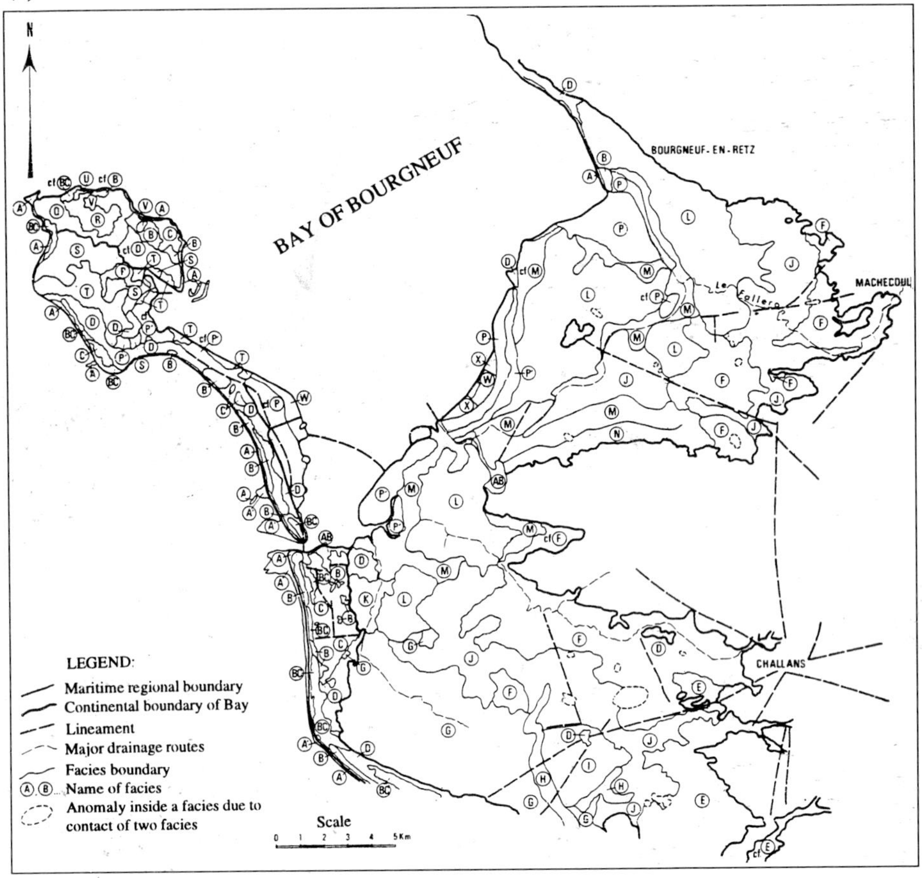

Plate 26. Interpretation of Landsat TM image of Breton bay region (after Minoux, 1986)

(b)

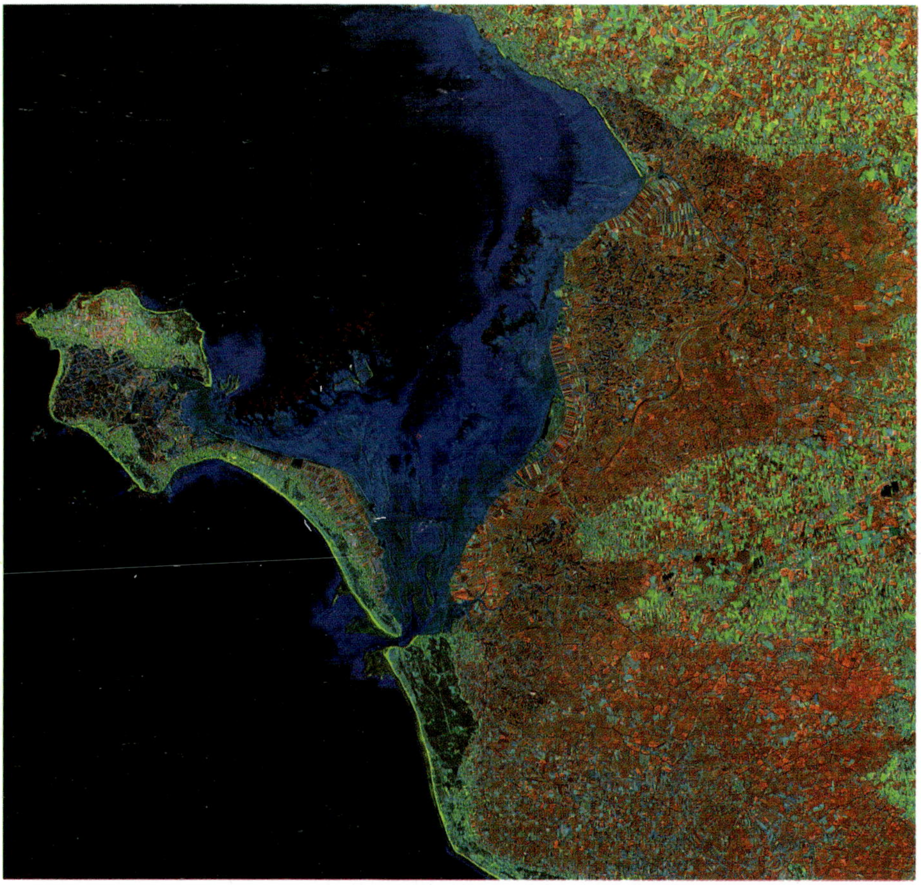

Plate 27. Landsat MSS image of Nafud At Thawayrat region (Saudi Arabia).

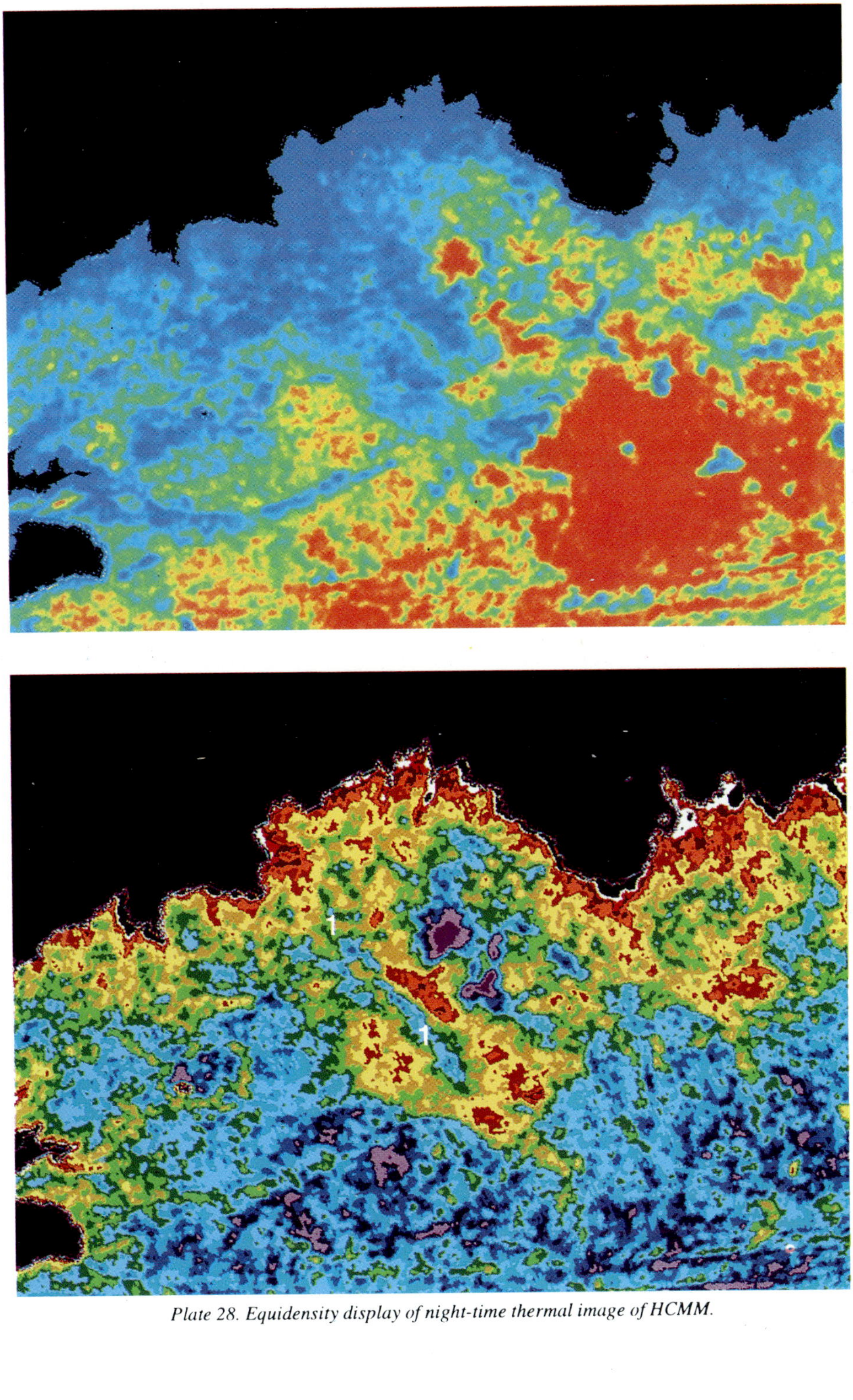

Plate 28. Equidensity display of night-time thermal image of HCMM.

Thermal image

Simulation from Spot bands

Plate 29. Thermal and visible (Daedalus) images of the Vans region (France). The thermal image reveals the existence of a clayey-limestone bed (cl) in the linestone sequence (L).

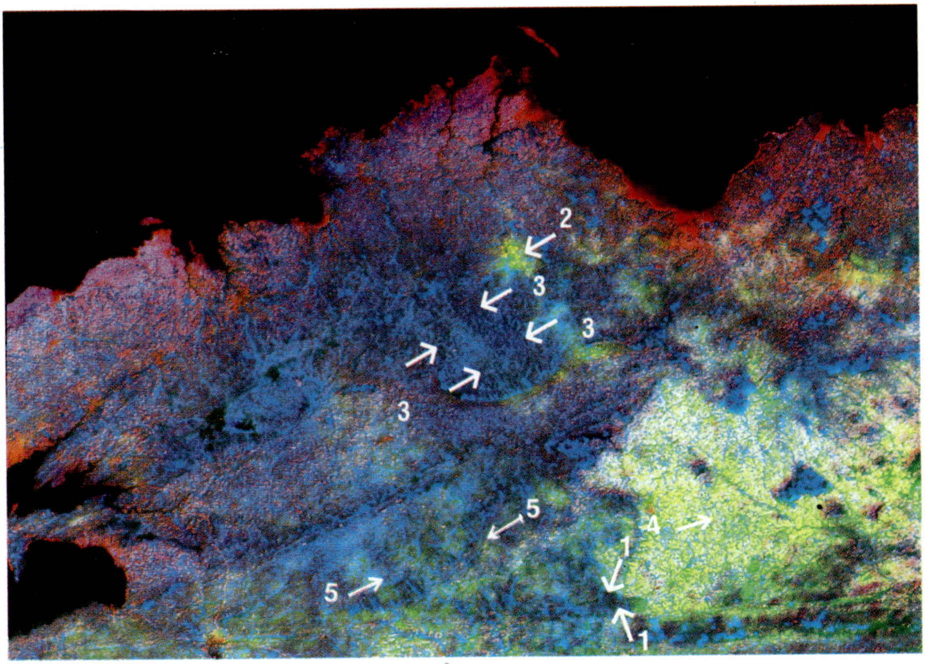

Plate 30. MSS-HCMM composite image of Bretagne (France) (also see Plate 10, points in which the two structures have a similar response).

1—Circular structures of East and West Baud; 2—Ploumagoar granite; 3—Quintin faults; 4—Palaeozoic formations; 5— Armorican Massif.

Plate 31. SIR A radar image of the Yambu region (Saudi Arabia) and the superimposed image of Spot and SIR A (CNES, 1986. Spot Image Distribution. BRGM Processing).

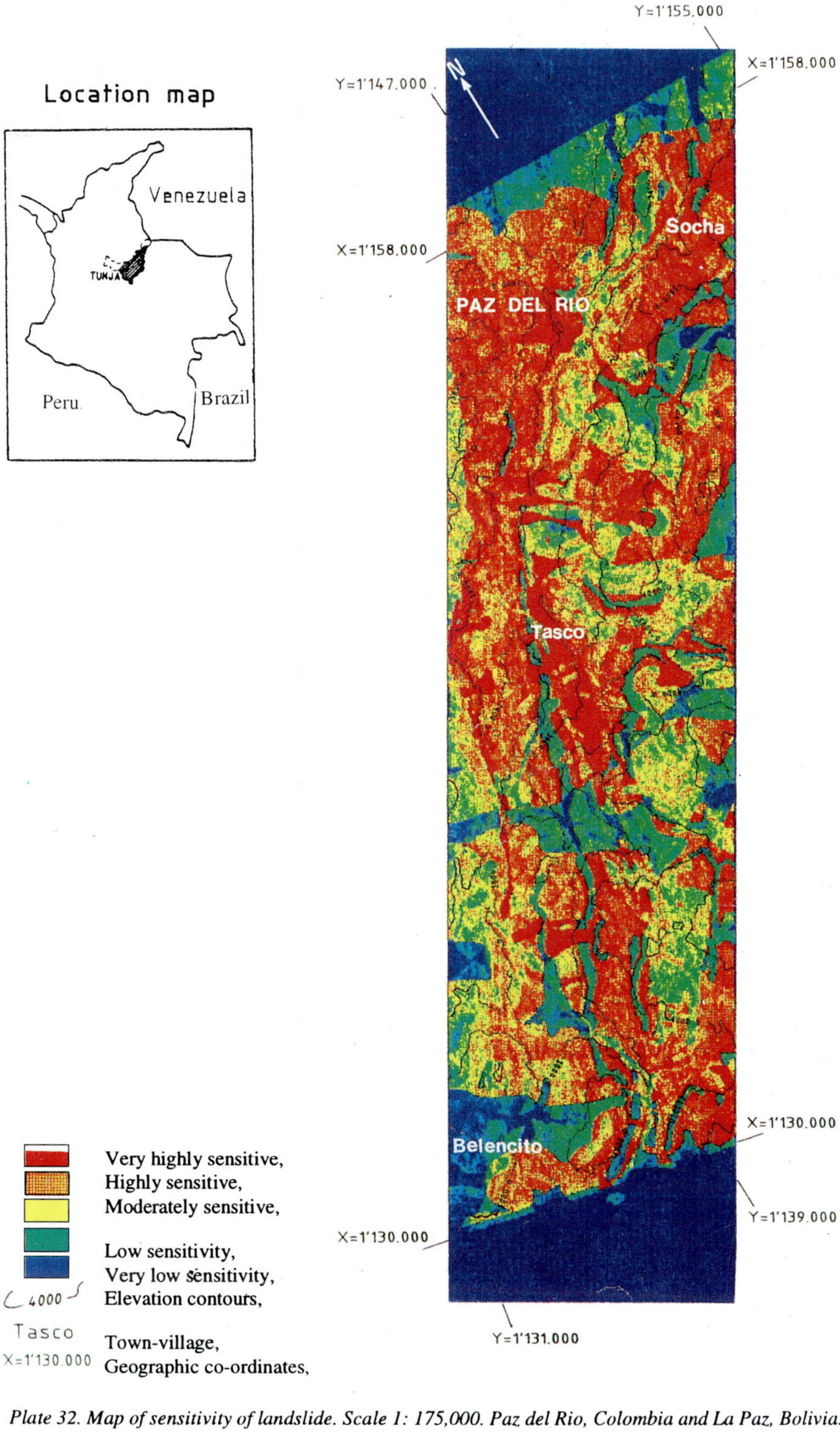

Plate 32. Map of sensitivity of landslide. Scale 1: 175,000. Paz del Rio, Colombia and La Paz, Bolivia.

Plate 33. Image of scintillometric data (Image Processing: Remote Sensing Department of BRGM. Goodyear radar data; co-ordination by Sodeteg. Processing of scintillometric data by CGG).

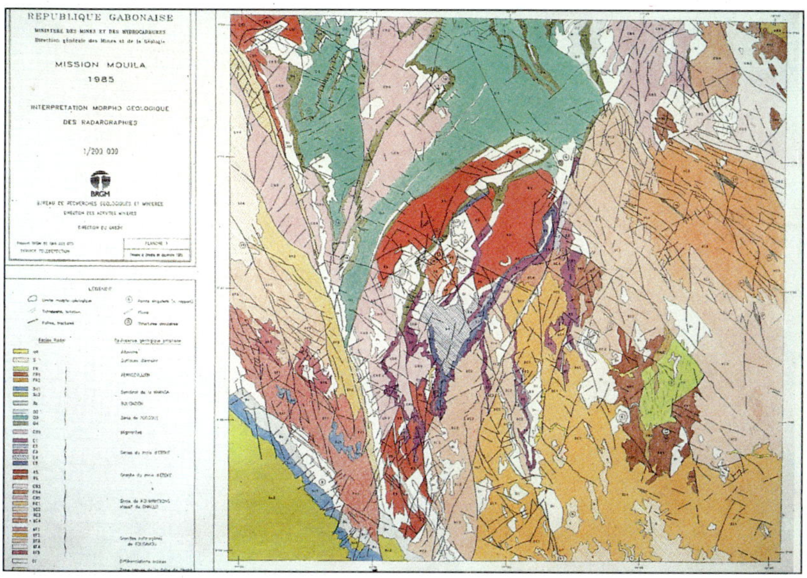

Plate 34. Goodyear radar image and its interpretation.

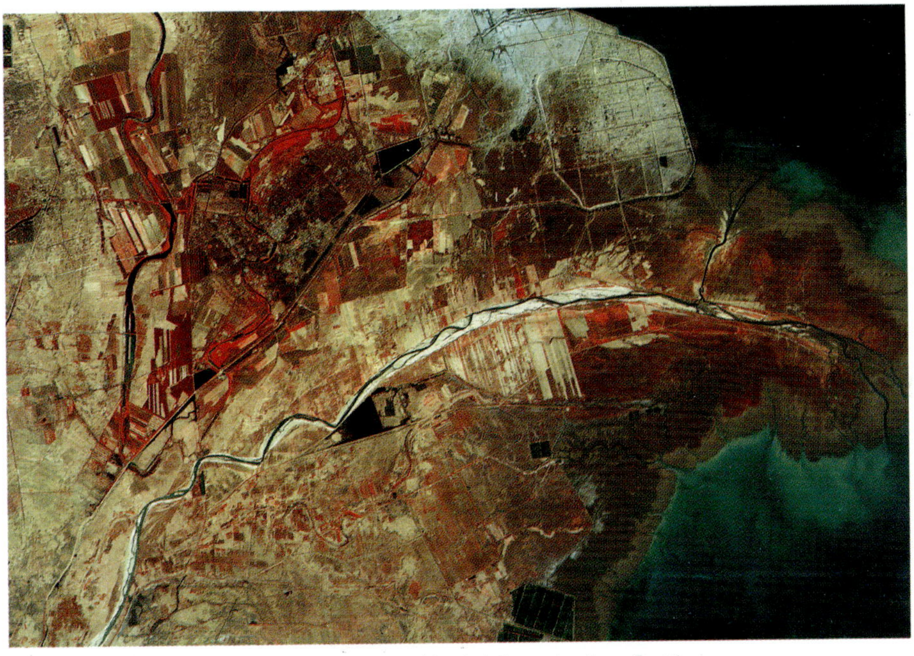

Plate 35. Remote sensing and basic information from Spot images
(CNES, 1988. Spot Image Distribution. BRGM Processing).
Top: erosion of the Loess plateau (China); Bottom: Yellow River delta (China).